Eco-Responsible Cities and the Global Ocean

Voula P. Mega

Eco-Responsible Cities and the Global Ocean

Geostrategic Shifts and the Sustainability Trilemma

Springer

Voula P. Mega
DG Research & Innovation
European Commission
Brussels, Belgium

ISBN 978-3-030-06705-2 ISBN 978-3-319-93680-2 (eBook)
https://doi.org/10.1007/978-3-319-93680-2

Softcover re-print of the Hardcover 1st edition 2019

Cover design: Serenissima Venice and the Bond of the Seas

This Springer imprint is published by the registered company Springer Nature Switzerland AG
The registered company address is: Gewerbestrasse 11, 6330 Cham, Switzerland

Disclaimer: "The views expressed are purely those of the writer and may not in any circumstances be regarded as stating an official position of the European Commission. The European Commission cannot be held liable for any use of the information contained in this document."

"I think of the ocean as the blue heart of the planet. Well, how much of your heart do you want to protect...?" … "With every drop of water you drink, every breath you take, you're connected to the sea. Health to the ocean means health for us"

Dr. Sylvia Earle, *Oceanographer and National Geographic Explorer-in-Residence*

"The Sea is the future of the Earth"

Ségolène Royal, *COP21, Paris 2015*

"Ignoranti quem portum petat nullus suus ventus est" (If you do not know to which port you are sailing, there is no such thing as a favourable wind)

Seneca, L. A., *Ad Lucilium Epistulae Morales. LXXI, 3*

"Cities also believe they are the work of the mind or of chance, but neither the one nor the other suffices to hold up their walls. You take delight not in a city's seven or seventy wonders, but in the answer it gives to a question of yours or to the question it asks you, forcing you to answer, like Thebes through the mouth of the Sphinx"

Italo Calvino, *"Invisible Cities," 1972*

"The sea is everything. It covers seven tenths of the terrestrial globe. Its breath is pure and healthy. It is an immense desert, where man is never lonely, for he feels life stirring on all sides. The sea is only the embodiment of a supernatural and wonderful existence. It is nothing but love and emotion; it is the Living Infinite"

Jules Verne, *"20,000 Leagues Under the Sea," 1872*

"In diversity there is beauty and there is strength… Diversity makes for a rich tapestry and all the threads of the tapestry are equal in value no matter their colour, equal in importance no matter their texture"…

M. Angelou, *"Rainbow in the Cloud," Virage Press, London, 2014*

Preface

These are fascinating and extraordinary times, uncertain and volatile, extreme and ambivalent, in terms of both risks and opportunities. The era of the Anthropocene is advancing as humanity is significantly changing the balance and the nature of Earth's ecosystems, including the largest of these, the Ocean. The impact of cities on the planet is huge and comparable to that of a force of nature. Cities are immense aggregators of people and resources, the places where tensions intensify, but where also countless and lofty opportunities are continuously co-created. These interconnected and influential ecosystems, the only human ones, have great potential as civic entities leading towards sustainable development.

As the second decade of the twenty-first century wanes, the world is in flux. Humanity in 2015, richer, better educated, and smarter than ever, declared its shared will to overcome the sustainability trilemma of deteriorating environment, unjust society, and slowing economy, and to embark on an invigorated shared journey with an ethical compass. The Sustainable Development Goals (SDGs) for 2030 constitute a universal call for action to end poverty, protect our blue planet, and ensure that all people enjoy peace and prosperity. Highly interlinked, the SDG represent a paradigm shift from traditional development assistance to a transformative agenda with epoch-making actions at the local, national, regional, and global level.

This is a pivotal moment for cities and oceans, subjects of dedicated goals among the 17 SDGs towards a shared future, during which the majority of wealth is increasingly produced in the developing world. The cities and oceans are also at the heart of the engaging Paris Agreement which provided a new impetus for the decades ahead. The universal landmarks for sustainable development brought a wave of responsibility towards addressing the challenges and the hopes that are on the rise. Once commitments were forged, new dynamics emerged from the rapid ratification of the Paris Agreement and the ongoing implementation of pledged action. But most authoritative thinkers believe that this is not enough... The road to sustainability will be an up-hill counterwave battle that asks for more radical decisive change.

This book argues that actions for sustainable cities and the global ocean, the most critical ecosystems, can make a huge difference for the future of humanity in the universal noble endeavour towards sustainability. It illumines the cutting-edge concepts and actions at the forefront by and for cities and the oceans, following the global engagements during the years 2015–2017. It proposes a journey around cities and seas and focuses on vigorous paradigm shifts to eco-design human systems and reconcile them with nature. The role of education, trust, and citizen empowerment can never be stressed sufficiently. Building on robust evidence, and the best transformational cases, this book tries to bring all messages together and provide structured advice for world leaders who responsibly and accountably strive for a better future.

* * *

The entry into force of the Paris Agreement on 4 November 2016, after its adoption by 194 countries and the EU, signalled the beginning of a new era for cities responsible for three fourths of the emissions. HABITAT III, the first high-level UN Conference following the adoption of the SDG 2030 by 193 countries, had taken place less than 1 month earlier, and the UN Ocean Conference "Our Ocean, Our Future," the first-ever conference on the implementation of a SDG, took place just a few months later. The years 2016–2017 have been critical for the future of cities and the seas, more than ever impacted by decisive economic, social, and environmental forces, and technological developments, in an uncertain, hyperconnected, and interdependent world.

The SDGs defied outdated linear economic, social, environmental, and political models. As several essential boundaries were irreversibly crossed, geostrategic resources came under reinforced stress, not least the nexus "water–food–energy." Biodiversity and ecosystems face hardened threats. Cities and the global ocean, intertwined with climate in ways still to be explored and understood, seriously suffer from pollution, unsustainable production and consumption, and forceful climate-related effects. However, the energy transition is irreversible, and cities are steadily heading towards the post-fossil age. Citizens have become more aware that responsible global governance needs to be more active on inclusive, creative communities and healthy ecosystems, especially the oceans. A dynamic balance between short-term action to address urgent extreme weather events and longer-term endeavours for sustainable development has to be carefully struck at all levels.

Cities tend to reinforce all dynamics that are boosted by physical proximity, from resource and energy efficiency to education and health service provision, but also pandemics and terrorism. By 2050, they could host 6.5 to 7 billion people, two thirds of all humanity, which means an additional urban population equal to the present populations of China and India combined. The ongoing rapid population attraction and growth of cities in the developing world is bolstering megacities, most of which lack adequate infrastructure and must still overcome unacceptable poverty and environmental deterioration. Sustainable cities strive to ensure access to safe and affordable housing and sanitation for all and to invest in public transport, green spaces, improved urban planning, and participatory, inclusive governance.

The global ocean is a major climate regulator and the largest carbon sink. The ocean has absorbed about one third of the carbon dioxide released by human activities since the beginning of the industrial revolution, mitigating climate change, but causing a 26% rise in ocean acidification. The sea of seas has intensely changed since Ferdinand Magellan undertook the first world circumnavigation in the sixteenth century. The ocean has become a connecting element among nations and people, and the human impact has dramatically increased. Marine and coastal biodiversity, on which more than 3 billion people depend for their livelihoods, has deteriorated dangerously during the past 40 years. Almost one third of our precious fisheries has been hugely harvested at levels above regeneration limits. The World Bank has highlighted that fully fished, overfished, or depleted stocks, or fisheries recovering from overfishing, increased from a little more than 60% in the mid-1970s to almost 90% in 2013; the sunken billions increased from $3 billion to $86 billion a year. Furthermore, marine pollution, an overwhelming majority of which comes from land-based sources and especially cities, is reaching unacceptable levels.

The scarcity of strategic resources, especially water, food, and energy, suffers from multiple inefficiencies that stem from outdated systems which produce unbearable amounts of waste, pollution, and emissions. Almost one third of the food or energy produced is wasted. Food that is not eaten requires almost 1.4 billion hectares of land, close to 30% of the world's agricultural land area, to produce. If food waste production were a country, it would be the third worst world emitter after China and the US. Humanity has to invent and thoroughly embrace better models to continue to thrive within the boundaries of the finite blue planet. Extreme events associated with climate change have reminded us of the fragility of the Earth and the seriousness of the turbulences. Cities based on proximity and interactions provide the ground to boost efficiencies and redesign production and consumption systems. Resilience is far more a dynamic process to be continuously steered than a state to be achieved.

The questions in the heart of the matter include these: What could the future of cities and the oceans look like, also in the context of action to implement the interconnected universal Sustainable Development Goals and especially the SDGs 11 and 14, and their precious links with all SDGs? What decisive steps, successes, threats, and tensions could shape trajectories towards the Sustainable Development Goals for 2030? What opportunities should not be missed to optimise prospects? How can the impact of advancing climate change and the depletion of strategic resources best be addressed? How can cities and the seas become more future proof? How can precious urban and ocean resources be co-optimised as valuable assets while being used sustainably as shared common goods? What knowledge, technology, and innovations can best support the burgeoning blue economy and the services provided by urban and marine ecosystems? How can world cities best grasp opportunities and manage risks, while engaging with their stakeholders and the world and promoting ocean sustainability? Which multi-win and multi-stakeholder alliances would likely bring the highest benefits? And, finally, what messages can be most enlightening for global leaders at all levels to decisively contribute to healthier seas and cities in a changing world?

* * *

What are the prospects for cities and the seas on this ageing, increasingly complex, uncertain, and interlinked urbanizing blue planet in the era of Anthropocene? The first chapter examines the megatrends that can impact the future of cities and seas as vibrant and vital dynamic places of changes and exchanges, in the aftermath of HABITAT III, the entry into force of the Paris Agreement, and the "Our Oceans, Our Future" high-level UN Conference. HABITAT III (Quito, October 2016), which convened for only the third time in 40 years, gave cities a dignified central place in the 2030 Sustainable Development Agenda. The Ocean Conference (New York, June 2017) concluded with a Call for Action for the implementation of the SDG14, and the UN declared in December 2017 the decade of ocean science for sustainable development to be 2021–2030. The interconnected human and physical ecosystems can and must have a major role in creating and distributing wealth and well-being in alliance with an inclusive, regenerated blue planet. Strong sustainability asks not only for the total capital, made from natural and physical, human, social, cultural and political, financial, built, and digital capital, to be preserved, but for each of its components to be strengthened and transmitted to future generations. If any of the precious forms of capital is unsustainably managed, benefits erode and progress towards achieving the SDGs is undermined.

Many interacting demographic, socioeconomic, and environmental trends, opportunities and risks, and political developments and technological breakthroughs impact the achievement of not only SGD11 and SDG14, focusing on cities and oceans, but in interdependence and cross-implementation with all other SDGs. This part presents the array of possible futures which invite to action for better lives to all citizens and powerful benefits across the globe. It highlights the role of the oceans, the "other 70%," the common heritage of humankind and immense but finite, over-exploited yet underexplored ecosystem, and clarifies their connections with cities and the responsibility to co-optimise eco-solutions across all parts of the trilemma, economy, environment, and society, at times of polycrises.

Humanity is already using the resources of the equivalent of almost 1.7 planets "like the Earth" to provide the goods and services needed each year, and this trend cannot lead to sustainable futures. In 2017, Overshoot Day was 2 August. In just 7 months the global community consumed all the resources that the planet can sustainably regenerate during an entire year. Authoritative studies highlight that the prevailing economic models led to overconsumption and resulted in disastrous effects for the environment and also the economy and society, if examined from a longer-term perspective, and undermined the future of humankind.

Cities and seas are vital ecosystems that can mobilise scarce resources, nutrients, and materials, ensure food security, and offer sustainable goods and services. But the fundamental resources of air, water, and soil face extreme tensions in many cities, and the global ocean suffers from pollution that often originates in cities. Urban air pollution continues to rise at an alarming rate and has become a critical risk for human health. Overfishing, pollution, and waste, especially plastics, put ocean health at risk. The essential nexus "water–food–energy," to which many add health, is under particular stress in cities, but oceans can extend possibilities for food, nutrition, and energy capacity (FAO 2013, 2016).

Resilience, the ability to rebound, is critical for cities, major concentrators of resources, food, materials, and products supporting many diverse human activities into a condensed space of intense interactions. Eco-responsible cities try to reduce their ecological debts on land and the sea and increase biocapacity, their ability to renew their assets. Some exemplary analyses and models of emission-neutral, waste-free, and ecological functional areas provide inspiration for reconciling the technosphere with the biosphere. They also highlight that communities and stakeholders have a to play cardinal role for the exploration, exploitation, and enhancement of their precious urban capital and the global commons, especially the oceans.

The third chapter argues that the battle for climate change will be won or lost in cities and the seas. Oceans and cities are the ecosystems that most affect, and are most affected by, climate. The oceans have a vital role in regulating the global climate system, generating oxygen and absorbing carbon dioxide; cities constitute the largest emitters and potentially most efficient energy users, but also the highest concentrated human resolve to change the status quo. Sea level rise, ocean warming, and acidification may have disastrous effects on the global ocean and thus on the future of the planet.

Cities impact climate not only because of their sheer size but also because of their democratic traditions and their global networks, which can play a great role for climate justice, and their power to disseminate influential patterns to the world. Climate change, subsidence, and sea level rise are important risks especially for low-elevated densely populated urban areas, but eco-responsible city developments can prevent disaster. Riverside, coastal, and deltaic cities are to bear the worst effects of the extreme events linked to climate change. Low-lying developing megacities are extremely vulnerable, but developed cities and assets are also at risk, and should share knowledge and technology to disseminate the best approaches.

Eco-responsible cities have to ensure access for all, at all times, to clean, secure, and affordable energy. Investments in energy efficiency and sustainably managed renewable energy resources can only have beneficial effects. Renewables, the central pillar of the energy transition, have overcome most cities of the world. In 2015, renewables globally surpassed coal to represent the largest source of installed power capacity, and then continuously broke records. Still, this change is not sufficient to cap temperature increase to less than 2 °C. Global emissions from energy reached their peak also in 2015, and the world inexorably entered an era of lower energy intensity. Offshore wind energy generation and ocean blue–green energy, including tidal and wave energy, can greatly extend the possibilities in the scene of the future.

Sustainable mobility and accessibility are fundamental for an equitable society and for linking the first and the last mile of all urban journeys to terrestrial and maritime networks and the broader world through integrated, smarter, and cleaner transport. The fourth chapter offers an insight into the evolving energy production and consumption patterns and the concerted efforts to raise the bar of sustainability excellence and overcome national and supranational energy targets. Intelligent mobility and connected and automated driving will be essential for the shift towards post-carbon cities, making the most of digital technologies and speeding up the deployment of zero-emission alternatives. Automated eco-ships may also be a near-future reality.

Cities are places of cross-pollination on a globalising chessboard, the very sites where many diverse economic flows interact and can be converted and intensified to better satisfy the expectations of citizens, businesses. and organisations. Brainports and well-functioning harbours can reinforce both local and global value chains and generate sustainable wealth. Their competitiveness depends on the strength of their functions, their links and networks, and their multiplier effects. Sustainable investments can help reconcile cities and oceans with people and the planet.

The evolving ability of cities to introduce new concepts and models and orchestrate the world economy is at the very heart of the fifth chapter, which also sheds light on the burgeoning ocean economy and its prospects for 2030. Generating and capturing synergies and bringing forth sustainability principles are essential for creating better assets, often from liabilities, and services that truly fulfil citizen needs. Eco-responsible blue–green businesses may provide a great contribution towards promoting an increasingly zero-waste, post-carbon circular economy and creating sustainable values from ethical citizen and enterprise values.

Fairness constitutes the ultimate ethos of the social architecture of sustainable cities that strive to offer citizens a fulfilling life that is worthy to live. Eco-responsible cities care much about the distribution of sustainable wealth across urban actors, places, and time. Interactions among diverse people and activities are invaluable assets for cities, schools for respecting differences and living harmoniously together. The sixth chapter focuses on the social capital of sustainable cities and their capacity for intergenerational and intercultural equity, public health, safety, and solidarity. Cities are themselves important direct and indirect sources of investment, education, and employment, the first factor of social integration, followed by housing. They can also give an impulse for local to global initiatives on land and in the sea. As migration rises on the global horizon, reinforcing inclusiveness is of increasing importance in cities that feel part of a universal community. No citizens should be left outside or behind; they should all have access to the conditions to enjoy a healthy, meaningful, and rewarding life, in harmony with the world commons.

Cities are extraordinary intellectual and cultural theatres at the crossroads of very diverse journeys, individual and collective, artistic and cultural, learning and creative, local and global. Port cities are reputed to be more open to innovation and trade, ideas, and exchanges. The urban and ocean heritage, including also the underwater cultural heritage, are exceptional forms of human expression. The sea has often inspired scientists and citizens to invest ideas and energy into new discoveries, entrepreneurial action, and cultural creation. Ocean-literate cities have a leading role in reinforcing the bond between cities and the blue planet.

A sustainable city has to cultivate the seeds of freedom for science and the arts to bloom and offer a space to all wishing to express, communicate, and interact. The seventh chapter presents a spectrum of actions to reinforce urban intellectual and cultural capital and make citizens proud of their immediate environment but also of the faraway ocean ecosystems that inspire them and ask them to act as responsible citizens. It also explores the principles of ecosystem-based strategic planning for better eco-designed policies in balance with the planet, on land, the coasts, and in the seas.

A shared vision for the future is the starting mobilising stage for transformational change. Strategic, holistic, transparent, ecosystem-based planning and management should strive to help realise the vision. Pathways to sustainability have to review many spatiotemporal patterns, both urban and marine. Time is a scarce and precious resource, which extends the limits of space and can serve as the litmus test of the well-being of societies. Urban and ocean observatories and indicators should take the pulse of local life and provide compasses in the journey towards sustainable development. Leadership should be science bound and citizen driven, striving for communities of best practice, continuously lifting the threshold of excellence and engagement.

Cities withstood time by promoting local democracies, which are reinvented every day. Democracy implies the power of citizens and the risk of entrusting citizens with the governance of all. As new models of deliberative democracy emerge, the eighth chapter clarifies the importance of principled governance, leadership at all levels, and citizenship in cities and links with oceans and the world. Citizen education, empowerment, and participation are cornerstones in democratic policy making. In an increasingly digital world, open data cities multiply, often in alliance with citizen scientists, offering invaluable insights for enlightened policy making, including that for the global ocean. Transparent institutional and energising partnerships, enriched with nudge interventions and quintuple helices, can maximize the potential of synergies and serve as catalysts of transformational change.

Coalitions and cross-networks allow overcoming national frontiers and the horizon. Interconnected alliances of democratic cities can lead to a global solidarity bond involving networks of cities and seas, with a major role in a world most noble partnership of the future. They can offer new democratic spaces between the micro-regulations of local communities and the macro-regulations of global governance.

Ideas are a vigorous global currency and matter most for challenges which are too big or too complex and have to be collectively addressed. Cities and the global ocean hold many hopes but require empowering bold actions by leaders able to understand and influence the interdependency of systems, build consensus, and mobilise stakeholders. What would be the most incisive insights for leaders at all governance levels to become champions of change? What are the priorities for cross-sector, cross-level, and cross-stakeholder engagement at a moment when humanity is on the verge of profound changes?

The final chapter brings together key insights and messages for the committed to lead change towards a radically better future. Purposeful cross-implementation of SDG11 and SDG14 can lead to multiple benefits at all levels. Sustainability asks for local, regional, national, and international actions to be fully coherent and aligned and bring a higher multiplier effect and impact. If all cities were adopting and protecting vulnerable marine ecosystems or protected areas, the overall effect could be greatly amplified. Significant data and insights are needed to address systemic stalemates so that obstacles are overcome, lofty opportunities are co-created, and new social contracts are sealed.

* * *

Cities and the ocean are extraordinary public goods. This book intends to offer an evidence-based, holistic, and integrated view of key urban and ocean sustainability issues at the horizon of 2030 and of 2050. All chapters trace the panorama of the most prominent issues at the heart of the matter and highlight systemic multi-stakeholder eco-solutions. The book tries to harvest the knowledge stemming from the evolving forefront of innovative responses and to propose key interactions stimulating new debates and partnerships on the oceans–cities nexus. Conceived as a starting point for greater engagement with the latest developments, it is illustrated with selected lighthouse projects which, if disseminated and scaled up, can make a big difference in addressing simultaneously all sustainability summits and dimensions of the 2030 SDGs agenda.

Exemplary pioneer initiatives can stimulate noble emulation and serve as torchbearers of actions for others to design for a better environment, a better society, and a better economy, optimised through higher citizen engagement. A concert of identified inspiring world cases and paradigm shifts, included in the chapters and reassembled in the Annex, provides a prelude for dialogue and exchange. The suggested references and websites offer readers the opportunity to go deeper into specific issues at stake and lead to unlimited ideas and innovations for the wonderful shared journey and meaningful progress for people and the Blue Planet.

If my previous book with Springer focused on coastal cities, the most intense ecosystems at the intersections of land, sea, air, and humans, this book invites going beyond the line of the horizon and raising the bar of excellence above the interconnected global commons. Trust, resilience, and inclusiveness may have never been so important as in the post-truth era in this closer-knit world, regulated by the force of oceans and powered by the will of humans.

Watercolour 1: Stockholm, first European Green Capital 2010

Brussels, Belgium
International Futures Day, 1 March 2018

Voula Mega

Acknowledgements

This book has been an extraordinary journey into cities and the ocean of creativity, and the world of the future that we all desire to be so much better. My heartfelt gratitude goes to all those who enhanced my ability for undertaking this journey with endless curiosity and passion, and all those who inspired, nurtured, and accompanied this wonderful navigation.

I also wish to genuinely acknowledge all those who taught me intellectual humility and courage and all those that made me wiser with meaningful and purposeful suggestions for what truly matters.

Happy, proud, and grateful for this fifth collaboration with Springer, I wish to express my warmest thanks to my editor and my known and unknown reviewers; it is immensely heartening to find such enthusiasm, knowledge, and commitment.

Watercolour 2: San Francisco, best positioned to attract global talent (Kearney 2017)

Last but not least, my special thanks go to the European Commission which granted me the authorisation to publish this book.

My overriding hope is that citizens, scholars, and decision makers will draw inspiration and insight from the heightening visions, mobilising concepts, and lighthouse actions that were identified, shaped, proposed, or thoroughly explored here, and will see cities and the seas through a different eye. The aim is that we all, the urban and ocean people, become more responsible citizens of cities and the seas.

Contents

Acronyms and Abbreviations

4IR	Fourth Industrial Revolution
100RC	100 Resilient Cities
ABNJ	Areas Beyond National Jurisdiction (Area)
AR5	Fifth Assessment Report (IPCC)
C40	Network of the World's Major Cities
CBD	Convention on Biological Diversity (UN)
CCS	Carbon Capture and Storage
CDP	Carbon Disclosure Programme
CEMR	Council of European Municipalities and Regions
CEO	Chief Executive Officer
CERES	Coalition of Environmentally Responsible Economies and Societies
CHP	Combined Heat and Power
CLIA	Cruise Lines International Association
CO	Carbon monoxide
CO_2	Carbon dioxide
COP	Conference of the Parties (UNFCCC), COP21 (Paris 2015), COP22 (Marrakech, 2016), COP23 (chaired by Fiji and hosted by the UNFCCC at Bonn, 2017)
CoR	Committee of the Regions
CSO	Civil Society Organisation
CSR	Corporate Social Responsibility
EBRD	European bank for Reconstruction and Development
ECA	European Council of Architects
ECF	European Cyclists' Federation
EDEN	European Destinations of Excellence
EDF	Electricité de France
EEA	European Environment Agency
EEZ	Exclusive Economic Zone
EFTA	European Free Trade Association (Iceland, Liechtenstein, Norway, Switzerland)
EFUS	European Forum for Urban Safety

EIU	Economist Intelligence Unit
EPA	Environmental Protection Agency (US)
EPR	Extended Producer Responsibility
ESI	Environmental Ship Index
EU or EU28	European Union (28 Member States since 1.7.2013: Austria, Belgium, Bulgaria, Croatia, Cyprus, Czech Republic, Denmark, Estonia, Finland, France, Germany, Greece, Hungary, Ireland, Italy, Latvia, Lithuania, Luxembourg, Malta, Netherlands, Poland, Portugal, Romania, Slovakia, Slovenia, Spain, Sweden, and the United Kingdom)
FAO	Food and Agriculture Organization (UN)
G7 countries	Canada, France, Germany, Italy, Japan, United Kingdom, and United States; EU always present
G8	The foregoing plus the Russian Federation
G20 countries	Argentina, Australia, Brazil, Canada, China, France, Germany, India, Indonesia, Italy, Japan, Mexico, Russia, Saudi Arabia, South Africa, South Korea, Turkey, United Kingdom, United States, and the European Union, represented by the Council, the Commission, and the European Central Bank
GCCA+	Global Climate Change Alliance Plus
GDP	Gross Domestic Product
GEF	Global Environmental Facility
GEO-6	Sixth Global Environmental Outlook
GHG	Greenhouse Gas [carbon dioxide (CO_2), methane (CH_4), nitrous oxide (N_2O), and the three main fluorinated gases)
GGBP	Greener, Greater Buildings Plan (New York)
GRI	Global Reporting Initiative
GOC	Global Ocean Commission
GPO	Global Partnership for Oceans (World Bank)
GVA	Gross Value Added
HDI	Human Development Index
IUUF	Illegal, Unreported, and Unregulated Fishing
IAPH	International Association of Ports and Harbours
ICAM	Integrated Coastal Area Management
ICAP	International Carbon Action Partnership
ICLEI	International Council for Local Environmental Initiatives
ICPC	International Crime Prevention Centre
ICT	Information and Communications Technologies
IDDRI	Institute for Sustainable Development and International Relations
IEA	International Energy Agency
ILO	International Labour Organization
IMF	International Monetary Fund
IMO	International Maritime Organization
IOC	Intergovernmental Oceanographic Commission (UNESCO)

IOM	International Organization for Migration
IoT	Internet of Things
IPBES	International Panel for Biodiversity and Ecosystem Services
IPCC	Intergovernmental Panel on Climate Change
ISA	International Seabed Authority
ISWA	International Solid Waste Association
ISOCARP	International Association of City and Regional Planners
ITF	International Transport Forum
IUA	International Union of Architects
IUCN	International Union for Conservation of Nature
IUU	Illegal, Unreported and Unregulated (fishing)
LDC	Least Developed Country
LED	Light-emitting Diode
LEED	Leadership in Energy and Environmental Design
LNG	Liquefied Natural Gas
MARPOL	International Convention for the Prevention of Pollution from Ships
MedPAN	Network of Marine Protected Areas in the Mediterranean
MIT	Massachusetts Institute of Technology
MPA	Marine Protected Area
MSC	Marine Stewardship Council
MSP	Marine Spatial Planning
NGO	Non-Governmental Organisation
NOAA	National Oceanic and Atmospheric Administration (USA)
NDC	Nationally Determined Contribution
NPCC	New York City Panel on Climate Change
NYC	New York City
NO_x	Nitrogen oxide
OECD	Organisation for Economic Co-operation and Development
OPRC	International Convention on Oil Pollution Preparedness, Response and Co-operation
OWHC	Organization of the World Heritage Cities
PM10 and PM2.5	Small and fine particulate matter
PSMA	Port State Measures Agreement
PV	Photovoltaics
SDG	Sustainable Development Goal
SEM	Southern and Eastern Mediterranean
SIDS	Small Island Developing States (21 in the Atlantic, 20 in the Pacific, and 9 in Caribbean, Africa, Indian Ocean, Mediterranean, and South China Sea)
SME	Small and medium-sized enterprise
SO_2	Sulfur dioxide
SOA	Sustainable Oceans Alliance
STEM	Science, technology, engineering, and mathematics

STOA	Science and Technology Options Assessment (European Parliament)
UCLG	United Cities and Local Governments
UfM	Union for the Mediterranean
UITP	Union Internationale des Transports Publics
UNCED	UN Conference on Environment and Development (Rio de Janeiro, 1992)
UNCLOS	UN Convention on the Law of the Sea
UNDP	UN Development Programme
UNEP	UN Environment Programme
UNESCO	UN Educational, Scientific and Cultural Organisation
UNFCCC	UN Framework Convention on Climate Change
UN-HABITAT	United Nations Human Settlements Programme
UNISDR	UN Office for Disaster Risk Reduction
UNPF	UN Population Fund
UNSDSN	UN Sustainable Development Solutions Network
UNWTO	UN World Tourism Organisation
WBCSD	World Business Council for Sustainable Development
WCCD	World Council on City Data
WCS	World Cities Summit
WEF	World Economic Forum
WFC	World Future Council
WHO	World Health Organisation
WMU	World Maritime University
WMO	World Meteorological Organisation
WOC	World Ocean Council
WPCI	World Port Climate Initiative
WRI	World Resources Institute
WTO	World Trade Organisation
WWF	World Wild Fund for Nature
Y4PT	Youth for Public Transport
ZEC	Zero-emission cities

List of Watercolours

Watercolour 3
New York, the City "Where the Future Comes to Rehearse" (M. Bloomberg)

Chapter 1
Planet Ocean, a World of Cities: A Journey Around the Shores of the Anthropocene

Abstract Oceans and cities are the most extraordinary public goods on this ageing, increasingly complex, uncertain and interlinked urbanising blue planet in the Anthropocene era. This chapter examines the megatrends which can impact the future of cities and seas as vibrant and vital dynamic places of changes and exchanges, in the aftermath of HABITAT III, the entry into force of the Paris Agreement, and the "Our Oceans, Our Future" high-level United Nations (UN) Conference. HABITAT III (Quito, October 2016), which convened for only the third time in 40 years, gave cities a dignified central place in the 2030 Sustainable Development Agenda, whereas the UN high-level conference "Our Oceans, Our Future" (New York, 5–9 June 2017) concluded with a Call for Action for the implementation of the SDG14, and the UN declared that 2021–2030 will be the decade of ocean science for sustainable development. The interconnected human ecosystems have had, and can continue to have, a major role in creating and distributing wealth and well-being in alliance with an inclusive regenerated blue planet. Strong sustainability asks not only for the total capital, composed of natural and physical, human, social, cultural and political, financial, built and digital capital, to be preserved, but for each of its components to be strengthened and transmitted to future generations. If any of the treasurable forms of capital is unsustainably managed, benefits are eroded and progress towards achieving the sustainable development goals (SDGs) is undermined.

This first part sets the global scene and reviews demographic, socioeconomic, and environmental trends, opportunities and risks, political developments and technological breakthroughs towards realising the vision of not only SGD11 and SDG14, focusing on cities and oceans, but in interdependence and cross-implementation with all other SDGs. It presents the array of possible futures that invite cities to action for better lives for all citizens and the oceans, the "other 70%," the common heritage of humankind and the immense but finite, overexploited yet underexplored ecosystem. It illuminates the 'cities–oceans' nexus and the responsibility to co-optimise eco-solutions across all parts of the trilemma of economy, environment, and the society, at times of polycrises.

V. P. Mega, *Eco-Responsible Cities and the Global Ocean*,
https://doi.org/10.1007/978-3-319-93680-2_1

1.1 Cities and the Seas: Megafacts, Trends, and Drivers of Change

1.1.1 On the Ageing, Increasingly Complex and Diverse, Uncertain and Interlinked, Urbanising Blue Planet

The impact of humankind on the planet has never been so intense. The twenty-first century brought a wind of acceleration and waves of manifold tensions on shrinking space and resources. The facts speak for themselves. In 2016, 54.5% of humanity lived in cities, which by 2050 could host 6.5–7 billion people, two thirds of humanity, which means an additional urban population equal to the combined population of China and India together. Almost all the urban expansion in the next decades will take place in the cities of the developing world, which promise many improvements for health and education but also may bring more slum populations and exacerbate social inequalities. Cities occupy just 3% of the Earth's surface, but account for 60–80% of energy consumption and 75% of carbon emissions. Rapid urbanisation is exerting strong pressure on fundamental resources, infrastructures, institutions, and public services. On the other hand, the high density of cities and the cross-pollination of exchanges can stimulate technological, social, and political innovation and progress (UN 2015a, b, 2016a, b).

Humanity is already using resources equivalent to those of 1.7 planets to provide the goods and services needed each year, and this trend cannot lead to sustainable futures on an increasingly multi-speed planet. The most critical world challenges also include inequality and instability and the global democratic governance deficit. Increasing inequality, of opportunity or of income, is a critical economic, social, and political challenge, and its significance has been amplified by the decline in public confidence and the insubstantial recovery from the 2008 crisis (Attali 2016; OECD 2017b).

The world's cities are growing in both number and magnitude. Cities increasingly face complex, interconnected, and rapidly evolving global challenges, such as climate change, migration flows, transnational governance, financial volatility, social inequalities, and environmental deterioration. Some of these problems have been aggravated as many cities have been hit hard by the crisis, in particular by unemployment and notably youth unemployment and social exclusion. Restoring trust, the highest form of human motivation, addressing diverse challenges, and seizing opportunities requires courage, ingenuity, and innovation, whether in education and research, or in investment and governance (EC 2016a).

Demography is a capital force. In 2017, the world population reached 7.6 billion. Humanity took more than a century to expand from 1 billion in 1804 to 2 billion in 1927 and more than 30 years to reach 3 billion in 1960. Fifteen years later, the world population reached 4 billion in 1975 and by 1987 and 1999 had gained 1 and 2 additional billion persons, respectively. In 2011, the world population hit the watershed of 7 billion people and then continued increasing at a slower pace. Annual growth rate reached its peak, of more than 2%, during the 1960s, and then started declining, but 83 million people are still added to the world population each year. The global landscape of demographic change seems very diverse. The population of developed regions is expected to slowly stabilise. Population is decreasing in Japan, and in Europe the demographic decline started in 2015. In contrast, the Least Developed Countries (LDC), the most rural and most rapidly urbanising, are projected to double in size from a population of around 900 million in 2013 to 1.8 billion in 2050 (UN 2015a, 2017a).

Continuing urbanisation, accompanied by lower fertility rates, imposes a brake on the global population explosion, which is expected to reach its apogee in the twenty-first century. Global population will have quadrupled during the period 1950 to 2100. The 2017 UN Prospects medium variant projections foresee a global population of 9.7 billion people in 2050 and more than 10 billion by the end of this century. Life expectancy is rising in both developed and developing countries. At the global level, it advanced from 52 years in 1960 to 71 years in 2015 and is projected to reach 76 years around 2045–2050 and 82 years in 2095–2100. By the end of this century, average life expectancy could be around 89 and 81 years in developed and developing countries, respectively. Healthy life-years will be the new fundamental measure for the well-being of societies (Kohli 2016).

Although the average number of children per woman has fallen quickly in large developing countries such as China, India, Indonesia, Brazil, and South Africa, rapid growth is expected to continue during the next decades in countries with high levels of fertility, such as Nigeria, Niger, the Democratic Republic of Congo, Ethiopia, Uganda, and Afghanistan. Nigeria, the country growing most rapidly in population, is expected to become the third largest country in the world shortly before 2050 (UN 2015a, 2017a).

1.1.2 The Ocean: The Contiguous Transboundary Body of Water, Earth's Ultimate Frontier

The oceans, the other 70% (UNEP 2011), form the largest physical ecosystem, including the Pacific, the Atlantic, the Indian, the Arctic, and the Southern Oceans. The ocean covers three quarters of the Earth's surface, contains more than 97% of the water on Earth, and represents 99% of the living space on the planet by volume. Only 10% of the ocean resources are known: there are nearly 200,000 identified species, but the actual numbers may lie in the millions. More than three billion people depend on marine and coastal biodiversity for their livelihoods in developing countries. Oceans constitute the kidneys of Earth, a great biological pump at the heart of water and nutrient cycles and global atmospheric and thermal regulation. But to perform these functions, the oceans must be healthy (UN 2016c).

The market value of marine and coastal resources and industries per year is about 5% of the global gross domestic product (GDP), but the oceans provide many ecosystem services free of charge. The oceans have absorbed about 30% of carbon dioxide produced by humans since the beginning of the industrial revolution, buffering the impacts of global warming. Sea rise, ocean acidification, and sea warming are the main effects of climate change on the global ocean, which also suffers from insufficient protection, unsustainable fisheries, and marine pollution (UN 2016a, b). The rampant effects of climate change require a 'whole ocean' strategy (IPCC 2014a).

The proximity of the sea offered humanity opportunities for fishing, which helped ensure a steady food supply. Coastal areas concentrate the most nutrients of all marine environments. Sunlight can penetrate the shallow waters above the continental shelves, which slope gently from the coastline of the continents to a depth of around 200 m and support complex marine ecosystems. Such ecosystems serve as nurseries for much of the biodiversity of the global ocean and protect coastal land from erosion caused by storm surges. Among coastal activities, harbours link local economies to regional and global markets. Maritime transport has provided vital economic links throughout history and made ports some of the most dynamic and visionary places in the world (Attali 2017; Port of Rotterdam 2012).

The ocean economy is an increasingly vibrant and vital part of the global economy, evaluated at $1.5 trillion in 2010, although marine ecosystem services are hugely underestimated. The global added values of ocean-based industries and ecosystem services are set to double by 2030 and provide 40 million direct jobs. As sea trade has increased much during the past decades and is expected to grow, port cities are likely to gain even more economic importance in the future (OECD 2013, 2016a).

All further ocean advances should be in line with sustainable development because the ocean is already under severe pressure. The oceans may be vast, but they are also fragile and vulnerable to pollution, overfishing, and climate change, which makes them warmer and more acidic, thus placing coral reefs, fish populations, and other marine life at risk. Nearly one third of the corals in Australia's Great Barrier Reef have disappeared. The rise in sea levels risks wiping entire ocean countries in the Pacific off the map. One in ten people live in the low-lying, most vulnerable coastal regions. Risks increase for the rich ocean diversity of plants, animals, and microbes, from the largest animals on the planet, the blue whales, to plankton and bacteria. The aesthetic, cultural, religious, and spiritual ecosystem services provided by the oceans are also threatened. Much marine pollution has come from cities that often overexploited the seas as a source of food and misused them as locations for waste disposal (UN 2016c).

Healthy oceans can greatly contribute to an inclusive sustainable future for small island developing states (SIDS) and coastal LDCs, as they are the fundamental base upon which the economies and identities of many of them are built, and they are also central to their contribution to the 2030 Agenda for Sustainable Development. A sustainable blue economy provides ocean countries with a basis to pursue a resource-conscious path to development while enhancing livelihoods and reducing poverty (World Bank and UN Department of Economic and Social Affairs 2017).

1.1.3 The Largest Physical and the Smartest, Only Human, Distributed Ecosystem, Critical for Sustainable Development

Cities and seas are subjects of dedicated Sustainable Development Goals (SDGs), the SDG11 and the SDG14, respectively, of the 2030 Sustainable Development Agenda, which initiated a new common responsibility era for humanity. Although nonbinding, the universal Sustainable Development Goals present major improvements on the Millennium Development Goals. They are universal, interlinked, and address key systemic barriers to sustainable development such as inequality, unsustainable consumption, weak institutional capacity, and environmental degradation. The first reports on their implementation highlight progress but also significant problems including methods for measuring success. Progress is uneven and has to be accelerated (UN 2016a, 2017b).

"If anything ought to be too big to fail, it is the ocean" stated a high-profile article entitled "The Ocean sustains humanity. Humanity treats it with contempt" (The Economist 2017) just before the opening of the UN high-level conference "Our Oceans, Our Future" (New York, 5–9 June 2017). The conference built on existing successful partnerships and stimulated new ones, bringing together all relevant stakeholders to assess challenges and opportunities. Four months later, the 4th edition of Our Ocean conference series, hosted for the first time from the European Union (EU), attracted €6 billion in the value of impressive commitments and assisted the generation of unique new partnerships between states, civil society, international organisations, and the private sector.

Cities and seas are public goods, enjoyed and shared by citizens in their pursuit of happiness. Cities are human nests and havens, places of social interactions, theatres of civilization, schools of abilities and values, and temples of culture and democracy. They have been defined as the human establishments that generate their economic growth through the "disordered order" of human interactions. A sustainable city is, first and foremost, a city which carries the seeds of progress in balance with nature, and sustains its ability to reinvent itself and grasp co-created opportunities for a brighter future for the present and the next generations. Innovation is a sine qua non condition for continuously reinventing, orchestrating, and optimising activities and stimulating synergies (Mega 2015).

Diversity is a decisive resource for cities, traditionally seen as propellers of innovation. Cross-pollination is a strong urban asset, closely linked to the proximity of very diverse people and activities. This closeness leads to a broad array of activities, which strengthens the resilience of places to shocks that affect particular types of economies. Cities of the same size and population have very different trajectories because of their composition of population and the ways they interact with their environment and create new opportunities for action and wealth

Urbanisation is increasingly occurring in lower-income developing countries. Many of the countries that are experiencing exploding urban growth are the most rural and located in Sub-Saharan Africa. This growth represents a major risk for urbanising poverty, which could undermine all opportunities that urbanisation brings for humans, including education and health services. Although the poverty rate is falling globally, the increasing proportion of urban poor is very disturbing.

The most intense world cities are coastal and port cities, at the crossroads of land, ocean, and atmospheric ecosystems. Their economies usually link to maritime routes, and their societies have to address interlinked tensions related to climate change, sea level rise and storm surge, inequalities and exclusion, migration and intercultural dialogues, and access to education, training, and innovation (Mega 2015). The rise in sea level is considered to be bringing one of the most serious megachanges (West 2016).

1.1.4 The Century of Africa, and Also India and China

In 2024, India, the first world democracy, could overpass China to become the most populous country in the world (UN 2017a). China will continue to be the powerhouse of Asia, but the spectacular growth rates of previous years are declining. As growth accelerates in India and radical transformations take place in its urban and rural economic landscapes, significant opportunities may arise, especially for eight high-performing states along with four very high performing city-states (MGI 2014). The faster progress of China and India ("Chindia"), which accounted for less than one half of G7 economies in 2010, implies that their wealth will account for more than the GDP of G7 by around 2025 and more than 1.5 fold larger by 2060. Such changes in the sharing of world wealth will be matched by a convergence of prosperity across countries, which, however, will still leave significant gaps in living standards between advanced and developing economies.

The Asian population is expected to peak at more than 5 billion around the middle of the century and then start a slow decline. Because of their sheer size and still rapid growth, India and China influence the key macroeconomic variables that are important for developing countries, including interest rates, prices of raw materials, and wage levels. They also have major impacts on global trading flows and investment patterns. Strengthened ties between developing countries, "South–South links" in trade, aid, and investment, are an increasingly important source of openness, knowledge, and finance for development (OECD 2017a).

If the twentieth century was the century of Asia, the twenty-first century is considered to be the century of Africa, the theatre of more than half of the global demographic growth by 2050. In 2015, 60% of Africa's population were children and young adults under the age of 25, whereas the median age is 20, less than half the median age of Europe (41 years). Africa is set to be the young continent of the world throughout the twenty-first century. It will be the only continent to continue growing in population, mostly urban, after 2050. If the world gains 8 billion citizens over the period 1950–2100, Africa could claim more than 3.4 billion and Asia nearly 4 billion. By 2050, urban Africa could gain more than 800 million citizens, more than Europe's entire present population (UN 2015a; Camdessus 2017).

Africa is the fastest urbanising continent. Asia and Africa together will account for nearly 90% of all increase in global urban population by 2050. Cities helped Africa to move from economic stagnation to greater than 5% GDP average annual growth and to weather several economic crises fairly well. The continent is the home of some of the most rapidly growing economies in the world, which urbanise rapidly. The most explosive urbanisation of the most rural continent should bring quantum leaps for education and health provision services, because concentrated populations are much easier to target, train, and service. Africa seems a land of opportunity for both citizens and investors. As in Asia, this growth is leading to a burgeoning urban middle class, which creates new markets for goods and services and promises more political stability, citizen participation, and democratic accountability.

1.1.5 A Maelstrom of Economic, Social, Environmental, Cultural, Political, and Technological Trends and Risks

Megatrends are powerful drivers of change and may exercise a profound and lasting influence on many, if not most, human activities, processes, and perceptions. Interpretation of possible opportunities and threats must, however, be done with caution, as these provisions may materialise under certain technology conditions and policy hypotheses. As the second decade of this century wanes, a more polycentric world and the earth's economic centre of gravity moving east and southwards appear to be structural global trends, accompanied by a shift of power from states towards private and civic actors. The middle of the decade experienced an historic reversal of the share of developing countries accounting 52% of the global wealth, exactly the same as the share of the developed countries in 1981. According to the Organisation for Economic Co-operation and Development (OECD), the economic and financial crisis accelerated this longer-term structural transformation in the global economy. Growth rates in the 2000s increased significantly for many developing countries, their numbers beginning to converge strongly with the affluent OECD countries, leaping from 12% to 65%. Dominant trends include the slowing growth in China, increasing difficulty in accessing global financial markets, demographic transitions, rapid urbanisation, unemployment and inequalities, climate change, and conflict and security (Camdessus 2017; OECD 2017a).

Inequality tensions continue. The prosperity gaps between rich and poor countries are shrinking, but widening within developing economies and in developed countries, which could face an uphill battle to sustain living standards. The richest 10% of the OECD population earn nearly ten times the income of the poorest 90%, compared with sevenfold in the 1980s. Since 1990, the number of world citizens living on less than a dollar a day has fallen by more than one quarter, approximately 500 million citizens, mainly in China. Other countries have made progress but at a pace insufficient to counter the effect of the escalating populations. Inequality and poverty still represent major challenges for sustainable development. A significant number of those entities whose incomes have not been increasing are losing faith in the global economic order and are developing negative opinions about openness to the world and immigration (OECD 2017d).

Global trends may signal the beginning of a demographic divide between countries with young populations and those with ageing populations, which nurture migration flows. The UN defines migrants as citizens residing in a foreign country for more than 1 year. From refugees and asylum seekers to expatriates, migration covers an increasingly diverse range of mobile citizens and life itineraries. According to the International Organisation of Migration, the movement of people is likely to become even more significant in the future, as a result of asymmetrical globalisation, demographic dynamics, and economic imbalances. The number of international migrants globally reached 258 million in 2017, about 3% of the world population: this figure includes almost 20 million refugees. India has the largest diaspora. Around 5 million people migrated permanently to OECD countries in 2016, well above the previous peak level that was observed in 2007 before the economic crisis. Cities are the main magnet for migrants. It is mainly in cities that newcomers seek economic, social, and cultural opportunities for the realisation of their potential in a world of solidarity and dignity (IOM 2015; OECD 2017d; UN 2017c).

As global population rises and life expectation increases, the pressures intensify on the natural and physical capital and life support systems. These tensions pose serious challenges to resource, water, food, and energy security, because climate change and depletion of natural resources are combined with the loss of land to urbanisation and degradation of terrestrial and marine ecosystems. The expanding world population and economy could cause further pressure on strategic resources. Water demand is expected to increase by 55% globally between 2000 and 2050, mostly for agriculture but also for manufacturing, electricity, and residential use. Groundwater depletion is a major threat to agriculture and urban water supplies in several regions. Deterioration of the quality of surface water in many countries has led to increased eutrophication, biodiversity loss, and disease. Almost 4 billion citizens, 40% of the global population, are likely to live under severe water stress by 2060 and 1.6 billion at risk of floods. Soil degradation will affect the amount of land available for productive agriculture: around half of the world's agricultural land is already affected by degradation, and the loss of yield may be as high as 50% in some African countries by 2050.

Despite the boom of renewable energies, fossil fuel combustion still dominates energy development and account for two thirds of global CO_2 emissions. Agriculture is a major emitter of the more powerful greenhouse gases of methane and nitrous oxide. The energy transition seems inevitable, and the share of fossil fuels in the global energy mix is set to decline because of the greater use of renewables. Growth in global demand is expected to slow from more than 2% per year in the past two decades to 1% per year after 2025. The future of nuclear power is uncertain, and although its capacity is set to rise considerably, its share in global electricity generation is expected to increase by only 1–12%. China will account for almost half the expected growth, overtaking the United States (US) by 2030 to become the largest producer of nuclear-based electricity. Almost 200 operational reactors of the 434 operating in 2013 will have been decommissioned by 2040, mostly in Europe, the US, Russia, and Japan (OECD 2016b; IEA 2017a).

Looking forward, at least two thirds of the growth in global final energy demand to 2050 will come from cities in emerging and developing economies. These cities are expected to be endowed with a large proportion of all new buildings by 2050, equivalent to 40% of the world's current building stock. They will also account for the great majority of the increase in urban passenger travel globally. Without change in current policies, that increased demand for energy services would double energy-related CO_2 emissions. As many of the urban areas in emerging economies are not yet fully built, their planning and construction should avoid the carbon-intensive old models of many cities in developed economies, while still offering citizens access to modern services of the same quality. International collaboration will therefore remain critical to ensure that cities around the world can draw on best practices, for example, that of urban buildings generating their own clean electricity. By 2050, rooftop solar could meet one-third of the electricity demands of cities (IEA 2016a).

If Europe gave the impulse to the rise of renewable energies, cities seem to take the lead in the transition to a post-carbon world. The International Energy Agency suggested that urban areas account for up to two-thirds of the potential to cost-effectively reduce global carbon emissions. Long-term technology pathways to limit the global temperature increase to no more than 2 °C indicate, as the most cost-effective approach, the deployment of low-carbon options in cities, especially in emerging and developing economies (IEA 2016b, 2017a).

Digital services will be instrumental for such a revolution. The Internet has been a great transformer of lives, and mobile services have been integrated into the evolving lifestyles of many citizens. The global network has been proposed as the guardian of a free society. The proliferation of social networks has spurred public expression and communication, but meaningful interactions or democratic expressions are questioned. The next wave of most potent technologies include the Internet of Things (IoT), big data analytics, artificial intelligence, neurotechnologies, robotics, micro- and nanosatellites, nanomaterials, additive manufacturing, advanced energy storage, synthetic biology, and blockchain (STOA 2015; WEF 2015). The IoT will build on advances in sensors and their connectivity through high-performance computing. Blockchains allow world citizens who do not know each other to organise into a network and keep trusted records. A new digital divide is looming on the horizon, expressed in number of connected devices per capita (12 for North America and 8 for Europe compared to 2.9 for Latin America and 1.4 for Africa by 2019). Artificial intelligence could drive the development of quantum computing, and quantum computing then could further drive the development of artificial intelligence, which may radically reduce the need for the labour force by as much as 50% in two decades. The State of the Future 2017 report warns that, as this mutual acceleration could grow beyond human control and understanding, science and policy leaders should further explore how to anticipate and manage this risk (The Millennium Project 2017).

Future societies will be further impacted by expanding mainstreamed digitalisation. Smart cities and communities gain political attention as highly interconnected, well-performing human and technical systems. During the next 15 years, firms will become predominantly digitalised, enabling product design, manufacturing, and delivery processes to be highly customised, integrated, and efficient. The costs of equipment will continue to fall, whereas the rise of open source practices will create further opportunities (OECD 2016b).

In depicting scenarios for the future, the 2017 edition of the State of the Future report, prepared by the Millennium Project, presents a compelling overview of humanity's situation, challenges, and opportunities, its potential for the future, and actions and policies that could lead to desired futures. This report suggests that the world is expected to continue improving during the next 10 years. World citizens will live longer and become healthier, wealthier, better educated, more urban, and increasingly mobile and connected. Life expectancy has increased by 10 years during the past 20 years to reach 72 years in 2017. Child mortality rate has dropped about 50% since 1990. Progress with citizen freedom and empowerment, women in Parliament, and literacy and education levels have demonstrated that humanity can and has still to overcome itself. The Internet, the global nervous system of connectivity, is on the road to completion, with more than 52% having mobile phones and more than 50% having smartphones.

Although extreme poverty fell from 51% in 1981 to 13% in 2012 and to less than 10% in 2017, the concentration of wealth is increasing, income gaps are widening, and low-employment economic growth seems to be the new norm. Urban infrastructures may become too complex and costly to manage. The long-term effects of global warming could produce massive and continual migration waves. The global health divide poses major concerns. Infectious diseases affect the developing world disproportionately, and their treatment will be further compromised by growing antibacterial resistance. Noncommunicable and neurological diseases are projected to increase sharply in line with demographic ageing and the globalisation of lifestyles linked to obesity (The Millennium Project 2017).

The World Economic Forum suggests that the five most likely risks in the next 10 years could include extreme weather events, natural disasters, cyberattacks, data fraud or theft, and climate failure. And the five most impactful risks could encapsulate weapons of mass destruction, extreme weather events, natural disasters, climate failure, and water crises. Environmental risks are an area of particular concern. 2017 has also been among the three hottest years on record and the hottest year ever without an El Niño event. Rising temperatures are a major risk for agricultural systems, in particular, for monoculture production. Pollution is responsible for one in ten deaths globally because 90% of the world's population lives in areas with pollutant levels exceeding the World Health Organization (WHO) guidelines. The creation of more megacities will only further advance this trend (WEF 2018).

In 2017, CO_2 emissions rose for the first time in 4 years as higher temperatures led to increased energy consumption and extreme droughts exacerbated a switch from hydro- to coal-fired power generation. The oceans continue to become warmer, and their capacity to absorb CO_2 may be declining. Many influential reports conclude that it is not too late to build a more resilient tomorrow, but we need to act with a stronger sense of urgency (OECD 2012b; UNEP 2016; WEF 2018).

Access to education and skills will become increasingly important, and gender divide is expected to shrink: however, women still constitute two thirds of the world's 700 million illiterate citizens. Growth in female enrolment at all levels of education will continue, with important implications for labour markets and family life. The growing internationalisation of university education is expected to continue, with education at all levels becoming more widespread. This evolution could create lofty opportunities for sustainable development and for engaging with climate action, healthy oceans, and inclusive communities (WEF 2016b).

1.1.6 The Rise of Megacities and the Global South: The Future Front?

In 2016, 23% of the global population lived in a city with at least 1 million inhabitants, and this percentage is expected to increase in all regions. The world hosted 31 megacities, officially defined as those with more than ten million people. In 2015, 45 cities had a population between five and ten million, and 10 of these are expected to be megacities by 2030. Between 1990 and 2030, the number of megacities will quadruple. Tokyo is always the largest city, and despite a projected decline is expected to remain the first in 2030, followed by Delhi. Twenty-four of the megacities are located in the global south. China alone hosts 6 megacities and India 5. At the other end of the spectrum, European cities gradually disappeared from the list of the top 20. Some of the megacities have existed for thousands years, whereas others such as the port cities of Guangzhou, Guangdong, and Shenzhen on the Pearl River Delta have sprung up almost overnight. Shenzhen, hosting the 8th largest stock exchange in the world, in 2017 entered the global top 20 of financial centres for the first time. Its port doubled its handling capacity in just 10 years and became the world's third busiest port in 2015. Other cities, such as Shanghai, went into a long evolution and dynamic reinvention; Shanghai became an urban regeneration and innovation hub and one of the major gateway cities into China (Arcadis 2016; UN 2016b).

Megacities, increasingly networks of networks, are the most complex social and techno-economic human ecosystems. They dominate world developments not only because of their sheer size but as the result of their influence, often strengthened through network interactions. Preventing and combatting poverty and environmental degradation are prime concerns. The endowment with and the management of expanding infrastructures, and the search of a more inclusive, intelligent, and effective metropolitan governance, are critical issues.

Megacities need to become better places in which to live and to develop in more harmonious symbiosis with the natural systems that support them. Urban sprawl is a major problem. As the world population grows, cities often expand far beyond their formal administrative boundaries. This pattern can be seen in many cities around the world, and not only in developing regions. Annually, Europe loses an area of the size of Berlin through urban and infrastructure expansion. From 2000 to 2015, the ratio of the land consumption rate to the population growth rate was the world's highest in Eastern Asia and Oceania, followed by developed regions. Unplanned urban developments undermine other drivers and determinants of sustainable development. For example, for every 10% increase in sprawl, there is a 5.7% increase in per capita carbon dioxide emissions and a 9.6% increase in per capita hazardous pollution (EC 2016a; UN 2016a, 2017b).

Exploding vibrant cities in the global south that are expected to experience the greatest increases in population have the lowest capacity and resources to provide access to fundamental services for all citizens. As much as 70% of city residents in the global south are underserved, lacking access to one or more core services including housing, water and sanitation, energy, and transport. For example, in 2012, more than 482 million urban residents lacked access to modern fuels and 131 million lacked access to electricity. In 2015, 140 million did not have clean water (WRI 2016).

Megacities must become smarter and enhance the intelligence of all their citizens and systems. Digitalisation has to be further harnessed and improve the ways in which cities and infrastructures function. In the future, citizens are expected to be linked by intelligent devises, robots, sensors, databases, and an infinite number of artefacts and interpersonal networks. Interconnection marked the rise of a middle class in developed countries during the twentieth century and is expected to be instrumental for the rise of a burgeoning middle class in many emerging countries during the twenty-first century. A wave of one billion additional consumers can boost competition for an array of products and services that will become affordable.

1.1.7 Global Interdependency and Pressures on Strategic Resources and Life Support Systems

Globalisation, propelled by technology and trade, has created a closer-knit world with higher interdependency among human settlements and sites of extraction or transportation of resources. Competition for resources is likely to exacerbate underlying tensions. That all may reap the full benefits of globalisation, many elements must be rebalanced, including access to quality education, social cohesion, efficient markets, policies for research, innovation, and development, and good public and corporate governance. Inequality is not inevitable, and globalisation can and must be equitable. Biodiversity losses and ecosystem crises may increase our consciousness of belonging to a universal community that is entering an age of scarcity. Visions for 2050 proposed by organisations and business associations underline that humankind can live well and within the limits of the planet, but our values and behaviour must change radically if we are to survive (WBCSD 2010a, b, 2014).

Responsible globalisation should not only be equitable but also should reinforce ethical principles. Without enhanced social justice, sustainability values, and democratic governance, the global crisis of 2008 could have an enduring impact. Terrorism and corruption are main hindrances to social inclusion and sustainable development.

Issues of trust are of crucial importance for navigating out of the crisis and into a more sustainable trajectory (OECD 2017b).

Humanity stands at a critical point, likely a point of no return. The multiplication of global actors and the diffusion of power could lead to a redefining moment. In 1972, the seminal report "Limits to Growth" by the Club of Rome suggested that humanity has 40 years to prevent the most negative consequences of resource over-consumption. Global population could peak at 8.1 billion people in 2042 because of rapid decline in urban fertility. Global wealth may grow, but much more slowly than generally expected because of slow productivity growth in mature economies. Global wealth may peak after 2052, and the investment share of the GDP may grow as society is gradually forced to handle issues of depletion, pollution, biodiversity decline, climate change, and inequity. Global energy use could reach a peak in 2040, because of a continued increase in energy efficiency. CO_2 emissions could peak in 2030, because of a shift towards renewable energy sources. Nevertheless, CO_2 concentrations may grow, and the global average temperature could exceed the threshold of +2 °C by 2050. This event could trigger self-reinforcing warming, with a possible systemic collapse in the second half of the twenty-first century (Club of Rome 2012).

Climate change is the most mobilising global threat over all life support systems. Climate change is already occurring in all continents and across all oceans (IPPC 2014). Most studies suggest that the universally accepted limit of 2 °C increase is very difficult to achieve given the present emissions gap (UNEP 2017). The focus of scientists and decision makers has expanded from mitigation to increasingly embrace adaptation. More than adaptation, the resilience of cities, regions, and coastal areas represents a strong goal reinforcing the immunity and robustness of places, communities, and functions.

Warmer temperatures, rising seas, and more frequent and severe extreme weather events threaten Earth's resilience. The complementary green, clean, and resilient agendas require balanced portfolios of policy mixes to provide multiple benefits for all world citizens. Such agendas combine long-term and immediate measures to halt environmental degradation leading to extreme disasters, including desertification, floods and storms, hurricanes, typhoons, and cyclones of increasing magnitude and frequency (World Bank 2012; OECD 2012a, 2016c).

Cities are concentrated demand and consumption centres of geostrategic resources comprising water, food, and energy. One sixth of the world's population is undernourished and lacks access to modern energy services. One third of the food and energy produced is wasted, unnecessarily polluting the planet further. Urban food and nutrition security is increasingly highlighted as a critical twenty-first century policy issue. Access to sufficient, nutritious, and affordable food is a basic human right and provides the foundation for citizens' ability to grow, learn, prosper, and reach their full potential. Cities are highly impacted by systems efficiency and sustainability, influencing the prevalence of quality and quantity of food, including availability, price, and behavioural patterns. As decisive actors in global and local distribution networks, cities can promote food and nutritional security, optimise flows, and boost food justice.

Food and nutrition security in cities calls for a multi-factor analysis for income and opportunity generation and distribution. Two thirds of the organic waste produced in the world originates in cities. Food systems for cities should be important for a sustainable urban metabolism. Urban lifestyles are directly connected with food systems functions. City residents are increasingly experiencing the 'triple burden' of undernourishment, micronutrient deficiencies, and obesity with the greatest impact observed among the urban poor, especially women and children. Last but not least, food security in urban areas is closely linked to urban poverty and access to goods and services. In the global south, a large majority of the urban poor rely on street food and informal labour.

Global energy demand could rise by more than one third by 2035, especially driven by emerging economies. Energy efficiency is highlighted as the cleanest and easiest way to save energy. Global demand for electricity is growing almost twice as fast as energy consumption, and huge investments are needed to satisfy the rising demand in developing cities and replace ageing energy infrastructures in developed ones. However, humanity has managed to decouple energy emissions from economic growth and in 2014–2016 celebrated, for the first time, 3 years in a row of flat emission levels in times of growth (IEA 2017a, b).

The changing global resource landscape suggests continued demand growth for food, energy, minerals, and timber resources. Tensions over scarce resources and raw materials could cause conflict and require new forms of crisis governance. Economic and environmental interconnections are extremely important in times of persistent low growth that sap collective ability to tackle environmental challenges. Many studies underline that having to choose between preserving the environment and developing the economy is an outdated dilemma. In the long term, socioeconomic development, on land or in the sea, is fragile without the sustainable management of resources (EEA 2015a, b; OECD 2016a; UNEP 2016).

The escalating pressure on natural resources during the past two decades has largely offset the benefits achieved through eco-efficient technology and practices. Research and development of new materials and energy options and technologies are of particular interest for cities striving to achieve emission reductions. Strong, light structural materials that reduce the energy costs of buildings and transport promise to increase urban resilience. The energy transition is expected to stimulate developments in generation of renewable electricity and new business models and behavioural patterns.

Extreme weather phenomena linked to climate change, natural disasters, irreversible biodiversity loss, energy disruptions, terrorism, pandemics, crime, and nuclear disasters are among the shocks regularly recorded by world organisations. The array of interrelated geopolitical, economic, technological, environmental, and societal risks, and their perceived likelihood, impact, and interconnectivity, are subjects of important analyses highlighting the role of cities as incubators of innovative responses of risk governance. Cities have to improve their ability to engage with uncertainty and address ocean-related threats involving significant and sudden, potentially high-impact, events such as storms, tsunamis, and floods, the timing and magnitude of which are very hard to foresee (WEF 2018).

1.2 The Urban and Ocean Capital in the Digital Age of Overexploitation

1.2.1 Increasing Attention on the Right to Healthy Cities and Seas

In 2015, sustainable development issues were receiving increasing attention, against a backdrop of environmental decline and persistent poverty, compounded by advancing climate change. The threat of insecurity of strategic resources has come back as the world is undergoing a period of uncertainty and intensified pressure on resources, driven in part by the scale and speed of demand from emerging cities. The horizon seems filled with supply disruptions, volatile prices, environmental degradation, and political tensions over resource sufficiency and access.

Cities and the seas are recognised as critical emerging issues for the global environment, almost universal in spatial scale and subject to new developments, scientific knowledge, scales, and impact. The UN Conference on Environment and Development (Rio de Janeiro, 1992) and Agenda 21 had already underlined the role of cities and local governments for sustainable development. Irresponsible land use planning, rigid zoning, and low-density peri-urban development into green areas have resulted in increased consumption of land, energy, and resources, pollution of air, soil, water, and noise, and greenhouse gas emissions. The impacts of climate change on oceans and the cities, the role of seas and green areas as carbon sinks, and the effects of compact cities versus urban sprawl are subjects of continuous research.

Twenty years after the ground-breaking UN Conference on Environment and Development, the Rio + 20 conferences, celebrated in the same emblematic city in 2012, tried to define “the future we want.” Cities and the oceans were two among the seven critical issues most likely to turn sustainable development from aspiration and irregular implementation into a systematic path to genuine prosperity for the present and the next generations. The Conference Rio + 20 pictured the world’s oceans and seas as vital conduits for health, food, energy, transport, and trade. Oceans serve as the world’s largest protein source, with more than 2.6 billion people depending on the sea for their primary source of protein. The crucial links among cities and the oceans were also reflected by the rest of the essential issues, including food, water, jobs, disasters, and energy. Governments and businesses underlined threats from climate change and ocean acidification, overfishing, illegal fishing, and subsidies that drive unsustainable consumption. They also debated the need to conserve and protect marine ecosystems to both restock the ocean and strengthen its resilience (UN Conference on Sustainable Development Rio + 20 2012; OECD-FAO 2017).

The massive urbanisation of the coasts of the emerging world should be seen as a golden opportunity to integrate ocean stewardship into the design of all urban functions, including housing and services, industrial spaces, and energy production and consumption systems. Cities generally have higher human development indexes, composed of life expectancy, literacy, education, and standards of living.

Since 1990, the human development index has shifted attention from income to well-being and people-centred policies, such as education and health services, which can better reach populations concentrated in cities. In line with SDG 2030, the 2016 Human Development Report suggests that human development is for everyone: in the journey to the future, no one should be left out. The profound shift in global dynamics is largely driven by the fast-rising emerging world and its megacities. Besides effective policies for reducing inherent vulnerabilities, cities must build capacities for disaster preparedness, physical and social resilience, and recovery from shocks (UNDP 2016).

Cities should offer access to public goods and expand rights to the commons for all—local citizens, migrants, and visitors. Cities are places that optimise the prospects for individual and collective well-being. The UN Human Settlements Programme (UN-HABITAT) suggests that urbanisation, if well harnessed, can foster development, social equality, economic prosperity, ecological security, and cultural vitality.

The Habitat III conference (Quito, October 2016) brought a unique consensus among all participating states that led to the adoption of the New Urban Agenda. This was a crucial step forward in the implementation of the 2030 Agenda for Sustainable Development, particularly SDG11 in its rich interconnections with all SDGs, and a cornerstone for other key agendas, particularly the Paris Agreement and the Sendai framework for Disaster Risk Reduction. The New Urban Agenda, which planned subsequent work for cities on five continents, is a call for an integrated, people-centred, and place-based action that takes into account the diversity of cities and their territorial context, while building on urban–rural linkages. Key commitments include the development of a harmonised definition of cities and the fostering of cooperation between cities to strengthen sustainable urban development.

The right of the city should be enjoyed by all, exactly as duties to cities apply to everyone. Since 2002, UN-HABITAT has convened the biennial World Urban Forum, bringing together all stakeholders working for better conditions in cities. Recognised as the foremost global arena for interaction among policymakers, local government leaders, nongovernmental organisations (NGOs), and expert practitioners in sustainable urban development, the forum focuses on burning urban issues, involving formal and informal dialogues, and concludes with action-oriented proposals. In 2018, the 9th World Urban Forum in Kuala Lumpur emphasised the implementation of the New Urban Agenda (UN-HABITAT 2016).

1.2.2 All Capitals Are Engaged in the Direction of Strong Sustainability

Sustainability involves a continuous renaissance and rebalance, resilience to threats, and the permanent reinvention of new and better opportunities. Qualitative leaps and paradigm shifts are crucial to maximise and optimise investments in capital, labour, and chance. Innovation and sustainability share a desire for permanent youth and grasping of new opportunities. In the long shadow of the global crisis, cities have to become smarter and regenerate their body and all its vital organs, but also their mind, with its ideas and concepts, and their soul, with its values and emotions. They have to reinforce their capacity to co-create the masterpieces of the future (EFLWC 1997).

Cities have the most complex capital, composed of human and social, natural, and environmental resources and cultural and political, technical and financial capital. All forms of capital are highly interconnected and interdependent, yet each is unique and needs particular attention. Human capital includes personal health and well-being assets, education and skills, and ethical values and lifestyles. Human and intellectual capital is identified by many city leaders as the most important urban asset, which can make a critical difference towards sustainable development.

The progress of a city can be evaluated through the evolution of its composite capital, forming the aggregate capital stock and its components. A key question concerns the extent to which the various components of wealth can be substituted for each other. For weak sustainability the heart of the matter is not whether a particular form of capital, such as fisheries, will be available infinitely, but whether human ingenuity can preserve and increase the global capital. An "integrated minimum" of the total urban capital has often been proposed as a nonnegotiable political line, allowing flexibility of substitution among the various forms of capital.

Strong sustainability implies that substitution of social and environmental capital is unacceptable, and that irreversible loss of any form of precious urban capital should absolutely be prevented. This is the plea of Agenda 21 and the Charter of European Cities and Towns Towards Sustainability, which state that cities should be founded on the carrying capacity of nature and advance towards social justice, prosperous economies, and environmental health. Equity is considered to be a precondition to sustainability, because the uneven distribution of wealth causes both unsustainable patterns and barriers to cultural change (Mega 2013).

Embarking on the journey of urban sustainability demands multiple efforts to prevent the depletion of the capital stock in all domains and at all levels, local, regional, national, and global. Cities have to manage their assets to address the specific challenges linked to each type of capital, without losing the overall sustainability perspective. Effectively managing the co-evolving dynamic capitals asks for a holistic, integrated approach, because all forms of urban capital are interdependent. Advancing towards strong sustainability through the simultaneous reinforcement of all urban capital requires strong political capital constructed of robust leadership and citizen participation. Beyond converging bottom-up and top-down approaches, nurturing "mutual trust" is essential for enduring change.

Ethical values are a critical element of good leadership, founded on collective and shared principles that are communicated clearly and can mobilise citizens. Public policy for sustainable development must orchestrate responses to the demands of many different groups and manage the allocation of resources between various, and often competing, claims. Developing the capabilities to ensure that cities manage their capitals effectively is essential to respond to citizen expectations. Responsible cities need to be sure that they can measure performance at all policy fronts and inject a culture of continuous improvement.

Most cities consider human capital to be their most precious resource. As cities offer conditions for better education and training, information and awareness, and health and well-being, they can reinforce it and create the preconditions for prosperity. The right to housing can enable the poor to create wealth. In the developing world, most slum dwellers live in extreme poverty. However, the social capital is often high. Squatter cities are vibrant self-constructed and self-organised places, with support communities and highly active groups. The multifactorial context, together with their predominant youth, makes slum dwellers one of the most vulnerable social groups. The social capital is a definitive feature of a city as a society, a community of shared aspirations and concerns. It expresses the richness of informal and formal relationships, the extent to which people feel connected, the dynamism of the local networks, the strength of the social bond. It comprises community traditions and patterns. Engaging all citizens is crucial for cities that wish to prevent crime, segregation, and exclusion (Hall 2013).

Natural capital embraces assets typically transacted in markets, whereas the physical or environmental resource capital includes the life support systems, offering their services on land and at sea. The aquatic capital includes not only water but life, the rich marine biodiversity. The natural and physical capital is being affected by the accumulating stress on ecological systems, terrestrial and marine, massive loss of biodiversity, and vulnerability to natural disasters, soil erosion, water stress, and desertification. Tensions over scarce resources and raw materials may also cause conflict and aggravate crisis.

Technical and digital capital includes the hardware and software of cities, infrastructures and buildings. It is harnessing the evolving possibilities of smart grids and services such as housing, water and sanitation, energy and transport, and access to world networks. Financial capital is vital for investments, and depends on pertinent framework conditions and the ingenuity of the private sector, especially in times of low growth trajectories (Khanna 2016; Piketty 2015).

The cultural and leisure capital depends on a complex array of attributes defining the distinct character of a city. Included are all cultural assets that form the visible essentials of a city and differentiate it from other cities on the world scene. Many cities try to develop a trademark that encapsulates their exceptional qualities and generates powerful and memorable positive associations. Strategies designed to develop cultural capital need to understand the public perception of cities by citizens, businesses, and visitors, and develop the steps to project cities forward into the experiences that they wish to offer.

The urban experience, an increasingly precious component of the cultural capital capturing all possible deeds to be lived in a city, often is linked to the aesthetic, religious, and spiritual ecosystem services provided by the oceans. It may spring from iconic scenes, for example, underwater monuments and sites, or the search of the ultimate discovery experience in aquaria and museums. The most visited cities are also leading advocates for improving quality of life and well-being. They can provide lessons for preserving and enhancing urban capital, with its tangible and intangible aspects (Mega 2013, 2015).

Last but not least, the political capital of a city depends on institutions and citizen participation, transparency, and accountability. Developing accountability is a key responsibility for cities wishing to engage their stakeholders. It involves the true implication of citizens in discussing possible futures and visions, defining objectives and fixing targets, and disclosing their performance against them, City leaders stress the importance of trust and engagement. In times of general distrust, many governments try to reinvigorate the participation of citizens in civic activities, increase the transparency and trustworthiness of decision making, and engage civil society in the conception, implementation, and evaluation of sustainable development policies (OECD 2017b; PwC 2016).

1.2.3 A Global eSociety in Digital Cities and Digital Oceans

Digitalisation is a poignant force for preserving and enhancing each part of the composite urban capital, and at the same time making it fit for the future, whereas innovations ask for exploiting new and different paths and destroying old and outdated patterns. From the initial invention to the final transformation, a chain of interactive processes could bring thorough changes that open up the range of opportunities and creates sustainable value. Trial and error are critical to better navigate failure and nurture the seeds of future successes. World smart cities can use cutting-edge digital technologies to enable and engage citizens. Collaborative technologies, bridging technological and social innovation, offer cities another way to make more effective use of resources, with smarter ways of collecting data and smarter ways to make decisions. A study suggests that successful smart cities of the future will combine the best aspects of technology infrastructure while enabling greater collaboration among urban communities and between citizens and city governments. Emerging investments in both smart technology and smart citizens include the collaborative and sharing economy, collective intelligence, crowd-sourcing, and crowd-funding.

The Fourth Industrial Revolution (4IR) and the worldwide digital interconnection of humanity give rise to a global society. By 2030, digitalisation is expected to have covered the entire planet. Virtual connectivity seems to complement, but not substitute for, the real bond and the relationships based on physical proximity. The Internet of Things promises a new age of ubiquitous connectivity and intelligence in which a set of smart components, robots, sensors, databases, products, services, and platforms connects, virtualizes, and integrates in a planet-wide network of networks and the "Internet of Everything." The connection of the real and digital world could lead to the formation of online cultures and lifestyles that transcend physical spaces and facilitate the integration of totally different spaces and flows. Understanding the policy implications in relationship to ethical values, privacy, security, and norms is essential (Gore 2013; WEF 2016c).

Science, technology, and innovation are pivotal in the search for substitutes for scarce and exhaustible resources and the adoption of new models of sustainable production and consumption. Knowledge is an inexhaustible resource, which can be enhanced and enriched through the density of urban networks and transactions. Urban possibilities to integrate functions, including product, process, and service innovations, social and public, and address threats are boundless. Cities can serve as the nexus for large-scale replicable innovations and support global innovation ecosystems. Open innovations offer the possibility for each city to build on other best achievements and capitalise on improvements as a network partner. Both incremental innovations, mainly bringing optimisation, and disruptive innovations bringing substitution, can constitute pathways for the future. Innovations must always be responsible, as, in a highly interlinked world with more value created online, important ethical questions arise.

Green technologies and eco-innovations must prevent the destruction of habitats and ecosystems and the depletion of natural resources. Anthropogenic degradation often comes through intensive use of resources and excessive production of wastes, such as overfishing and marine debris. Environmental degradation has strong effects on health and on the availability of resources. As strings of extreme natural phenomena are developing into major catastrophes, affecting citizens and infrastructures, human livelihoods and economic systems, understanding and measuring the impacts of climate change through real-time data gathering and networks of sensors becomes urgent.

Public health is a critical urban system constantly challenged and evolving. Advanced automation can provide multiple improvements in care and contribute to the general adaptation of technology and infrastructures to support of ageing populations, including performance-enhancing technology and treatments. Antibacterial and antimicrobial compounds (antibiotics) may soon be less readily available, because every dose of antibiotics creates selective evolutionary pressures as new antibiotics have been continuously developed to replace older, increasingly ineffective ones. The rich marine biodiversity holds many promises for new medicines to be developed.

Humanity is on a digital journey with new game rules. For cities and the oceans, the fourth industrial revolution is first and foremost a data revolution. Increasingly tightly integrated, highly interactive streams of information are translated into new understanding and insights. The emerging digital ocean and cities can enable and accelerate solutions, empower citizens, and foster accountability. Blockchains provide new business directions and models announcing digital coins and the Internet of Values (WEF 2017; Tapscott 2016).

Digitalisation and the development of circular or spiral economy models can foster the sustainable use of renewable resources, reduce primary demand for materials, and limit waste to the bare minimum, thanks to prevention, reuse, and continuous upcycling. As mass production is being replaced by on-demand, user-centred manufacturing, new opportunities arise. Additive manufacturing, also known as 3D printing, and more recently 4D printing, could revolutionise the productive web of cities. Each home could become not only energy independent but also auto-sufficient in food, materials, and products. If the first industrial revolution promoted standardisation, the 4IR is bringing customisation (EC 2016a).

Governments, enterprises, and citizens continuously ask for more cost-effective technology and more powerful devices and applications. The supply chains to satisfy these demands lead to explosive growth in data and analytics and to new competition in almost every scientific and industrial domain. Advances in genomics, biotechnology, neuroscience, and life sciences are radically transforming the fields of medicine, agriculture, and molecular science. The possibilities leading towards artificial land and even artificial life multiply, and the era of post-humanity seems close.

Interactions between the physical and the virtual world are being massively facilitated by advances in materials and biological sciences. Human bodies could further benefit from the production of soft robots with organic tissues. Artificial and human intelligence may compete, or coexist and harmoniously reinforce each other. Cyborgs could perform complex tasks as do humans and in all economic activities, and may be particularly effective for working in risky open seas conditions.

Data become a form of nonexhaustible resource, and technology can democratise access for all citizens. The Internet has been a great transformer of lives, and mobile services are being integrated into the shifting lifestyles of many citizens. The global network of networks is being proposed as the guardian of a free society, and the proliferation of social networks has spurred public expression and communication. The deluge of information accessible to citizens, unprecedented in human history, continues to grow exponentially and requires better judgement capacities. Large data analytics are expected to accelerate multiplication of knowledge, foster new technological advances, and enable management of increasing complexity and diversity.

Smart grids and their underwater components enable cities to enhance scientific and industrial capabilities and take advantage of transformations in manufacturing, infrastructures, and services including those from the ocean. Future smart cities may have to strive to balance sustainable infrastructure with the provision of smarter services to citizens, including those from the global ocean. We may advance towards an integrated supply and demand structure that provides food, energy, materials, security, mobility, and communication more effectively and efficiently.

Public acceptance and willingness to engage are capital for the further development of digital services. Studies conducted by the European Parliament research service concluded that some digital services, such as electronic procurement, met with early-stage uncertainties and resistance. The interest of public authorities in conducting cross-border, fully electronic procurements did not appear to be sufficiently strong. Online health applications met with problems of public acceptance. Last but not least, the creation of a comprehensive system for electronic voting in Europe may still meet with important resistance (STOA 2011, 2013).

Inclusiveness, transparency, and accountability are crucial for the electronic participation of citizens in public decisions. From eLearning or eHealth to eGovernance and eBusiness, the most challenging goal is to build digital ecosystems for knowledge sharing and public services. Advancing from the simple digital reproduction of administrative procedures to citizen-friendly constant and uninterrupted provision of online services and on mobile devices implies a true shift in culture from the use of technologies to support public operations to the integration of digital technologies in shaping public service reform. Mobile social media leapt to almost two billion people or 27% of global population in 2015 but interactive eGovernment services were still used only by 26% of the EU population. Still, only one in three public websites is mobile friendly (EC 2016a).

The natural world, including the largely unexplored marine world, reflects the immense potential inherent in the genetic code of all living organisms. Progress in synthetic biology and metabolic engineering capitalises on this potential in unprecedented ways, enabling the development of new biological processes and organisms to serve specific purposes of personalised medicine, nutrition, and disease prevention. Developments are laying the foundation for a revolution in healthcare and more effective well-being.

Public transparency and trust in the ability of technological systems to prevent negative surprises are important issues for society, the ultimate frontier, this being particularly important at the time when humanity needs huge infrastructural investments. Urban infrastructures in emerging countries require the creation of whole functional districts. In industrialised cities, infrastructures are ageing and need renewal to deliver services such as water and energy distribution and public transport. Widespread sensor networks are needed to help manage the new infrastructures and allow the plethora of applications. The development of personalised services may be followed by another wave of innovations aimed at further tailoring and targeting services, from manufacturing to leisure.

The substitution of scarce and depleted resources is a key issue for the green and blue revolution 2.0, which has been stimulated by the rising global demand by more empowered citizens and consumers for healthy environments and nutritious food and fresh water for all. Progress across the biological and physical sciences and innovation hold the promise of further reinforcing optimisation of resources, as in achieving maximum sustainable yields in fisheries, preserving biodiversity, and decreasing the carbon and environmental footprints of economic activities.

Since the discovery of scuba diving by J. Cousteau and advances with satellite observation, marine and maritime research and innovation, together with enabling technologies, have achieved extraordinary progress. The exploration of space offered the first images of Earth, the overwhelmingly blue planet, and drew attention to the oceans (UNEP 2011). Since then, ocean observation has progressed rapidly, through more advanced observation satellites and in situ seaborne sensors, mapping, and forecasting. Ocean science to understand the diverse socioecological systems and services is "big science," complex and multidisciplinary, expanding in both magnitude and scope, and requires multidimensional observations and multi-stakeholder engagement. The exploration of the unfathomable diversity of ocean resources asks for submersible technologies, robotics and video-surveillance, and high-performance equipment for marine operations. World infrastructures include 784 marine stations, 325 research vessels, and more than 3800 Argo floats. New opportunities are associated with hyperconnectivity, such as big data analytics. The major technical challenge is the complexity of the data and analytics across disciplines rather than the amounts of data per se (UNESCO/IOC 2017).

The UN Decade of Ocean Science (2020–2030) aims at filling knowledge gaps. UNESCO/IOC reminds us that, for 99% of habitable marine areas, the basic biodiversity knowledge required for effective management is missing; only 5% of the ocean floor has been mapped and only 1% of the mapped seafloor has been gridded at high resolution: 103 million square miles of the deep sea exist in perpetual darkness and up to one million marine species could still be unknown to science. Only three humans have explored the deepest known point of the ocean. The possibilities are endless.

The vision for the digital ocean involves networks of sensors, manned and unmanned systems, and satellites, connected to provide instant, affordable public access to data, information, and services. Progress towards the digital ocean is advancing with a mounting number of increasingly smarter sensors and devices interconnected to produce near real time information and integrated marine technologies and cloud services. The implications of the Fourth Industrial Revolution for the oceans may challenge humanity's ability to advance in ocean governance and finance. The ocean has no voice, is weakly governed, and is poorly served by markets. More data is being generated every week than in the last millennia, and approximately 50% of all research data and experiments are considered not reproducible. Experiments with blue clouds progress towards free access to a commons of scientific data. Smart grids supporting the development of smart cities rely on ocean infrastructure, which may face conflicts among diverse users. Submarine cables have an important role in all forms of communication through the Internet, and increasing demand is expected to result in more capacity. Although cables and their protective corridors cover only very narrow strips of the seabed, they introduce a dividing line across the open ocean seabed that may cause conflicts with other activities (UN 2016c; WEF 2017).

1.3 Cities and Oceans in the Aftermath of Historic Commitments

1.3.1 From COP21 to HABITAT III: A Floor for Progress, Not a Ceiling

The 17 Sustainable Development Goals, adopted by the 70th General Assembly of the United Nations in 2015, constitute a universal call for action to end poverty, protect the planet, and ensure that all people enjoy peace and prosperity. The SDGs represent a paradigm shift from traditional development assistance into a transformative agenda with actions driven as much at local, national, and regional levels as at the global level. The first Sustainable Development Goals annual report suggested that this monumental journey will require all orchestras of the world to work together. It proposed a benchmark for the 15-year implementation period of the 2030 Agenda, providing an evaluation of where the world stands, including that on cities and on oceans, and highlighting the most significant gaps and challenges (UN 2016a).

Cities and oceans were among the issues of major attention with fully dedicated SDGs interacting among them and the array of the rest of SDGs. Cities and oceans also received much attention few months later at COP21. Traditionally, discussions on climate policy have largely portrayed cities as the source of problems, painting them as foremost polluters and intensive resource consumers. The race for limiting climate change initiated a new era for cities and oceans as critical sources and spaces for solutions.

The Paris Agreement commits all countries to holding global temperature rise to "well below 2 °C." This goal will require economy-wide global, national, regional, and local transformations. A robust, multi-sector response will be a key component of this desirable equation to pivot societies towards a more sustainable future. Leadership, mobilisation, technological change, economic competition, social innovation, and influential public–private multi-stakeholder alliances are crucial for climate action. Climate change requires much more ambitious greenhouse gas (GHG) emission targets and policy portfolios including both short term and for the longer term. Stringent mitigation leading to CO_2-equivalent concentrations of about 450 parts per million in 2100 would meet the 2 °C target. The 2 °C scenario characterised by 40–70% reductions in global GHG emissions by 2050, compared with 2010, asks for an increase in renewable electricity supply from a share of approximately 30% to more than 80% by 2050. Energy performance and mobilisation of cities will be most decisive.

COP21 has been a forum for cities to share actions, to create a more structured dialogue regarding local solutions, and to commit to new steps, building upon and scaling up transformative actions that have already been taken. City leaders across the globe have already demonstrated their resolve to develop and implement innovative solutions that not only address climate change but also make their cities better places to invest, work, live, and visit. Cities have established themselves as laboratories of climate change solutions, working towards multi-integrated strategies that also improve livelihoods and well-being (C40 and Arup 2015, 2016).

The significant mobilisation of cities at COP21 signalled a paradigm shift in political perceptions on the role of cities. During the conference, more than 400 mayors signed on to the Compact of Mayors, the UN initiative to reduce emissions in cities around the world, which later united with the European Covenant of Mayors and became the Global Covenant of Mayors for Climate & Energy. The Paris City Hall provided a beautiful backdrop to the energy-conscious cities events at COP21, a manifestation of bold, local climate actions. And the momentum continued with the "Climate Action 2016," held in Washington, DC, in May 2016, a multi-stakeholder summit that deepened and expanded the coalitions of government, business, finance, philanthropy, civil society, and academic leaders. The search focussed on near-term actions and long-term needs and created a springboard for climate implementation in the pre-2020 period.

Action against climate change has been a crucial call for cities to become smarter. Already in 2012, Amsterdam received the first world award designed to identify the cities, initiatives, and solutions that have contributed the most to developing the Smart City concept. With its Open Data project in urban mobility, the city department on infrastructure and transport has provided the public with data with the aim of stimulating better mobile services for citizens. The ultimate goal is to develop the best possible application that provides multimodal urban transport information and improve mobility. Among the C40, Yokohama has been praised in 2016 for its Smart City Project to more effectively manage energy use and mitigate climate change by engaging citizens and stakeholders.

The alliance C40 presented a roadmap to assist cities progressing on the goals of the Paris Agreement. According to a study conducted with Arup, the next years are critical to move from initiatives in planning and pilot stages to fully transformative, city-wide measures. The mayors of Paris, Mexico City, Madrid, and Athens committed to improve air quality in the four cities through banning the use of all diesel-powered cars and trucks by 2025, incentivising alternative vehicles, and promoting walking and cycling infrastructure (C40 and Arup 2016).

Habitat III, which took place in Quito in 2016, in line with the bi-decennial cycle that begun in Vancouver in 1976, and continued in Istanbul in 1996, offered a generational opportunity to reinvigorate the global commitment to sustainable urbanisation and adopting a New Urban Agenda. In Quito the world asserted itself as overwhelmingly urban, against 38% urban at the time of the Vancouver conference and 45% at the time of the Istanbul conference. The Conference aimed to promote renewed political commitment for sustainable development, assess accomplishments versus objectives, tackle urban poverty, and identify emerging challenges and shared visions. As the first UN global summit after the adoption of the Sustainable Development Agenda 2030, it offered a unique opportunity to discuss the ways that cities are preparing to fulfil their role in the climate change era and the goals for sustainable development (UN-HABITAT 2016).

In Habitat III, cities reconfirmed their determination to have a crucial role in the transition to the civilisation of sustainability, as highly engaged interconnected cooperating superpartners rather than competing superpowers. The cities have always been the very places where people concentrate and interact to create and seize opportunities. Already in the seventh century BC, Alcaeus had suggested that "cities are made out of men trying to grasp opportunities"—for cities can multiply chances and opportunities through proximity and interaction. As biological species come together in mutually supportive environments, humans come together to optimise their chances for prosperity. Synergies have always been generated and captured through a spectrum of activities, ranging from defence and religion to innovation and politics (Gore 2013; Mega 2013, 2015; NYC 2015a, b).

The New Urban Agenda is an action-oriented consensus that sets global standards of achievement in sustainable urban development and should help meeting the needs of the urban underserved and more prosperous, sustainable, and equitable cities for all. Despite considerable calls to action in recent years, the enormous challenge of rapid urbanisation in developing countries exceeds by far their current capacity to address the expanding urban footprint, secure and affordable shelter for all, regional water risk management, and universal access to fresh water and to clean, affordable, reliable energy and urban transport (WRI 2016; UN-HABITAT 2016). The European Union, reputed to have the most sustainable cities, tries to lead world developments with its Urban Agenda initiated with the Pact of Amsterdam in 2016. Cooperation among Member States, cities, the European Commission, and other stakeholders is critical to stimulate growth, liveability, and innovation in the cities of Europe and provides a model of multi-level governance (EC 2017b).

1.3.2 Waves of Commitments for the Global Ocean

The future of humankind is inextricably linked to the future of the ocean. An ocean threatened by unsustainable and illegal fishing, marine pollution, and climate-related impacts cannot be auspicious for humanity. Since 2014, the first three editions of "Our Ocean" conferences have generated commitments valued at more than \$9.2 billion to protect 9.9 million km^2 of ocean, an area equal to that of the US. In 2016, following The Netherlands and Chile, the US hosted the third "Our Ocean" Conference in Washington, DC, to catalyse actions and empower a new generation to lead the way towards a healthy and sustainable ocean. The third conference generated 136 new initiatives on marine conservation and new commitments for the protection of almost 4 million km^2 of the ocean.

Ocean acidification, warming, and sea level rise threaten economic livelihoods, marine resources, and the integrity of coastal areas, tourism, and recreation. The ocean has absorbed more than 90% of the increased heat in the Earth system since the 1970s. This capacity has helped limit global average temperature rise, but at a high price. Sea ice and glaciers are melting around the world, exacerbating sea level rise and endangering marine life. By the end of the twenty-first century, sea levels could rise at least another 1–4 feet, affecting the lives and livelihoods of coastal communities. The pace of these changes is increasing, making it difficult for coastal and marine communities to adapt. Coral reefs and polar ecosystems are especially vulnerable.

The ocean, the world's largest carbon sink, has absorbed about one third of the carbon dioxide released by human activities since the beginning of the industrial revolution, mitigating climate change. But this mitigation comes at an exorbitant price, because carbon dioxide when dissolved in seawater lowers the pH level of the oceans, thereby increasing its acidity and changing the biogeochemical carbonate balance. Concerns about ocean acidification, first expressed in the early 1980s, have been confirmed, and the extent of its impact on marine ecosystems is being investigated.

Ocean acidification is one of the most critical aspects of the impact of the increasing CO_2 emissions on ocean ecosystems. Ocean acidity has increased by about 30% globally since the industrial revolution and is projected to increase by 170% by the end of the twenty-first century, as compared to the pre-industrial level, unless global carbon emissions are significantly curtailed. This is "the other CO_2 problem," and it could cause irreversible damage to marine species and ecosystems. Long-term observations within ocean, coastal, and coral reef environments are essential to better assess the impacts (UNEP 2011; IPCC 2014b).

The impacts of these changes are already clear, such as coral bleaching around the world. The protection and restoration of marine ecosystems and coastal habitats, such as mangroves, seagrass, and wetlands, one third of which have been destroyed by heedless development, can both sequester additional CO_2 and increase resilience to rising temperatures and extreme weather events. Coastal zone planning can help reduce the expected effects of sea level rise, and careful fisheries management can adapt to shifts in species populations and ranges. To be effective, these responses require coordinated action and commitments from governments and stakeholders around the globe (GOI 2013).

Ocean degradation has a direct impact on climate, and climate change affects marine and coastal ecosystems and their dependent populations. COP21 hosted a key event on oceans, the Oceans Day (Paris, 4 December 2015). Ocean and climate are inseparable, in ways that are still to be explored and understood. The Ocean and Climate Platform, created by representatives of civil society and of the research community in 2014, aimed at contributing to the general mobilisation for a universal, ambitious, and binding climate agreement, and ensuring that the ocean is fully integrated in the future climate regime. The "Because the Ocean" declaration, signed by 22 countries, called for a special report of the Intergovernmental Panel on Climate Change on the ocean, a UN conference on Ocean and Seas to establish a regular review and benchmarking of SDG14, and the development of an ocean action plan under the UNFCCC. The world system was never so quick to integrate at least the two first proposals (GOC 2016; UN 2017a).

The third 2016 "Our Ocean" Conference highlighted best practices and innovative efforts to address marine pollution, mostly originating on land. Nutrient pollution comes from diverse sources, including agricultural runoff and sewage and wastewater discharges. It overloads marine environments with high concentrations of nitrogen, phosphorus, and other nutrients, which can produce large algal blooms. The decomposition of these algae after they die consumes oxygen, creating hypoxic, or oxygen-depleted, 'dead zones' where fish and other marine life cannot thrive. An estimated 500 dead zones exist in the world, and many more areas suffer the adverse effects of high nutrient pollution.

A parallel "Our Ocean, One Future: Leadership Summit" to engage the next generation of the committed was co-hosted by the U.S. Department of State, Georgetown University's Edmund A. Walsh School of Foreign Service, and the Sustainable Oceans Alliance. High-level ocean leaders joined a selected group of 150 university students aspiring to build leadership and spur action. The Summit brought together political leaders, entrepreneurs, scientists, NGOs, and citizens to identify high-impact actions.

The fourth 2017 "Our Ocean" Conference, hosted by the European Union in Malta in 2017, saw, for the first time in history at such a scale, the public and the private sector coming together to form a true "ocean community." More than 400 new commitments were made in support of safe, sustainable, and healthy oceans, worth more than 7 billion euros (€): this figure includes €1.5 billion for maritime security and €2.7 billion to support a sustainable blue economy. This European edition also took up the challenge of engaging with all stakeholders, governments, foundations, and civil society. More than 125 businesses, from blue chips to start-ups, announced 100 commitments in all six areas of action. The European Parliament proposed a pilot project to boost ocean literacy in Europe and to consolidate understanding of the seas and oceans at all levels, from children at primary school up to CEOs and policy makers. Our Ocean 2017 was not only about making many more pledges, but also tracking progress against 700 commitments, with the announcement of a fund to support the tracing of achievements that was broadly welcomed.

Another important series of ocean summits is the series of the World Ocean Summits initiated by The Economist Group beginning in 2012, which held the fifth edition in Mexico in 2018. Each World Ocean Summit convened diverse stakeholders and bridged the perspectives of business, government, academia, and civil society on an array of fundamental issues. The Summits aim to accelerate the transition to sustainable use of the oceans, given the general consensus that the current status is unsustainable. Across the world, humans have inflicted dramatic changes on the seas. Yet the ocean remains a vital resource with the potential to generate enormous economic growth. Balancing human activity with the long-term health of the oceans is essential (EIU 2015).

1.3.3 Accountable Governance and Anticipatory Democracy

Urban democracy is a historic conquest of cities and a major asset for rising in the world scene and leading sustainability-oriented action. Anticipatory democracy is vital for building momentum towards desired futures. Globalisation is at the crossroads of trust and rebalance, as, in an increasingly polycentric world, wealth and power shift from west to east and north to south, from local to global and back; emerging G20 economies drive more growth and south–south exchanges are intensified. Although poverty is decreasing, debt and economic insecurity are mounting and income disparity is widening. Government indebtedness has increased greatly in OECD countries, and the spectrum of multiple inequalities appears everywhere. A more conscious search for sustainable development could bring a better future for all (UNDP 2016; The Millennium Project 2017).

Progress towards sustainable development can only be achieved through active citizen participation and local democratic processes. Democracy is the very essence of citizenship and collective life in cities. There is a feeling of urgency to restore the healthy functioning of representative democracy at all levels and make political systems sustainable. Democratic governance is critical in directing the flows of power and shaping the interrelated global and local patterns; it can prevent power imbalances and social inequalities interacting with environmental distress and resulting in even more worrisome situations. Global governance often weakened the voices of emerging world cities and excluded marginalised groups. Approaches emboldening cities to bring about socioeconomic and political change hold enormous promise for human-centred policies at the local and global level. Improvements to the democratic process include greater use of foresight and deliberative participative interactions.

Enlightened and empowered citizens can have a major role in moving cities into sustainable trajectories and reinventing a new world order, nurtured with the growing sense of belonging to a human world community. Universalism is vital for human development (UNDP 2016). Citizens are increasingly interconnected by extensive networks and interpersonal flows. The emergence of a global conscience is increasing momentum for sustainable development, and shifts of power away from states towards local, civic, and private actors are often presented as compelling drivers of change. Greater awareness of the global nature of the citizens' most important concerns can bring a higher conviction of the potential of shared solutions (EUISS 2012; NIC 2017).

The near-universal access to education, the smarter use of enabling information and communications technology, and the importance of lifelong education, together with the evolution in the status of women in most countries, are promising elements for global governance. The process toward gender political-economic equality seems irreversible. Even if humanity may still wait for more than 80 years for gender parity in the workplace, women are becoming empowered throughout the world and increasingly engaged in professional life and representative democracy at all governance levels. The Global Gender Gap Report, by the World Economic Forum, offers a framework for capturing the magnitude and scope of gender-based disparities in key areas such as access to healthcare, education, political participation, and economic equality. It seems that there is a strong correlation between those countries that are most successful at closing the gender gap and the most competitive economies. The top performing countries according to the latest Global Competitiveness Index also rank best according to the Global Gender Gap Index (WEF 2016a, b).

Accountable governance will be critical for improving living conditions in the world's top cities. Megacities attract huge investments for infrastructures, roadways and rail structures, ports, power grids, and telecommunications that need to be efficiently and effectively managed. Business and civil society networks are also expected to fill a critical role. Their influence could become greater than that of many states and lead to new forms of governance and civic action. Trust is essential for public authorities to orchestrate the transition to sustainability (OECD 2017a, b).

The scale and speed of growth have profound implications for multi-level, multi-stakeholder governance, because the emerging profile of the future cities confronts national structures. Strong and interconnected local leaders play an increasing role in international politics. The rising power of intermediate cities could signal important growth and innovation coming from cities at a faster pace than at present. Furthermore, intense networks, alliances and coalitions of cities can be much more agile, responsive, and effective than international established organisations (EC 2016b).

The contribution of the civil society is expanding, fuelled by the technological revolution, the end of national governments' monopoly on information, the increasing complexity of policy problems, the swelling size of government, the crisis of confidence in elected officials, and the effects of globalisation. State fragility from inequality and a massive increase in the urban population are expected to reinforce megacities and metropolitan areas, both in relationship to their regions and as interconnected partners in a more diffused power landscape. By 2030, the 50 largest world cities could concentrate more resources than most medium-income states and have a more prominent international role (EC 2015).

Cities could instill a new sustainability ethos in the global debate as leading interconnected democratic actors in an era of citizen disengagement and mistrust. The transition from the nation-state system to a multipolar set of actors outside traditional government structures asks for strategic coalitions bringing together governments, corporations, academic institutions, and civil society organisations. Democratic cities, networks of networks, could lead in networking with their citizens and partner cities. As no single power is able to play a leading role in the search for shared solutions to the global ocean, strengthened links among all levels of governance, from local to global, and between state and non-state actors or among cities and civil society, seem to be decisive for the future of humanity. Pioneering cities able to promote a transformative agenda for the global ocean need first to be committed sustainability players on their territory and able to motivate their interconnected partners.

1.4 The Bond Between Cities and the Oceans: Expectations from the Sustainable Development Goals 2030

1.4.1 Achieving the Interconnected Universal SDG11 and SDG14 and Others

The 2017 Sustainable Development Goals Progress Report provided an overview of the world's implementation efforts to date and suggests that although progress has been made across all areas of development, the pace of progress has been insufficient, and advances have been uneven to fully meet the implementation of the SDGs and ensure that nobody is left behind. The report suggests that if the world is to eradicate poverty, address climate change, and build peaceful, inclusive societies for all by 2030, key stakeholders, including governments, must accelerate the implementation of the SDGs. Economic losses from natural hazards are reaching an average of $250 to $300 billion a year, with a disproportionate impact on Small Island Developing States, Apologies (SDIS). Although the global unemployment rate decreased from 6.1% in 2010 to 5.7% in 2016, youth were nearly three times more likely than adults to be without a job. In 2015, 85% of the urban population used safely managed drinking water services, compared to only 55% of rural populations (UN 2017b).

Effectively tracing progress on the SDGs requires robust, accessible, reliable, timely, and disaggregated data at all levels, which poses a major challenge to national and international authorities (OECD 2017c). Although data availability and quality have steadily improved over the years, statistical capacity still needs strengthening. Nearly 1 billion people have escaped extreme poverty since 1999, but about 767 million remained destitute in 2013, most of whom live in precarious situations. An alarmingly high number of children less than 5 years of age are still affected by malnutrition. In 2016, an estimated 155 million children under 5 were stunted, with low height for their age. The distance to sustainable energy is still huge. Access to clean fuels and technology for cooking increased from 50% in 2000 to 57% in 2014, but more than 3 billion people lacked access to clean cooking fuels and technologies, and this led to an estimated 4.3 million deaths in 2012.

Oceans are particularly crucial for people living in coastal communities. The sustainable use and preservation of marine and coastal ecosystems and their biological diversity is essential to achieve the 2030 Agenda, in particular for Small Island Developing States (SIDS) fully dependent on oceans. SIDS have to address particular structural challenges to advance towards sustainable development, given their unique vulnerabilities including small size, remoteness, narrow resource and export base, and exposure to global environmental, ocean, and climate risks.

Fisheries contribute significantly to global food security and livelihoods, but overfishing impairs the functioning of ecosystems and reduces biodiversity. Fish stocks must be strictly maintained within biologically sustainable limits, at or above the abundance level that can produce maximum sustainable yields. Biodiverse marine sites require effective sustainable safeguarding. And the momentum is building. Since 2000, the proportion of marine sites that are designated as key biodiversity areas and are completely covered by protected areas increased from 15% to 19%. The process, led by national governments, enjoys broad participation from civil society stakeholders.

Sustainable Development Goal 11 asks for cities and human settlements to become inclusive, safe, resilient, and sustainable. By 2030, they have to ensure access for all to adequate, safe, and affordable housing and basic services, provide access to safe, affordable, accessible, and sustainable energy and transport for all, and especially for the most vulnerable, enhance inclusive and sustainable urbanisation and capacity for participatory, integrated, and sustainable planning and management. In many parts of the developing world, the demographic transition to urban areas, rampant poverty, threatened ecosystems, massive construction of infrastructure, and lack of inclusive and accountable governance have resulted in multiple weaknesses. Explosive urbanisation has led to the rise of mega-slums. Africa is home to nearly two thirds of the global slum community. More than 60% of Sub-Saharan Africa's urban population lacks adequate access to clean water, nutritious and affordable food, adequate sanitation, and modern energy services.

The quest for sustainable and coordinated urban development starts with national policies and regional development plans. As of 2015, 142 countries, accounting for 75% of the world's urban population, had a national urban policy in place or under development. Cities also have to safeguard the world's cultural and natural heritage and organise their protection against disasters. Well-managed cities and other human settlements striving for sustainability can be incubators for innovation and ingenuity and key drivers of sustainable development (UN 2017b).

Sustainable development requires the production of goods and services that improve the quality of life in harmony with the planet. The production and consumption system must minimise the natural resources and toxic materials used and the waste and pollutants generated, throughout the entire process. The plastics economy, flourishing in cities and accounting for about 6% of global oil consumption, equivalent to the consumption of the global aviation sector, came with a price for the seas. Oceans may contain more plastics than fish by weight by 2050 (WEF, Ellen McArthur Foundation, and McKinsey 2016).

The UN Ocean Conference (New York, June 2017) was the first conference in a series planned to address the implementation of SDGs. The Conference adopted by consensus a concise, focused, intergovernmentally agreed declaration in the form of a "Call for Action" to support the implementation of SDG14 and a report containing the co-chairs' summaries of the partnership dialogues, as well as a list of extraordinary voluntary commitments for the achievement of the Goal and targets. The conference outcomes establish the norm for subsequent conferences in terms of stringency in the Call for Action and the follow-up mechanisms including monitoring, reporting, and verification. Local initiatives are decisive to the fulfilment of the goal. The City of Malmö organised an international conference on the local implementation of the Goal in collaboration with the World Maritime University.

The conference harvested 1328 commitments, 45% of them by governments, and demonstrated the collective will and mobilisation. NGOs were the second committing entity with 20% of the commitments. Among the oceans, the North Atlantic attracted the lion's share (37%). The target 14.2 (By 2020, sustainably manage and protect marine and coastal ecosystems to avoid significant adverse impacts, including by strengthening their resilience, and take action for their restoration in order to achieve healthy and productive oceans) concentrated the majority of commitments, followed by 14.a (Increase scientific knowledge, develop research capacity and transfer marine technology, taking into account the IOC Criteria and Guidelines on the Transfer of Marine Technology, in order to improve ocean health and to enhance the contribution of marine biodiversity to development) and 14.1 (By 2025, prevent and significantly reduce marine pollution of all kinds, in particular from land-based activities, including marine debris and nutrient pollution). Cities are involved in many of the commitments.

1.4.2 Sustainable Cities, Seas of Opportunity

Cities are ecosystems of power which can be viewed from various complementary angles. A kaleidoscope of global rankings sheds a fascinating light on the ability of cities to lead into a sustainable world (Mori Memorial Foundation 2017). The 2016 Arcadis Sustainable Cities Index ranking 100 global cities on the three key dimensions of sustainability—people, planet, and prosperity—indicates potential leaders for the global commons. The ranked cities have very different median ages, spanning from 18.7 years in Nairobi to 46.6 years in Tokyo. Evidence suggests that cities have to balance their performance across these three dimensions. Cities leading in one dimension often underperform in another dimension or all the dimensions. Zurich tops the overall ranking and the planet sub-ranking. It performs well concerning profit but occupies only the 27th place in relationship to people. European cities dominate the ranking with 13 European cities in the top 15. Singapore in 2nd place and Seoul at 7th place are also among the most sustainable cities. Vancouver appeared as the highest ranking city in North America and New York the most sustainable city in the US. Australian cities are in the top half of the ranking, whereas rapidly growing emerging cities sit mostly in the fourth quartile (Arcadis 2016).

The capacity of cities to grasp opportunity is critical for their strength to protect the global commons and ensure a better future for humanity. The PricewaterhouseCoopers Cities of Opportunity series provide another unique private sector-led ranking of global smart cities. London, New York, Tokyo, Amsterdam, and Paris are ranked at the top according to the criteria of intellectual capital and innovation, technology readiness, and city gateway. The intellectual capital, the most precious element of the urban ecosystem, does not just happen, but can be developed. Similarly, innovation is not a commodity, but a human process arising from creativity and an environment sparking ideas and supporting interactions. Cities can catalyse multiple chain reactions cross-pollinating intellectual capital. The 2016 top ten cities in intellectual capital and innovation include London, San Francisco, Paris, Amsterdam and Toronto tied in fourth place, New York, Los Angeles, Tokyo, Sydney, and Stockholm and Chicago sharing tenth place (PwC 2016).

Technology readiness is a criterion that frames the technological potential of a really smart city that uses technology to improve citizen life. This indicator has undergone substantial revision, enhancement, and expansion to include broadband quality. The score measures quality or connection reliability, speed, and value. New variables include mobile broadband speed, which complements the broadband quality score, digital performance, and security. The top ten cities include Singapore, London, Amsterdam, New York, Stockholm, Hong Kong, San Francisco, Tokyo, Paris, and Toronto. The third criterion tried to take the pulse of a city's social, economic, and cultural magnetism and connections. London remains first when comparing cities as gateways to the world. The French capital also performs very well on airport connectivity, and its role is to be further reinforced with the 2024 Olympic Games.

Singapore, the leading island state, is a major sociocultural and commercial hub, the fourth global financial centre, for decades the top port of the world, and always one of the five busiest ports. At the crossroads of the seas, Singapore has a vibrant multicultural population, and its world-class diversified economy depends on trade, especially manufacturing, representing more than one fourth of prosperity. The city-state has the third highest income per capita in the world and also ranks high in relationship to education, healthcare, and economic competitiveness, and also demonstrates a gold standard for water management (WEF 2016a).

In China, Shanghai, the top world port city, hosts more than 18 million inhabitants in search of harmonious coexistence between local and post-modern realities in the port, the old city, and futuristic Pudong. The city invests in education and innovation, favours the bicycle as a means of sustainable mobility, and capitalises on digital assets, while respecting traditions. In 2010, Shanghai was the first emerging metropolis to host a world exhibition on "Better City, Better Life," which invited the world to advance towards "Cities of Harmony." The Shanghai declaration highlights that citizen understanding and pursuit of a better life are both the foundations and the engines of urban development.

1.4.3 Blue Coalitions of the Willing for the Global Commons and the Longer Term

Strategic cities of power and harmony can lead by example and by network emulation to prevent, in extremis, the "tragedy of the public commons," resulting from the overexploitation of precious environmental resources, both terrestrial and marine. The interconnected local democratic leadership can be instrumental for global awareness and radical change in direction. Local short-term profits should not prevail but rather serve global collective long-term efforts. The joint consequences could be highlighted through the well-known metaphor of the 'prisoner's dilemma,' which advocates for the art of cooperation, principled negotiation, and concerted action.

Cities can do much for the global ocean, starting with significantly reducing their emissions and waste and cultivating a public consciousness of connectivity and multi-stakeholder engagement with the global commons. Climate action gave a new impetus to coalitions and alliances and multi-stakeholder forums, increasingly involving citizens and the civil society. Political short-termism and human greed have been identified as the main causes for the outgrowth of humanity. Bold leaders at all levels call for a radical shake-up in politics and business to advance on climate change, reduce inequality, and improve individual behaviour and corporate practices (WWF 2015a, b, 2016).

More and stronger coalitions could accelerate action on climate change and the ocean, through initiatives for energy-efficient buildings, faster market penetration of efficient technologies, and reduction of emissions. Thinking and acting "Now for the Long Term," the Oxford Martin School suggested a world pro-climate coalition composed of G20 countries, 30 companies, and 40 cities. Cities can also help end discrimination against future generations by revising discounting methods and adjusting them to take account of the evolving uncertainties, risks, and ethical implications for the following generations (Oxford Martin School 2013).

The Global Covenant of Mayors for Climate & Energy is a great example of a strong global network building forward momentum. In June 2017, 7451 cities, representing 674,790,274 citizens worldwide and 9.32% of the total global population, engaged with the Covenant, bridging two major EU and UN initiatives. Urban diplomacy is a rising function reinforcing engaged local leadership. Cities can work together to change the fundamental conditions that gave rise to the accumulating natural and marine problems and make a systemic difference of amplified significance.

Cities can lead alliances for a sustainable blue economy and the health of the marine resources that underpin it. Blue growth comprises a broad array of activities related to food, jobs, and opportunities provided by the oceans and coastal assets. The concept emphasises conservation of ecosystems and habitats and sustainable management of aquatic resources and equitable benefits to coastal communities which are directly dependent on them. If the marine ecosystems are in good environmental status, responsibly sharing a space as immense as the 70% of the globe would bring multiple benefits to all and everyone (EC 2017a).

Thirty-five years after the signature of the UN Convention of the Law of the Sea (UNCLOS), known as the Constitution for the Oceans, which came into force in 1994, ocean governance is recognised as a major issue. Advanced technologies such as robotics, marine biotechnologies, or underwater systems call for the regulation of new activities, especially in deep sea waters, in areas beyond national jurisdiction. Better coordination is needed for marine protected areas, environmental impact assessments, capacity building, and rules on the transfer of marine technology, genetic resources, and benefit sharing. The private sector, operating in transport, oil and gas, wind and ocean energy, fisheries, aquaculture, or coastal tourism, is an important player, and marine social society organisations are active drivers of change (EC 2016c).

Engaged eco-responsible cities could go far beyond their immediate horizons and lead concerts of human blue alliances for healthy and productive oceans. About one third of the global ocean is controlled and managed by governments. But in the high seas, Earth's last precious commons, care and monitoring are limited. High seas are no longer protected by their relative inaccessibility. Their complex ecosystems are subject to negative impacts from human activities ranging from shipping to deep sea overfishing, all compounded by a lack of comprehensive knowledge, effective instruments, and coherent governance. A landmark resolution, adopted at the UN General Assembly on 24 December 2017, marked the beginning of formal negotiations for an international treaty to conserve and sustainably use the high seas. The decision paves the way for a range of measures, such as a much-needed system of global marine protected areas to sustain ocean life. United in their engagement to help reverse ocean degradation, cities could be key advocates (OECD 2017c).

Eco-responsible interconnected cities can bring a major contribution to the protection and enhancement of the world blue commons, following, for example, the recommendations by the Global Ocean Commission (GOC), which, in 2014, proposed closing seas, ports, and markets to illegal, unreported, and unregulated fishing. Since then, many cities have joined the Port State Measures Agreement (PSMA), which was entered into force in 2016 as a binding, international treaty designed to combat illegal, unreported, and unregulated (IUU) fishing. By denying such unscrupulous fishers safe haven and access to markets, the PSMA is expected to improve seafood practices and drive cities toward fairer trade and greater sustainability. Through their networks and communication with their peers, cities can help win the fight against IUU fishing. The EU Fair Trade City award can inspire empowering measures. No port city can be proud to be targeted by illegal operators as a shelter for unsustainable practices regarding any commodity, not least the most emblematic fisheries (GOC 2014, 2016).

Cities could play an influential role in a multi-stakeholder breakthrough coalition for the seas to enhance global ocean accountability and demonstrate that they do care not only about their immediate territories but the whole planet. Together with their industries, universities, and civil society associations, they can engage in the protection of watersheds and ocean biodiversity, including care and recovery for the high seas, the ending of overfishing and harmful subsidies, the closing of ports and markets to illegally harvested seafood, and the prevention of pernicious marine litter, especially plastics. Having more plastics than fish in the oceans by 2050 is a fatality that cannot be permitted (WEF, Ellen McArthur Foundation and McKinsey 2016).

References

Arcadis (2016) Sustainable cities index 2016. Putting people at the heart of city sustainability, Amsterdam

Attali J (2016) Vivement après demain. Fayard, Paris

Attali J (2017) Histoire des Mers. Fayard, Paris

C40 and Arup (2015) Powering climate action: cities as global changemakers, New York

C40 and Arup (2016) Deadline 2020: how cities will get the job done, New York

Camdessus M (2017) Le Monde vers 2050. Flammarion, Paris

City of New York (NYC) (The) (2015a) One New York. The plan for a strong and just city. New York

City of New York (NYC) (The) (2015b) Building a Smart + Equitable City. New York

Club of Rome (2012) 2012–2052 – a global forecast for the next forty years, Rome

Economist (The) (2017) The ocean sustains humanity. Humanity treats it with contempt, London

Economist (The) Intelligence Unit (2015) Ocean world summit. Executive summary

European Commission (EC) (2015) Global trends to 2030: can the EU meet the challenges ahead? ESPAS report. Brussels

EC (2016a) Opportunity now! Europe's mission to innovate, Brussels

EC (2016b) Shaping the future. Thoughts on the future of society and governance. European Strategy and Policy Analysis System, Brussels

EC (2016c) Global ocean governance. Joint Communication, Brussels

EC (2017a) Report on the blue growth strategy. Towards more sustainable growth and jobs in the blue economy, Brussels

EC (2017b) Report from the commission to the Council on the Urban Agenda for the EU, Brussels

EEA (2015a) The European environment. State and outlook 2015. Synthesis report, Copenhagen

EEA (2015b) The European environment. Assessment of Global Megatrends, Copenhagen
European Foundation for the Improvement of Living and Working Conditions (EFILWC) (1997) European cities in search of sustainability. A panorama of urban innovations in the European Union, Dublin
European Union Institute for Security Studies (EUISS) (2012) Global trends 2030 citizens in an interconnected and polycentric world, Paris
FAO (2016) The state of world fisheries and aquaculture. Rome
Global Ocean Commission (GOC) (2014) From decline to recovery. A rescue package for the Global Ocean, Oxford
GOC (2016) The future of our ocean. Next steps and priorities, Oxford
Gore A (2013) The future: six drivers of global change. Random House, London
Gulbenkian Ocean Initiative (GOI) (2013) Blue natural capital: towards a new economy, Lisbon
Hall P (2013) Good cities, better lives: how Europe discovered the lost art of urbanism. Routledge, London
International Energy Agency (IEA) (2016a) World energy outlook. Paris
IEA (2016b) Energy technology perspectives 2016. Towards sustainable urban energy systems, Paris
IEA (2017a) Energy technology perspectives 2017, Paris
IEA (2017b) World energy outlook, Paris
International Organization for Migration (IOM) (2015) World migration report: migrants and cities, new partnerships to manage mobility, Geneva
International Panel of Climate Change (IPCC) (2014a) Climate change 2014: impacts, adaptation and vulnerability. Bonn
IPCC (2014b) Climate change fifth assessment report, Bonn
IPPC (2014) Climate mitigation, Bonn
Khanna P (2016) Connectography: mapping the future of global civilisation. Random House, London
Kohli HS (ed) (2016) The world in 2050. Oxford University Press, Oxford
Mega V (2013) Quintessential cities, accountable to the future. Innovation, sustainability and leadership. Springer, New York
Mega V (2015) Conscious coastal cities. Sustainability, blue green growth and the politics of imagination. Springer, Cham, Switzerland
MGI (2014) India's economic geography in 2025: states, clusters, and cities
Millennium Project (The) (2017) 2017 State of the future report and index, Washington, DC
Mori Memorial Foundation (The) (2017) Institute for Urban Strategies Global power city index, Tokyo
National Intelligence Council (NIC) (2017) The paradox of progress: global trends 2035, Washington, DC
Organisation for Economic Co-operation and Development (OECD) (2012a) Green cities: new approaches to confronting climate change. Paris: OECD
OECD (2012b) Environmental outlook to 2050: the consequences to inaction. OECD, Paris
OECD (2013) The competitiveness of global port cities. Synthesis report. OECD, Paris
OECD (2016a) The ocean economy 2030
OECD (2016b) Horizon scan of megatrends and technology trends in the context of future research policy. OECD publishing, Paris
OECD (2016c) Making cities work for all, data and actions for inclusive growth. OECD publishing, Paris
OECD (2017a) Beyond shifting wealth: perspectives on development risks and opportunities from the global south. OECD publishing, Paris
OECD (2017b) Trust and public policy: how better governance can help rebuild public trust. OECD publishing, Paris
OECD (2017c) Measuring distance to the SDGs targets; an assessment of where OECD countries stand. OECD publishing, Paris

OECD (2017d) International migration outlook. OECD, Paris
OECD–FAO (2017) Agriculture outlook 2017–2026, Paris
Oxford Martin School (2013) Now for the long term, Oxford
Piketty T (2015) Capital in the twenty-first century. Harvard University Press, Cambridge
Port of Rotterdam (2012) Port vision 2030, Rotterdam
PricewaterhouseCoopers (PwC) (2016) Cities of opportunity 7: the living city, London
STOA (2011) E-public, e-participation and e-voting in Europe: prospects and challenges. European Parliament Research, Brussels
STOA (2013) Security of e-Government: E-passport, E-procurement, E-Health. European Parliament Research, Brussels
STOA (2015) Ten technologies which could change our lives. European Parliament Research Service, Brussels
Tapscott (2016) Blockchain Revolution How the Technology Behind Bitcoin and Other Cryptocurrencies Is Changing the World. New York: Penguin Random House
United Nations (UN) (2015a) World population prospects: 2015 revision, New York
UN (2015b) Global sustainable development report, 2015th edn, New York
UN (2016a) Progress towards the sustainable development goals 2016, New York
UN (2016b) The world's cities in 2016. Data booklet, New York
UN (2016c) First ocean assessment, New York
UN (2017a) UN Population prospects. The 2017 revision, New York
UN (2017b) Progress towards the sustainable development goals 2017. Report to the UN Secretary-General. New York
UN (2017c) Our ocean: call for action, New York
UN Conference on Sustainable Development Rio+20 (2012) Future perfect, Rio de Janeiro
UNDP (2016) Human development for everyone, New York
UN Environmental Programme (UNEP) (2011) The other 70%. Paris
UNEP (2016) Global environmental outlook, 6th edn, Paris
UNEP (2017) The emissions gap report 2017. A UN environment synthesis report, Paris
UNESCO/IOC (2017) Global ocean science report, Paris
UN-HABITAT (2016) The new urbana agenda, Nairobi
West FM (2016) Megachange: economic disruption, political upheaval, and social strife in the 21st century. Brookings Institution Press, Washington, DC
World Bank (2012) Inclusive green growth. The pathway to sustainable development. World Bank, Washington, DC
World Bank and United Nations Department of Economic and Social Affairs (2017) The potential of the blue economy: increasing long-term benefits of the sustainable use of marine resources for Small Island developing states and coastal least developed countries. World Bank, Washington, DC
World Business Council on Sustainable Development (WBCSD) (2010a) Vision 2050. Geneva
WBCSD (2010b) A world of sustainable cities, Geneva
WBCSD (2014) The urban infrastructures initiative. Final report, Geneva
World Economic Forum (WEF) (2015) Top ten urban innovations. Geneva
WEF (2016a) The global competitiveness report, 36th edn, Geneva
WEF (2016b) The global gender gap report, 11th edn, Geneva
WEF (2016c) Forth industrial revolution, Geneva
WEF (2017) Harnessing the fourth industrial revolution for oceans, Geneva
WEF (2018) The global risks report, 13th edn, Geneva
WEF, Ellen McArthur Foundation, McKinsey and Co (2016) The new plastics economy: rethinking the future of plastics, Geneva
World Resources Institute (2016) Towards a more equal city: framing the challenges and opportunities, Washington, DC
World Wide Fund (WWF) (2015a) Living blue planet report. Species, habitats and human well-being
WWF (2015b) Reviving the ocean economy: the case for action
WWF (2016) Living planet report

Selected Websites and Social Media

#TAaward18
#BasqueDeclaration
#oceansummit
www.c40cities.org
www.atkearney.com/global-cities
www.deltacities.com
www.citymayors.com
www.csiro.au
www.eea.org
www.en.unesco.org/ocean-decade
www.espas.eu
www.europarl.europa.eu/stoa/
www.habitat3.org/
www.gov.uk/government/collections/future-of-cities
www.globalcovenantofmayors.org
www.goc.org
www.iom.int
www.lincolninst.edu
www.metropolis.org
www.millennium-project.org
www.nesta.org.uk
www.nexcities.org
www.nyc.gov/forward
www.oecd.org
www.ourocean2017.org/commitments
www.panda.org
www.sdg.iisd.org/
www.unhabitat.org
www.undp.org
www.unep.org
www.unfpa.org
www.un.org/sustainabledevelopment/sustainable-development-goals/
http://unsdsn.org/
www.unesco.org/ioc
www.wef.org
www.wri.org

Watercolour 4
Tokyo, the Largest and Among the Most Powerful Global Cities

Chapter 2
Threatened Urban and Ocean Biodiversity: The Imperative of Resilience

Abstract Humanity is already using the natural resources of almost 1.7 planets like the "Earth" to provide the goods and services needed each year, and this trend cannot lead to sustainable futures. The 2017 Earth Overshoot day was 2 August. In just 7 months the global community consumed all the resources that the planet can sustainably regenerate over an entire year. Authoritative studies highlight that the prevailing economic models led to overconsumption in multiple forms, resulting in disastrous effects for the environment and also for the economy and society, if examined from a longer-term perspective, and undermined the future of humankind.

Cities and seas are vital ecosystems able to mobilise scarce resources, nutrients, and materials, to ensure food security, and to offer sustainable goods and services. But the fundamental resources of air, water, and soil are under extreme pressure in many cities, whereas the global ocean suffers from pollution that often originates in cities. Urban air pollution continues to rise at an alarming rate and has become a critical risk for human health. Overfishing, pollution, and waste, especially plastics, put ocean health at risk. The essential nexus "water–food–energy," to which many add health, is under particular stress in cities. Healthy oceans can contribute with seafood, renewable energy, and ecosystem services.

Resilience, the capacity to rebound, is critical for cities, major concentrators of resources, food, materials, and products supporting many diverse human activities in a condensed space of intense interactions. Eco-responsible cities try to reduce their ecological debts on land and in the sea and increase biocapacity, their ability to renew their assets. Some exemplary analyses and models of emission-neutral, waste-free, and ecological functional areas provide inspiration for reconciling the technosphere with the biosphere. They also highlight that communities and stakeholders take a cardinal role for the exploration, exploitation, and enhancement of their precious urban capital and the global commons, especially oceans.

V. P. Mega, *Eco-Responsible Cities and the Global Ocean*,
https://doi.org/10.1007/978-3-319-93680-2_2

2.1 The Impacts of Excessive Consumption, Pollution, and Overfishing

2.1.1 Cities and the Seas, Global Assets at Risk: Is There Still Time to Reverse Impacts?

Cities must offer citizens the possibility to lead a fulfilling life in balance with nature and within the boundaries of this bountiful planet. They have to provide clean air and fresh water, healthy food, shelter, energy and transport, materials, services, and opportunities for education, employment, and leisure. For this, cities interact with their surroundings and impact a broader territory. Their ecological footprints reflect their resource consumption patterns and the production of waste and emissions over multiple value chains. Cities can collectively influence more than 70% of the global ecological footprint (Gore 2013; Mega 2013).

The global ocean produces almost half the oxygen supporting life on Earth and absorbs more than a quarter of the carbon dioxide emitted into the atmosphere. It has been described as the kidney of the planet or as a huge biological pump at the heart of global atmospheric and thermal regulation and water and nutrient cycles. But the ability of the ocean to continue providing these essential ecosystem services is being compromised by marine pollution, environmental degradation, overfishing, rising temperatures, higher sea levels, and acidification and dead zones, the largest of which at the mouth of the Mississippi River reached a size equivalent to half the area of Switzerland (GOC 2016; West 2016).

Humanity consumes far more than the ecological resources and services that the planet can annually regenerate. Earlier and earlier each year, the Global Footprint Network reports the Earth Overshoot Day, when demand for natural resources and ecological services starts surpassing global supply. In 2017, the Network reported that Earth Overshoot Day, the day humanity has used all the natural resources that the Earth could renew during the full year 2017, was the 2nd of August: this means that humanity is consuming resources almost as great as those of 1.7 equivalent planets. In the wealthier countries, the Earth Overshoot Day happened earlier. In Portugal, the 2017 Overshoot Day coincided with World Environment Day, on June 5. It sounded a clarion call, as it indicated the date that the global demand for natural resources would exceed nature's annual budget if everyone on the planet lived as an average Portuguese citizen.

The ecological footprint provides a measure of the bioproductive land and sea area required to provide the essential resources and absorb the resulting waste and emissions. A number of city footprint assessments highlight that cities tend to have larger footprints than countries, on average. An assessment of the ecological footprint of 19 coastal cities in the Mediterranean region suggests that Valletta, Athens, and Genoa are cities with the highest per capita ecological footprint, almost double the per capita ecological footprint of the region. Food and transport are the main footprint drivers in cities from the south and the north rim of the Mediterranean, respectively (Baabou et al. 2017).

The ecological footprint per capita in high-income countries is in average almost three times that of middle-income countries and about six times that of low-income countries. Carbon emissions make up more than half the footprint of high- and middle-income countries; biomass-based components still represent the largest footprint share for low-income countries. A discussion organised by the Global Footprint Network in Guimarães, during the city's Green Week festival, explored the multiple consequences and initiated action. Ten Portuguese cities and regions analysed their footprints.

There is urgency to move the date of Earth Overshoot Day back on the calendar. Humanity is jeopardising its very future, on land and the sea. A critical threshold was already crossed in the 1970s when global demand started outstripping the planet's biocapacity. Since then, ecological overspending has entered a vicious cycle. As resource deficits become larger and resource prices remain high, pressures become unbearable and the accumulation of debt weakens the ability of communities to initiate virtual cycles. Unbridled and unconscious consumption has led to the depletion of many finite resources, and the consumption of renewable resources at a pace more rapid than their rate of regeneration (WWF 2015a, 2016).

The major sixth Global Environmental Outlook (GEO-6) published by the UN Environment Programme (UNEP) in 2016 concludes that environmental threats intensify across the planet, but there is still a window of opportunity to act and prevent disaster. A series of regional reports on the state of the planet's health signal that environmental deterioration is occurring much faster than previously thought and that action is more urgently needed than already had been advocated (UNEP 2016a; IPBES 2018).

The GEO-6 regional assessments indicate that the global community shares common environmental risks in many parts of the world. In almost every region, population growth, urbanisation, rising levels of consumption, desertification, land degradation, and climate change have led to severe water scarcity. The assessments suggest that there is still time to tackle many of the worst impacts of environmental decline, such as the damage to marine ecosystems and the rising level of air pollution, but urgent action is needed.

2.1.2 Air Pollution in Cities, the World's Most Widespread Public Health Risk

Air pollution critically affects humans, in both the natural and the built environment. Poor air quality, both ambient and indoor, can cause serious health problems. Exposure to air pollution is largely a multi-pollutant process. The primary source of air pollution is the combustion of fossil fuel in energy generation, industrial processes, and transport. Contaminants include nitrogen oxides (NO_X), particularly nitrogen dioxide (NO_2), known to cause specific damage to lung tissues and to contribute to acidification, eutrophication, and photochemical smog. Most NO_X emissions are produced by energy production and consumption. Despite cuts in nitrogen oxides and non-methane volatile organic compounds, air pollution caused by ozone persists. Heat waves could further boost ozone exposure.

Urban air pollution continues to rise at an alarming rate and significantly affects human health, causing respiratory and cardiovascular-related illnesses in particular. Around 6.5 million deaths are attributed each year to poor air quality, making this the world's fourth-largest threat to human health, after high blood pressure, dietary risks, and smoking. At the same time, awareness is mounting and more cities are monitoring their air quality. The World Health Organisation (WHO) database included, in 2016, 3000 cities in 103 countries, almost double that of 2 years earlier, because of the widening recognition of the associated health impacts, risk of stroke, heart disease, lung cancer, and chronic and acute respiratory diseases, including asthma. High concentrations of small and fine particulate matter (PM10 and PM2.5), including pollutants such as sulfate, nitrates, and black carbon, represent the greatest risk for public health.

Cities can easily become pollution hotspots, as they concentrate people, resources consumption, construction activity, and traffic. Most sources of urban outdoor air pollution are well beyond the control of individuals and demand action by cities, as well as national and international policy makers to promote cleaner and more efficient energy production, sustainable transport, and sound waste management. Reducing industrial emissions, increasing use of renewable power sources and clean transport, such as walking and cycling, are among the most common policy measures.

Air quality guidelines offer global direction on thresholds and limits for key air pollutants that pose health risks. The WHO guidelines indicate that by reducing particulate matter pollution from 70 to 20 micrograms per cubic metre ($\mu g/m^3$), air pollution-related deaths could be reduced by roughly 15%. Many cities monitor ambient air quality and witness concentrations of particulate matter above the limits indicated by the WHO. All regions of the world are affected, but populations in low-income cities are the most impacted. An analysis of a total of 795 cities in 67 countries and their levels of small and fine particulate matter, during the period 2008–2013, provided some interesting insights. Even if the developed world cities are less concerned that the cities of the developing world, many of them do not offer acceptable quality air to their citizens (Friends of Europe 2013; WHO 2017).

Global urban air pollution levels increased by an average of 8%, during the period 2008–2013, despite improvements in some regions. In general, urban air pollution levels were lowest in high-income countries, with lower levels most prevalent in Europe, the Americas, and the Western Pacific Region. The highest urban air pollution levels were experienced in low- and middle-income countries in Eastern Mediterranean and Southeast Asia, with annual average levels often exceeding five to ten times the recommended WHO limits. In Africa, urban air pollution data are more than imperfect; however, available data have revealed particulate matter levels above the average. More than half the monitored cities in high-income countries and more than one third in low- and middle-income countries reduced their air pollution levels by more than 5% in 5 years.

The energy sector is responsible for a great part of urban air pollution. Without changes to the way that the world produces and uses energy, the toll from air pollu tion on human life is certain to rise. An IEA 2016 World Energy Outlook Special Report has been dedicated, for the first time, to the links between energy, air pollution, and health. The report sets out the scale, causes, and effects of the problem and the ways in which the energy sector can contribute to solutions. Energy production and use, mostly from unregulated, poorly regulated, or inefficient fuel combustion, are the single most important sources of air pollutant emissions, the majority (85%) of particulate matter, and almost all the sulfur oxides and nitrogen oxides. Coal is responsible for about 60% of global combustion-related SO_2 emissions, a source of respiratory diseases and a precursor of acid rain. Fuels used for transport, and especially diesel, generate more than half the nitrogen oxides emitted globally, which can cause respiratory problems and the formation of other hazardous particles and pollutants, including ozone (IEA 2016).

Even in the European Union (EU), the air pollution problem is still far from being solved, despite falling emission levels and reductions of some air pollutant concentrations in recent decades. Governments increasingly try to involve stakeholders through platform debates and make citizens more aware of air quality through interactive information services, such as "Dublinked." Measurements from five air monitoring sites around Dublin city provide hourly results for SO_2, NO_2, carbon monoxide (CO), and particular matter. Black smoke monitoring is also carried out, although this source has been radically reduced since the introduction of the coal ban in heating in the early 1990s.

In 2016, the Brussels government adopted the Plan "Air-Climate-Energy" to significantly reduce emissions and improve quality of life in a fair "burden-sharing" agreement with the regions. Air pollution is mainly caused by traffic and marked by repeated violation of NO_2 levels, notably from diesel. The target is to decrease greenhouse gas (GHG) emissions by 30% by 2025. The plan followed public consultation and comprises 64 measures including the creation of permanent low-emission zones and the promotion of exemplary clean buildings. The concept of low-emissions zones has already been implemented in Sweden since 1996 and in Germany and the UK since 2008. Other measures include cogeneration of electricity and heat, with an increasing share of renewables, and digital solutions for the smart city.

2.1.3 Journey in Regional Environments on Land and in the Sea

In the Asia Pacific region, unprecedented economic growth and unsustainable consumption patterns have led to worsening air quality, water scarcity, and waste generation, threatening human and environmental health. the increased demand for natural resources, extensive agriculture, palm oil, illegal logging, unsustainable aquaculture, and the illegal wildlife trade are causing environmental degradation and biodiversity loss.

The Asia Pacific region continues to be the world's most disaster-prone region. About 41% of all natural disasters reported during the past two decades occurred in the region, which also accounted for 91% of the world's deaths attributable to natural disasters in the last century. By the 2070s, top Asian cities in terms of population exposure to coastal flooding could include Bangkok, Dhaka, Guangzhou, Kolkata, Mumbai, and Shanghai, threatening hundreds of millions of people. Deforestation is a major environmental problem in the region and depletes another critical carbon sink. In Southeast Asia, the average area deforested annually between 2005 and 2015 was more than 1 million hectares, resulting in the release of hundreds of millions of tonnes of carbon dioxide every year.

Exploding demographic and economic growth, a rising middle class, and intensified industrialisation has led to increasingly unhealthy, polluting, and carbon-intensive lifestyles in many Asian cities. Unsafe sanitation, disposal of untreated wastewater, and runoff of agrochemicals are responsible for a rise in water-borne diseases, especially in population-dense urban areas. Uncontrolled dumping, which is still the main solid waste management method in the region, is also a major source of disease. In Mumbai, for example, about 12% of total municipal solid waste is still burned either openly on the streets or in landfills (UNEP 2016a).

Organic waste flowing through cities is also significant and has a huge untapped potential. Municipal organic waste, from solid waste streams and wastewater from sewage systems, is traditionally seen as a costly economic and environmental problem. The World Economic Forum estimates that potential global revenues from the biomass value chain, comprising the production of agricultural inputs, biomass trading, and biorefinery outputs, could be as high as $295 billion by 2020. The bio-cycle share in the economy is higher in developing countries, where the share of waste is also higher. Case studies suggest that shifting the system towards one aligned with circular economy principles and designing more effective recovery and processing systems can turn organic waste into a source of revenue and help to restore natural capital and prevent streams ending in the ocean (Ellen MacArthur Foundation 2017).

The expanding global middle class is the main driver for accelerating domestic consumption. The size of the global middle class is projected to increase from 1.8 billion (2009) to 4.9 billion in 2030, with most of this growth coming from Asia. The OECD predicts that global spending by the middle class will grow from $21 trillion in 2016 to $56 trillion by 2030 and that more than 80% of this increase is expected to come from Asia and the Pacific.

In West Asia, a rise in the amount of degraded land and the spread of desertification are having profound economic and environmental impacts. The region is suffering from overexploitation of groundwater resources and deteriorating water quality, as well as unsustainable patterns of consumption. High population growth and rolling conflicts mean that the carrying capacity of the land has become too low to support people with freshwater and food. Only 4 of 12 countries in West Asia are above the water scarcity limit of 1000 cubic metres (m^3) per person per year (UNEP 2016a).

Continuous conflict and the mass displacement of people throughout the region are also triggering severe environmental impacts that are endangering people's health. Heavy metals from explosive munitions and radiation from missiles have leached into the environment as a result of the region's conflicts. The 2.97 million refugees in Lebanon, Jordan, Yemen, and Iraq are placing an immense environmental burden on the region, overwhelming governments, and increasing the risk of disease outbreaks.

In Africa, the fundamental resources of land, water, and air are threatened from pollution and inadequate infrastructures, especially in bourgeoning cities. Soil erosion, salinisation, pollution, and deforestation led to land degradation damaging agricultural productivity, nutrition, and human health, especially in sprawling urban areas. A growing population and a rise in the demand for firewood could impact the forest cover in Africa, which is likely to continue shrinking, declining to less than 600 million hectares by 2050. Many of the region's fisheries, both inland and marine, face overexploitation from illegal, underreported, and unregulated fishing. The continent's large young population has an opportunity to drive growth. Low-carbon, climate-resilient choices can develop the continent's infrastructure, accelerate industrialisation, increase energy and food production, and promote sustainable natural resource governance. Indoor air pollution is responsible for 600,000 premature deaths every year in Africa, given the continent's reliance on the use of biomass for cooking, lighting, and heating.

The proportion of the population with access to clean water is increasing, from 64% in 2005 to 68% in 2012, although absolute numbers remain high. More than half of the population in Sub-Saharan Africa still does not have any access to improved sanitation, compared to 90% coverage in North Africa, with a vast difference between urban and rural areas. African megacities such as Cairo, Kinshasa, and Lagos, and emerging megacities such as Dar es Salaam, Johannesburg, and Luanda, face challenges from poor management of sanitation services because of inadequate infrastructure. Land-based activities causing pollution of freshwater bodies ultimately impact coastal and marine resources.

In North America, environmental conditions have improved thanks to policies, institutions, data collection and assessment, and regulatory frameworks. Air quality, in particular, continues to improve in response to concerted action and favourable trends in technology and energy markets. Drinking water quality is extremely good, although water scarcity is of increasing concern. A valuable network of well-managed protected areas is helping to conserve biological diversity. However, environmental challenges have emerged in recent years. Climate change is generating impacts across the region, and aggressive hydrocarbon extraction methods could bring increased emissions, water use, and induced seismicity. The coastal and marine environment is under increasing threat from nutrient loads, ocean acidification, ocean warming, sea level rise, and marine debris (UNEP 2016a, b; UNEP/GRID-Arendal 2016).

Climate change has damaging effects on the environment, on human health and well-being, and, in some cases, on human security. In 2012, Hurricane Sandy was directly responsible for approximately 150 deaths and $70 billion in asset losses. The 30 centimetres (cm) of sea level rise off New York City since 1900 likely expanded Hurricane Sandy's flood area by approximately 65 square kilometres (km), flooding the homes of more than 80,000 additional people in New York and

New Jersey (WRI 2012). A 5-year drought around the state of Texas ended in 2015 with devastating floods. Findings suggest that global warming exacerbated the drought by approximately 15% to 20%.

The Arctic is experiencing a profound transformation largely driven by interacting forces of climate change and increased human activities. As one of the first areas of the world to experience the most severe impacts of climate change, the Arctic region serves as a barometer of change for the rest of the world. Warming in the region has increased at twice the global average since 1980 and, during the past 20 years, there has been a progressive and dramatic decrease in the extent of summer sea ice, which has led to an increased surface area of blue water during the summer months. The largest contributions to global glacier ice loss during the early twentieth-first century were from glaciers in Alaska, the Canadian Arctic, and the Greenland ice sheet, as well as in the Southern Andes and Asian mountains. Together these areas account for more than 80% of the total ice loss. The melting of sea ice has created new expanses of open ocean, allowing large populations of phytoplankton to bloom, altering the marine food chain (UNEP 2016a).

2.1.4 *The Fisheries Crime and the Sunken Eco-Billions*

Fisheries constitute an emblematic resource and a pillar of the global economy. Heavily subsidised overfishing has often been described as a major crime. Global fisheries are in crisis as their productivity has decreased tremendously, evidenced by the fact that catches did not increase as rapidly as the global level of effort. Uncertainties abound, including mounting impacts of climate change, sea level rise, increasing ocean temperatures, acidification, and changes in patterns of the currents. The Code of Conduct for Sustainable Fisheries, adopted by FAO in 1995, after 2 years of discussions, highlights the limits of voluntary versus compulsory obligations.

Biological overfishing has led to economic overfishing and urgency for reduction of global fishing activities to allow fish stocks to recover from overexploitation. The World Bank study on the Sunken Billions, revisited in 2017, suggests that fisheries that were fully fished, overfished, depleted, or recovering from overfishing increased from just over 60% in the mid-1970s to about 75% in 2005 and almost 90% in 2013: this has serious consequences and costs. Reducing the global fishing effort would allow fish stocks to recover from overexploitation, boosting the profitability of the sector from an estimated $3 billion to $86 billion per year (World Bank 2017).

The concept of maximum sustainable yield, embedded in the United Nations Convention on the Law of the Sea, is instrumental for the restoration of fisheries stocks. It is at the heart of the EU common fisheries policy and has already given results in EU seas, especially the Atlantic Ocean and North Sea. Managing capture fisheries according to maximum sustainable yield has been successfully implemented in many single-species fisheries. Sustainable management, taking account of fishery and ecological interactions among different fish stocks predating on each other or competing for food, is only beginning.

Eco-responsible cities should pioneer ethical and sustainable fishing practices. They should close their markets to Illegal, Unregulated and Unreported (IUU) Fisheries combated through initiatives such as the Safe Ocean Network. One of the most effective means to address IUU fishing is by implementing the FAO's Agreement on Port State Measures to Prevent, Deter, and Eliminate IUU fishing. The Port State Measures Agreement requires Parties to refuse port entry or use by fishing vessels from another country that is known or suspected to have engaged in IUU fishing. Strong and effective implementation of the Agreement will ensure consistent enforcement measures and improve coordination among countries in preventing loopholes and miscommunications that can favour illegal fishers.

Cities can introduce awareness campaigns, shed light on the role of seafood in healthy diet, and shift fish consumption patterns, which concentrate on a small number of species. Sustainable seafood education efforts bring together cities, schools, organisations, and programmes such as the Monterey Bay Aquarium's Seafood Watch Programme, to promote nutrition security. Cities could also bring attention to the assessment of the ecological footprint of food that gives seafood a priority place (WRI 2016).

Policies for fisheries and aquaculture should systematically identify risk and build responsive processes. Fish have an important role in global food security and nutrition and can also disproportionately impact rural and coastal areas, especially in less developed countries. Securing a future with healthy fish is a necessary part of feeding a growing population and providing inclusive economic opportunities (OECD 2015).

As many fisheries are overcapitalised and suffer from low and limited returns, prospects are uncertain. Forward assessments suggest that the world would gain 13% more fish in volume and at least 50 billion per year if all depleted stocks were recovered. There is clear evidence of biological and economic recovery of fish stocks that are managed in a sound way. Twenty-five years ago, Canada decided to protect the Northern cod stocks on the Grand Banks by closing a fishery that had all but collapsed from overfishing: this move ended more than 400 years of fishing tradition and caused 30,000 persons to become unemployed overnight. The bold and painful decision had positive results, and the cod stocks are slowly increasing. This example demonstrates that, given a chance, the ocean can recover its health and generate employment and growth (OECD-FAO 2017).

Aquaculture, seen as the future of fish production, had a spectacular development, from a niche market into the most important source of fish products for world consumers, and can grow by more than a third during the next 10 years. Average annual aquaculture growth in OECD countries has accelerated and averaged 2.1% per year in 2017. Globally, it is more spectacularly rapid, at 6% per year. Moreover, average prices of aquaculture products are also increasing, and efforts for ethical aquaculture are leading to better products and acceptance (OECD 2017b).

Cities can take measures and encourage fisheries and aquaculture to embrace reform that will put them on the path to greener growth and sustainable development. Global governance, markets, and trade can create a sounder frame for fisheries. The near-term vision for aquaculture is for the sector to produce more and better in smaller places and be less dependent on wild feedstock for feed, improve its environmental performance, and better interact with other coastal activities (OECD 2015).

Aquaculture and capture fisheries are co-dependent, as feed for many high-value farmed fish is provided in large part from capture fisheries, as is also highlighted in the "Inseparables" campaign of the European Commission. They are also competitors for space, markets, and potentially for labour and governmental support. Marked progress has been made in replacing feed sources from capture fisheries with other products, and especially waste from land-based agriculture and forestry could be instrumental.

The regulatory landscape is an important factor for the future growth of aquaculture: this includes national development plans, supranational policies, for example, the EU common fisheries policy asking for maximum sustainable yield and the respect of the landing principle, institutional innovation, certification, and spatial planning. Economic measures could send the right price signals to all actors along the value chain and help the most efficient producers to expand. A broader perspective and the involvement of all stakeholders can yield better informed and more effective policies (EC 2017a).

2.2 Towards Resourceful Urban and Marine Ecosystems

2.2.1 A Living Ocean, Seventh World Economy, at the Brink of Collapse

In 2015, the World Wide Fund for Nature conservatively estimated the value of ocean assets to be at least \$24 trillion. The ocean's wealth rivals the size of the world's leading economies and could rank seventh in global economy with an annual value of goods and services of \$2.5 trillion, but this precious capital is being seriously eroded. It seems that the ocean is changing more rapidly now than at any other moment in history. Research suggests that oceans are under such severe pressure that essential natural systems may simply stop functioning (OECD 2016; WWF 2015b).

The bi-annual report Living Planet alarmed us about the state of the natural capital upon which our societies and economies depend. Global populations of fish, birds, mammals, amphibians, and reptiles had already declined by 58% between 1970 and 2012, and this puts the world on a trajectory of a potential two-thirds decline within a span of the half century ending in 2020. The Living Planet Index, which measures trends in 10,380 populations of 3,038 vertebrate species, declined 52% between 1970 and 2010 (WWF 2016).

The Living Blue Planet report suggested that almost half of all marine vertebrate populations collapsed between 1970 and 2012. Although the most rapid decline was observed between 1976 and 1990, there is currently no sign of overall recovery at the global level. Beyond this generic trend, some fish species of importance for regional economies, livelihoods, and food declined even more dramatically. Around one in four species of sharks, rays, and skates is threatened with extinction, primarily the result of overfishing. Tropical reefs have lost more than half their reef-building corals during the past 30 years, and nearly 20% of global mangrove cover was lost between 1980 and 2005. Only a small part of the ocean is protected, and only a portion of this area is effectively managed. Increasing marine protected area coverage to 30% could generate up to $920 billion during 2015–2050 (WWF 2015a).

The global index masks considerable regional variations. Some marine populations have been recovering from previously depleted levels in northern latitudes while failing in tropical and subtropical regions. There are also seas with an unacceptable lack of data. The marine environment is largely unexplored and is far from being comprehensively monitored. The Zoological Society of London has created indices for a number of ecosystems and species types, providing sufficient robust data available from published reports, journal articles, and fisheries research.

Humanity is rapidly approaching unprecedented tipping points that, if unheeded, may inexorably lead to systemic collapse. Pollution and overfishing have already exceeded unacceptable levels. In the future, climate change is expected to have the greatest impact on the ocean capacity to provide its ecosystem services. The World Ocean Day, celebrated each year on 8 June, emphasises individual and collective responsibilities in sustainably managing ocean resources. Canada proposed World Oceans Day in 1992 as a country surrounded by three oceans and the longest coastline in the world (202,080 km), responding to a call to action by the Brundtland Report,"Our Common Future," in 1987. That groundbreaking document identified the need for global cooperation to mitigate the growing threats to the ocean. In 2006, Ocean Networks Canada became the world's first interactive ocean portal providing real-time deep sea and coastal data from all three coasts. In 2017, UNESCO organised the first ever Ocean's 8 Celebration, which honours national champions in ocean science, and launched the first Global Ocean Science Report, highlighting the importance of international cooperation to build capacity in global ocean science (UNESCO/IOC 2017).

The future role of ecosystems on land and in the sea for human well-being depends increasingly on their health and self-regenerating capacity. At the horizon of cities, many of them started introducing conscious resource-based management, recognising that humans are an integral part of the ecosystems, which are vital in supporting human life. The process emphasises the necessity of sound scientific understanding and strong citizen participation to address complex and often controversial issues. Policy makers at all levels and practitioners from across the public, private, philanthropic, and civil society sectors suggest that, given the urgent resource constraints, there is a need for a dramatic resource optimisation. Many cities try to optimise synergies between sectors and prevent losses through innovative integrated and cost-effective planning, and participative governance, without forgetting the global commons (GIZ and ICLEI 2014).

2.2.2 Urban Natural and Marine Capital in the Decade of Biodiversity

Marine resources are often perceived as "less finite" resources than those terrestrial. The deterioration of marine ecosystems has long been hidden by the apparent immensity, mobility and relative, until recently, inaccessibility of large parts of the global ocean. In 2002, the UN signatory countries of the Convention on Biological Diversity (CBD) committed to achieve, by 2010, a significant reduction in biodiversity loss including that in the ocean. That year of high and noble intent was defined as the International Year of Biodiversity and expanded later to the Decade of Biodiversity to frame efforts with a longer perspective (2011–2020).

However, in spite of hundreds of internationally agreed goals and voluntary commitments, the world continues on an unsustainable path. The OECD Environmental Outlook to 2050 suggested that the potential cost of inaction on climate change could be as high as 14% of average world consumption per capita in 2050. Continued degradation and erosion of the natural environmental capital, together with irreversible effects, could endanger two centuries of improving living conditions. Multiple-benefit best cases confirm that green growth is not prohibitively expensive and can help humanity to strike a renewed green deal (ICLEI 2012, LSE and ICLEI 2012; OECD 2012a).

The Cities and Biodiversity Outlook, the world's first global analysis of the links and opportunities between urbanisation and biodiversity, highlights a wide range of worthy initiatives by cities, local authorities, and sub-national governments in both developed and emerging countries. The assessment underlined the potential benefits from taking into full account terrestrial, freshwater, and marine ecosystem dynamics in urban planning and policy: this should be seen as a major opportunity to make giant leaps for the reduction of biodiversity loss and improvement of the quality of life. The global initiative "Cities in Biodiversity Hotspots," involving more than 250 cities located in and around the world's biodiversity hotspots, most of them in tropical environments, provides a platform for action and mutual learning (UN CBD, Stockholm Resilience Centre and ICLEI 2012).

Urbanisation is heavily drawing on natural resources on a global scale, with serious effects on biodiversity and ecosystem services. Urban regions must take increased responsibility for conceiving, designing, and implementing actions that take into account their profound connections with and impacts on the rest of the planet. Urban areas are expected to continue to expand faster than urban populations. The total urban area is likely to triple between 2000 and 2030, and urban populations are expected to nearly double in the same timeframe. Food, materials, and energy are thus drawn in great quantities from all parts of the world, despite the sustainability calls for giving priority to local and seasonal resources.

Dynamic world cities emerge as both the main drivers and beneficiaries of a paradigm shift towards green growth, proposed as a shortcut to sustainable development. Inclusive green growth brings opportunities for new environment-friendly businesses and jobs, while managing the structural changes for the transition to a greener society. Greening growth is necessary and can be efficient and affordable (World Bank 2012). A concert of international reports on green growth suggests that the way forward requires integrated innovative solutions to tackle political constraints, change deeply entrenched behaviours, and develop the right instruments to monitor progress (ICLEI 2012).

The sustainable management of natural resources, including marine resources and ecosystem services in an increasingly urbanised world, is a major challenge for cities, long typified as concrete jungles inhospitable to flora and fauna, ecosystems, and habitats. Depending on their values, performance, and governance, cities can be an opportunity or a threat for biodiversity. Seizing the opportunity asks for a mix of high-quality green and blue commons in dialectic symbiosis with dense and compact built-up areas.

Many cities together with their main stakeholders have declared their will and readiness to act and correct their enormous ecological footprints. Urban design and infrastructures have an important role. Population density, a compact form, and public transport significantly reduce the per capita ecological footprint. European cities such as Amsterdam and Copenhagen are witness of decades of investments in population densification, compact developments, and cycling.

Vancouver declared its ambition to become the greenest city of the world by 2020. The Greenest City Initiative, involving all levels of governance, citizens, and business in a range of projects, aligned all efforts towards this direction. The city, which never agreed to be crossed by a motorway, has the lowest carbon footprint among major North American cities through a dense built environment, limited urban sprawl, sustainable public transport, and many green and open public spaces. Long-term planning, investment in multimodal transport, and an open attitude towards migrants made the city environmentally friendly and culturally attractive.

A global system of cities wishing to lead the world towards sustainable futures has to closely cooperate with other actors such as national, regional, and local governments, multinational corporations, and civil society. Each of these actors has important roles in sharing and managing local and planetary resources. Actions by a consortium of municipalities or state governments operating at larger scales are likely to accomplish even more in addressing challenges of ocean sustainability. Partnerships across urban and rural communities are also expanding to address systemic sustainability issues and the interconnections and impacts on broader variable geometry territorial scales.

Effective stewardship of ecosystem services links cities to ecosystems outside their boundaries, and the multitude of actors and factors that shape and sustain the resource flows. This responsibility involves implementing the ecosystem approach of the Convention on Biological Diversity and supporting local governments in addressing the Aichi Biodiversity Targets, the 20 ambitious goals of the CBD's Strategic Plan for Biodiversity 2011–2020, many of which have been artfully reintegrated in the set of targets under SDGs. The targets provide a framework of action for all stakeholders. The first target suggests that by 2020, at the latest, people are aware of the values of biodiversity and the steps they can take to conserve and manage it sustainably (UN CBD 2010).

The "Fibre City: Tokyo 2050" is a vision for the future of the first world metropolis aiming at a radically new balance between the natural and built environment. It responds to four urban challenges, including a shrinking population, an ageing society, environmental crisis, and the risk of earthquakes. While unprecedented population declines are expected in Japan, the metropolis could benefit from preserving the freed-up land as a precious resource. The concept of the Fibre City 2050 focuses on the ways in which urban fibres can be used to construct an alternative image of the metropolis. As an imaginative view of Tokyo, from macro to micro, Fibre City intends to inspire global cities retrofitting for greater harmony with nature, better access to public transport and services, and enhanced liveability with rediscovery of historic features such as the Edo canals and bridges.

Offering citizens easy access to high-quality wetlands and local nature reserves is one of the great attributes of cities. Singapore and Perth have built elaborate urban forest canopy walks. More cities have designed imaginative parks in aquatic and marine settings. Central Park in New York City, Park Guell in Barcelona, the Pharo gardens in Marseilles, Griffith Park in Los Angeles, and Stanley Park in Vancouver are prime examples of accessible green urban oases. Stanley Park, in the heart of the built urban landscape, includes 400-hectare natural West Coast rainforest and enjoys scenic views of water, mountains, sky, and trees. Since its opening, in 1888, the City and the Park Board work together to preserve and protect the ecology of the park and ensure its sustainability, while enhancing Vancouver's liveability (Mega 2015; NYC 2015a, b).

Sydney has planned a large-scale scheme to ensure that every resident be within a 250-m walk of continuous green links that connect to major city parks. For several years Lisbon has been developing green areas to protect the environment and to improve the quality of everyday life. One of the conscious eco-projets, the Vale de Alcântara green corridor, is expected to become a structuring park of the Portuguese capital. This 3-km corridor, lined with tree-populated green areas and irrigated with recycled water, will be entirely crossed and enjoyed by bicycle or foot, from the city's major forests to the Tagus River.

In Dublin, more than 250 species of birds have been recorded and most of the observation coastal sites are easily accessible from the city centre by public transport. In Porto, the Passeio Alegre on the waterfront, a most spectacular landscape expanding 83 hectares along the river and the ocean, is teeming with happy pedestrians and visitors enjoying its services. The urban park is a meandering labyrinth of winding paths that extends over 10 km, covered by trees, shrubs, and aquatic plant species. From physical exercise to kite flying, the park supports a very diverse array of activities bridging the city with the Atlantic coast. Several themed gardens, including the garden of feelings, the garden of aromatic plants, the garden of roses, and the apple tree estate, offer oasis environments for citizens and visitors.

The City of Amsterdam hosts a vast network of rivers and lakes, as well as urban beaches and marine ecosystems with hundreds of species. From a small fishing village in the twelfth century to a world-class city in the twenty-first century, Amsterdam solemnly declared sustainable development its utmost goal. Although highly urbanised, the city has a wide diversity of flora and fauna and a unique network of urban canals and parks which provide recreational ecosystem services. The city's environmental policy plan, reviewed every 4 years, ensures that biodiversity is given priority, well integrated with its economic and social development. Just 6 km from Amsterdam's central station, the Ilperveld fen is part of the Natura 2000 sites in Europe, the larger world coordinated network of protected areas, established according to the Birds and the Habitats Directives of the European Union. The site offers breathtaking views of large flat fens, criss-crossed by a myriad of ditches and canals. Originally used for hay-growing and pasture, it became a municipal dump in the 1980s and was abandoned by the thousands of meadow birds. A new life has been injected into the site since its inclusion in the Natura 2000 framework and a large restoration has been undertaken. Farming and cattle grazing has also been reintroduced to maintain the wet meadows in good condition for the recovery of the ecosystems. The area is also very popular with birdwatchers, nature lovers, and school children.

Water parks are amongst the most vital urban commons and can integrate beautifully with green parks. Biodiversity and ecosystem services are global public goods, but local and regional authorities have the legal power to designate conservation areas and the ethical obligation to protect them through incorporation of biodiversity concerns into their spatial planning processes. Public commitment is essential for sustainable communities, which identify biodiversity as a precondition for resilient cities. The designation of marine protected areas and the development of green infrastructures offers cities the opportunity to integrate biodiversity and resilience into local and regional plans taking into account the land and the sea dimensions (UN Convention of Biological Diversity (CBD), Stockholm Resilience Centre and ICLEI 2012).

Attractive green areas and blue waterfronts are essential for populations and nature in cities to thrive. Copenhagen provides a prime example. An ancient Viking fishing village, strategically located at the entrance to the Baltic Sea, the Danish capital / European Green Capital in 2014 made the Blue and Green City one of its priorities. Swimming races and cliff-diving contests with athletes jumping into the harbour from the roof of the opera pay tribute to the city's long symbiosis with the sea. Copenhagen's ambitious goal of being carbon neutral by 2025 and the eco-metropolis pathway are among the best in the world.

Urban farming helped Detroit overcome the spiral of decline, construct social links, and create a new future through producing local and biotic features at the scale of a large city. Many more cities could become fertile and work to transform urban grey to urban blue–green spaces. Urban agriculture has the potential to become pervasive within cities. All roofs could host urban micro-farms and all walls could support vertical gardens and micro-farms. Vertical farming in tall buildings could replace less productive single-story greenhouses as the source of all city-grown produce. Food labs and experiments suggest that 'slow food' can have a long future (Gilsoul 2011).

Cities and agriculture are responsible for nutrients and waste, especially plastic packaging, entering the marine environment. The very qualities that make plastic useful, including its unmatched strength, light weight, and low cost, make it problematic once discarded. Plastic waste is making its way into the world's oceans, where it can take centuries or longer to completely decompose. Once in the ocean, plastics interact with the biosphere. It has been documented that plastics affect nearly 700 species, from plankton to whales. Apparently plastics not only harm animals externally, but their ingestion affects growth and make them less successful in reproduction and less able to detect and evade predators. The potential impact from plastics moving up the food chain is an important question that is actively being intensively researched.

Furthermore, many hundreds of fishing nets are lost or abandoned every year, often because of conflicts between trawlers. The problem of lost or abandoned fishing nets is severe because of the nondegradable netting materials, which can maintain 10% of their catching capacity once lost. Much waste comes from food packaging, which does not always meet the performance standards. Various research and innovation projects try to develop sustainable, eco-friendly food packaging coating products made from seaweed extracts and starch and applied to paper and cardboard in the form of a spray. This is a prime field for cities and businesses to join efforts (EEA 2013, 2014).

2.2.3 *Zero Urban Waste and Zero Marine Debris*

Urban and marine litter are prime symbols of resource inefficiency and inaction on land and in the sea. Marine debris comes mainly from land activities, and only one fifth results from shipping. Only 15% of the debris remains on the surface; 70% goes to the seafloor. The debris pollutes beaches, harms ecosystems, undermines the recreational value of the sites, and prevents the optimal enhancing of marine resources and the development of the blue economy (GOC 2014, 2016). If no urgent action is taken, marine litter is projected to increase by 44% by 2030.

Plastics are by far the most prevalent component, contributing an estimated 60–80% of the total marine debris. It has been estimated that 8 million tonnes of plastic waste enter the world's oceans every year (Jambeck et al. 2015). At least 5.25 trillion pieces of plastic have already accumulated in the oceans. The worst offenders are in Asia, with China responsible for 28% of all plastic in the ocean, followed by Indonesia with 10%. The global plastics sector, of which bio-plastics make up only 1% of the market, is a powerful sector resulting in large amounts of nonbiodegradable plastic waste. Plastic marine debris contains highly toxic substances in concentrations that are more than a hundred times higher than those normally found in ocean waters, affecting marine life and ecosystems.

Marine litter affects all marine habitats, from densely to remote populated regions. The highest quantities have been found in the deep Mediterranean Sea, but even at the Arctic margins deep-sea littering was found to increase considerably in the past decade. At least 1 million seabirds and 100,000 marine mammals die each year because of plastic pollution. Plastic pollution is a carrier of invasive species, threatening native ecosystems. Toxic chemicals are adsorbed by the plastic, increasing the concentrations a million fold. If such pieces of plastics are ingested, high concentrations of the toxins enter the food chain. Synthetic polymers in the ocean should be regarded as hazardous waste (Eriksen et al. 2014).

It has been estimated that the average density of marine debris varies between 13,000 and 18,000 pieces per square kilometre, with far higher concentrations, more than 200,000 pieces per square kilometre, in the convergence zones between two or more ocean currents. Computer model simulations, based on data from about 12,000 satellite-tracked floats deployed since the early 1990s as part of the Global Ocean Drifter Program, confirm that debris is transported by ocean currents and tends to accumulate in a limited number of subtropical zones where currents converge, the gyres.

In the dynamic ocean environments, the accumulation of floating garbage has led to potentially five garbage patches scattered globally, located in the North and South Pacific Ocean, North and South Atlantic Ocean, and Indian Ocean. The total amount of plastic garbage, estimated around 36,000 tonnes, is unequally dispersed, and twice as much is in the North Pacific than in the North Atlantic. Economic impacts, estimated after beach cleanup costs, tourism losses, and damages to fishing and aquaculture industries, are also very important (Sesini 2011).

In the North Pacific, the immense cauldron known as the Great Pacific Garbage Patch contains an estimated 100 million tonnes of plastic waste and chemical sludge, much of which is produced by cities. The Pacific Garbage Patch, also known as the 'eighth continent,' is made up of two main collections of waste at opposite ends of the Northern Pacific Ocean, one off the Japanese coast (Western Garbage Patch) and the other off the western US (Eastern Garbage Patch). Around 80% of the island is made up of plastics originating from land-based activities in North America and Asia; the other 20% is primarily fishing nets from trawlers and waste from cargo ships.

Cities can make innovative alliances and engage in jointly addressing this major challenge. Plastic takes thousands of years to degrade, unless biodegradable, and when exposed to the sun it gradually breaks down into tiny particles or microplastics, often mistaken for food by sea animals and birds. In addition, chemicals and colorants from plastics leach into the water, potentially causing harm to human and marine life. Marine species and seabirds often ingest plastic. It is estimated that, by 2050, 99% of seabirds will have ingested plastic and can become entangled in plastic waste items (IUCN 2017).

Public awareness of marine litter increased rapidly during past years, which brought many initiatives to tackle this devastating transboundary issue at international and intergovernmental levels and also at the national, regional, and local scale. UNEP #CleanSeas campaign, launched at the 2017 Economist World Ocean Summit in Bali, is urging governments to promote plastic reduction policies, industry to minimise plastic packaging and redesign products, and urging consumers to change their throwaway habits. The campaign attracted far-reaching pledges to turn the plastic tide. Indonesia committed to reduce its marine litter by a massive 70% by 2025 and Costa Rica invests in measures to dramatically reduce single-use plastic through better waste management and education.

In the EU, the 7th EU Environmental Action Programme (2014–2020) calls for a quantitative waste reduction headline target, supported by source-based measures. Research projects try to better shed light on not just reducing waste from marine activities but using it for new products, for example, bio-based polymers from shell waste from the fishing industry. In 2018, the adoption of the first-ever EU-wide plastics strategy aims at all plastic packaging on the EU market to be recyclable by 2030, the consumption of single-use plastics to be reduced, and the intentional use of micro-plastics to be restricted. Every year, Europeans generate 25 million tonnes of plastic waste, but less than 30% is collected for recycling. According to estimates, 95% of the value of plastic packaging material is lost after a very short first use, whereas unrecycled plastic takes hundreds of years to break down. There is a strong business case for transforming the way products are designed, produced, used, and recycled in the EU in the context of the circular economy (WBCSD 2017).

The EU-funded Ocean Sampling Day is another inspiring initiative which took place on the midsummer solstice day, the longest day of 2014. Science teams around the world with the help and involvement of citizen scientists gave the oceans a health check. Schools are often associated with such initiatives, and cities can organise their regular health checks while promoting science for youths. The UN Environment and "Think Beyond Plastic" launched a worldwide innovation challenge for university students to fight plastic in the oceans, with winners invited to attend the Sixth International Marine Debris Conference and receive mentorship to develop their ideas into viable projects.

In the framework of the EU Regional Seas Conventions, action plans on marine litter focus on its prevention or reduction and address both land- and sea-based sources of marine litter, through a range of actions at the national or regional level such as improved waste and wastewater management, port reception facilities, targeted fishing for litter, education, awareness raising, and outreach activities. The European Environment Agency has developed the application "Marine Litter Watch," a citizen science-based tool that can help fill data gaps relevant for policy, while raising awareness about the problem of litter and the policy actions to prevent and address it.

Plastic trash is everywhere, and awareness of it has mounted sharply during past years. The environmental charity Plastic Oceans Foundation paired up with media and initiated in 2017 a campaign to have the giant mass of trash officially recognised by the UN as the globe's 196th country, the emerging nation "Trash Isles," with its dedicated currency, flag, citizenship, passport, and stamps. Al Gore is the nation's first honorary citizen. The campaign also has a call to action, issued as The Trash Isles Manifesto and aiming at developing biodegradable materials, introducing the carbon tax, and promoting recycling.

The German Presidency of G7 has initiated an awareness campaign on marine litter in 2016. Inspired researchers and innovators joined forces to clean the oceans. Already in 2012, the young inventor B. Slat designed a system, the Ocean Cleanup, for concentrating and catching plastic debris driven by ocean currents. The system has drawn much attention and has also been the subject of a successful crowd funding campaign in 2014. By acting as an artificial coastline, the system uses the ocean currents to concentrate the plastic by orders of magnitude. A barrier of solid screens catches the floating plastic, but allows sea life to pass underneath the barrier with the current. A central collection point extracts and buffers the debris, to allow for its shipment to land. By recycling the debris and selling the semi-finished product directly to interested companies, the Ocean Cleanup aims to make the operation self-sustainable. The first prototype system was deployed in June 2016, and the first full-scale deployment is planned for 2020 in the Pacific.

In 2017, aquariums around the world have joined forces to fight marine litter, as part of a global awareness-raising campaign launched by the European Commission, a contribution to the UN #CleanSeas campaign. Aquariums offer a television screen to the ocean, with millions of visitors across the world and a great multiplier potential. The "World Aquariums against Marine Litter" campaign started in summer 2017, in the run-up to the fourth international "Our Ocean" conference that the EU hosted, for the first time, in Malta in October 2017. The chain of events started at Monaco's Oceanographic Museum and involved about 150 aquariums from 38 countries on five continents.

Recognising the global nature of the challenge, the G20 decided to work together to promote and initiate measures and actions at local, national, and regional levels to prevent and reduce marine litter. Taking into account the need for comprehensive multi-stakeholder involvement, the G20 acknowledged the role of non-state actors and encouraged private sector engagement. Knowledge gaps were judged important. However, a lack of certainty in scientific evidence can no longer be accepted as an excuse for inaction. Humanity has knowledge enough to act (UN 2016).

The G20 suggested that the tools to reduce marine litter have to be as diverse as the challenge of marine litter. There is no 'one size fits all' solution. Upstream prevention is far preferable to downstream removal, through the '3Rs' (reduce, reuse, recycle) approach, which implies that waste prevention should be the first option, followed by reuse and recycling. To support implementation of the activities, the G20 launched a voluntary Global Network of the Committed, a platform addressing marine litter linked to the UNEP's Global Partnership on Marine Litter to offer information exchange and sharing, dissemination and transfer of standards, experiences, and knowledge.

Moving from the international recognition of the problem to concrete actions at local, regional, and national levels is essential and linked to the opportunity offered by the circular economy. Major producers should also be more involved in managing the entire life cycle of their products. The policy approach of the Extended Producer Responsibility (EPR) brought more possibilities for coaction. With EPR, producers are given a responsibility for funding or managing the recycling, treatment, or responsible disposal of their products in the post-consumption stage. Extended producers responsibility systems had spurred a wave of innovation and entrepreneurship in the field of waste management and recycling, leading to a flourishing industry with 2 million job opportunities and an EU-wide average recycling rate of 65% for all household and industrial packaging.

The world's cities produce up to 7–10 billion tonnes of urban waste each year. Three billion world citizens lack access to controlled waste disposal facilities. Holistic waste management could prevent costs and reduce greenhouse gas emissions by as much as 20% annually, as was suggested by UNEP and the 2015 Global Waste Management Outlook, in association with the International Solid Waste

Association (ISWA). Volumes of waste are likely to even double in lower-income African and Asian cities by 2030. An urgent response to the world's mounting waste problem is not only a public health and environmental necessity but also a sound economic investment. The outlook suggests that inaction is costing countries five to ten times more than investments in proper waste management.

This integrated global overview to the waste problem encourages a shift in thinking about the broader concept of resource management and calls for immediate improvement of waste prevention, collection, and recycling of resources. It demonstrates that wise management of resources, including waste, can cut critical costs and bring benefits from recovered materials. The Belgian region of Flanders achieved the highest waste diversion rate in Europe, growing from nearly zero in the 1980s to more than 70% in 2013. This stage was possible with a smart mix of social, fiscal, and legal policies, such as prevention education, establishing reuse centres, and a "pay as you throw" taxation system based on the 'polluter pays' principle. Lighthouse stories also come from developing countries. For example, the inclusion of informal recyclers into the municipal waste management system in Bogota, Colombia, is diverting 1200 tonnes of waste daily from the landfill and employing more than 8000 people (UNEP/ISWA 2015).

Zero-waste cities have to show the way forward. San Francisco has been a front-runner in drastically reducing waste by deploying a demanding solid waste management programme. Zero waste means that products are conceived, designed, and used according to the principle of highest and best possible use, and that discarded materials cascade through reduce, reuse, and then recycle or compost. San Francisco developed a long-term strategy and a threefold approach to address the legal, administrative, and social dimensions of waste management. It enacted strong waste reduction policies and partnered with Recology, a materials management company, to innovate and promote a culture of recycling and composting. Policies embraced a wide range of actions, from production and packaging to consumption, public and private management, and government procurement. The city recycles or composts 77% of its waste, the highest rate of any major US city.

At the other end of the spectrum, difficulties seem insurmountable in the unplanned peripheries of the developing metropolises. Kolkata's Solid Waste Management Improvement Project won the C40 2016 award. Kolkata's climate change risks have been exacerbated by unsanitary disposal and waste dumping of an average of around 450–500 g municipal solid waste per capita every day. Problems included the absence of source segregation arrangements and only limited (60%) house-to-house collection using 50–55% open vats. The operational efficiency of the Kolkata Municipal Corporation transport system was about 50%, with a fleet composed of about 30–35% old vehicles. The waste collected has a low energy value with high moisture and inert content. Open dumping and the threat of groundwater pollution, as well as saturation of an existing landfill site, have been the most pressing problems. The Kolkata Environmental Improvement Project emphasises segregation at the source, investment in disposal arrangements, and optimised transport. Forward looking, the project aims to eradicate open dumping and burning of waste and to limit the concentration of methane gas generated in landfill sites (UNEP 2016b).

The initiative "Closed Loop Oceans" is designed to fund waste management and recycling solutions in Southeast Asia, with investments addressing the root causes of plastics in the ocean by investing in the systems and technologies to capture and transform waste into valuable assets before reaching marine environments. Nearly half the plastic that flows into the ocean every year escapes from waste streams in just five rapidly developing Asian economies, including China, Thailand, Vietnam, Indonesia, and the Philippines. The market-based investment strategy and supporting mechanism aims to catalyse investments from larger, less nimble actors such as development finance institutions and expects to build recycling and waste management infrastructure and maximise recycling profitability.

In 2016, a network of 90 NGOs from around the world including Greenpeace, Oceana, the Story of Stuff Project, 5Gyres, and Clean Water Action have come together under the banner "Break Free From Plastics" and launched a massive global movement to achieve a future free from plastic pollution. In Europe, Rethink Plastic is an alliance of leading European NGOs, including Seas at Risk and Surfrider Foundation, fighting against plastic pollution. With many active groups, supporters and citizens in every EU Member State, the alliance played a role for the adoption of the 2018 EU Plastics strategy. And this is only the beginning.

2.3 The Over-Stressed Geostrategic Nexus: "Water-Food-Energy"

2.3.1 Water, the Blue Gold, an Irreplaceable, Vital, and Scarce Resource

Fresh water, the blue gold, is a vital and scarce natural resource, which, unlike oil, cannot be replaced. Access to clean water is regarded as a fundamental human right. Intelligent infrastructure and surveillance systems to limit leakage, which often reaches 30%, are being implemented or planned in many cities. Leakage and risk detection are increasingly parts of integrated management and early warning systems. Many cities introduced strategies for the wiser use of water resources, facilitated by digital installations and smart meters. The Tokyo system for identifying leaks is considered to be exemplary; Stockholm and Singapore are well known for their proactive water management policies.

Major world cities literally suck their surrounding areas to satisfy the water needs of their populations. A global survey suggests that large cities occupy only 1% of the Earth's surface but the watershed providing their water covers 41% of the land surface. Cities with at least 750,000 inhabitants move 504 billion litres (l) of water each day in an accumulative distance of 27,000 km. Los Angeles, Boston, Mumbai, Karachi, Hong Kong, Tokyo, New York, Tel Aviv, Sydney, and Athens top the list of the cities that literally suck their surrounding regions (McDonald 2014).

Stockholm, the first 2010 European Green Capital, is a pioneer city in water protection and management. The Swedish capital stands out for its high quality of environment and life, and demonstrates strong public health performance, high educational attainment, and social protection levels. The municipality has done much to improve wastewater treatment and reduce the impact from storm water. The Stockholm Water Company has improved the water quality by radically cutting down the discharge of phosphorus and nitrogen. The sensitive Stockholm archipelago has a high water and environmental quality and great recreational value.

Australia, essentially an ocean country, is the driest populated land. Sydney faces particular challenges as an additional 1.3 million people are predicted to move into the city during the next decade. Urban renewal projects are planned around corridors and hubs that have access to public transport and major job and housing developments in Western Sydney. As a result, Sydney Water is exploring options for future supply models that complement the existing centralised system to provide intelligent and affordable options. A study, led by Arup and Sydney Water, tried to map future water supply in Sydney and other major cities in the face of population growth, increasing scarcity of water, and continued budget pressures. The study suggests that water authorities should migrate towards a more hybrid model which incorporates greater decentralisation and autonomous management of water supply, increased participation of additional service providers, and smarter management of the water grid (Arup and Sydney Water 2014).

Water is visible and valued in many initiatives introduced by cities that had to learn about living with water. The initiative Canadian Water Sustainable City of the Near Future is built on the expectations of key stakeholders. Canada faces a water and wastewater infrastructure deficit that is expected to grow. The Blue Economy Initiative sought to inspire trust and dialogue among Canadian decision makers and influential actors on the opportunities and benefits of preserving water, and the severe risks of inaction. Four broad areas were identified to make tangible progress towards sustainability, including financial responsibility, progressive regulation and governance, customer-oriented information, and cutting-edge technology.

The "water-energy" nexus merits special attention. The energy sector already accounts for 15% of the world's total water use. As energy needs are set to grow, water becomes an increasingly important criterion for assessing the sustainability of energy projects. Expanding power generation could lead to an 85% increase in the volume of water that is reused.

Sustainability asks for water to be respected throughout the life cycle. A US overview of water recycling applications, including drinking water, non-drinking urban and industrial uses, irrigation, groundwater recharge, and ecological enhancement, shed light on various municipal water reuse projects, such as irrigating parks or providing industrial cooling water. Drinkable water projects account for only a small fraction of the water being recycled. However, many drinking water treatment plants draw water from a source that contains wastewater discharged by a community located upstream (NRC 2012a).

Wastewater treatment technologies processes are easily adapted to reclamation plants to meet the quality requirements of intended future applications in the water cycle. The concentrations of chemicals and microbial contaminants in reuse projects designed to provide drinking water supplies can be comparable to or lower than those commonly present in many supplies. Although the costs of reclaimed water are often higher than current water sources, external costs and benefits, for example, seasonal peak demands on the drinking water system, can strike a difference. Depending on the specific requirements, recycling and reuse projects could also have a larger or smaller carbon footprint than alternative options (NRC 2012a).

Sustainable wastewater management is one of the most widespread challenges and is linked to climate action and urban resilience. Green infrastructure captures and filters pollutants. The green alternative is often less expensive than structural engineering options and can help reduce costs of flooding. Green roofs, permeable materials, alternative designs for streets and buildings, trees, gardens, and parks, and rain harvesting systems are increasingly supplementing aquatic or substituting hard infrastructure investments.

Seawater and saltwater promise alternative water resources in the face of freshwater insecurity. Clean seawater can become a precious drinking resource. Coastal aquifers may contain seawater and salty groundwater from ancient times. State-of-the-art desalination technologies can remove salt from both seawater and brackish water and provide a new source of freshwater. Techniques used to desalt water are also useful for removing other impurities common to contaminated source waters. The process still tends to be very costly and energy intensive. Desalination has to be managed according to strict sustainability principles. Research suggests that combining alternative energy sources with desalination technology can lead to more economical and environmentally friendly systems. Desalination is especially appealing to water-stressed Middle East and Arab countries, such as Kuwait, Saudi Arabia, and Israel. The plants use reverse osmosis, a process that orients water through a membrane with tiny pores that retain most of the dissolved salts as well as most organic compounds and microbiological contaminants.

In Europe, Spain is a desalination leader. The Barcelona desalination plant, located next to the mouth of River Llobregat, is the largest reverse osmosis-based desalination plant and was awarded the "Desalination Plant of the Year" award by the Global Water Awards 2010. Created by the public utility responsible for the supply of water to the city of Barcelona and the surrounding eight districts, the plant aimed to increase water security and its availability in the metropolitan region. Inaugurated in July 2009, it is supplying drinking water to about 1.3 million inhabitants in the region, almost 20% of the population.

The captured seawater is screened and disinfected. Its coagulation is conducted in two chambers, and the sludge is collected at the surface by overflow. The pre-treated water is fed to the reverse osmosis membranes. The process used at the plant is based on a partial dual-pass method. In the first pass, the pre-treated water is pressurised and fed into 16,800 membranes fitted in ten racks in a single row. To optimise energy consumption and maximise water production concurrently, the membranes are set up in a hybrid system. The second pass helps to keep the boron content below the 1 mg/l limit, imposed by the European regulation. The treated water has a low pH value and reduced salt concentration. It can become drinkable, if re-mineralised by lime filters and disinfected by adding chlorine dioxide.

Singapore's water supply and sanitation utility received the Stockholm Industry Water Award for its holistic approach to aquatic resources management. The city is ranking high in water management in the challenging environment of a densely populated island. Access to water is universal, affordable, and efficient. Innovative integrated water management approaches such as the reuse of reclaimed water, the establishment of protected areas in urban rainwater catchments, and the use of estuaries as freshwater reservoirs have been introduced along with seawater desalination to reduce dependency on imported water. Singapore does not rely only on physical infrastructure, but also emphasises proper legislation and enforcement, water pricing, public education, and heightened awareness, as well as investment in research and development.

Singapore is a water-scarce country because land to collect and store rainwater is limited. For this reason, it invested in a diversified and robust supply of water through the Four National Taps, namely, local catchment water, imported water, NEWater, high-grade ultra-clean reclaimed water, and desalinated water. Since 2011, the total water catchment area has been increased from half to two thirds of Singapore's land surface with the completion of three reservoirs in urbanised areas. By 2060, Singapore's water use is expected to more than double. To strengthen water security and drought resilience, the city-state continues to build up the capacities of NEWater and desalinated water, which are weather-resilient sources. Currently, NEWater and desalination capacity are able to meet up to 30% and 25% of water demand, respectively, and this could reach 55% and 30%, respectively, of the 2060 water demand (National Environment Agency 2016).

Sand, the second most important natural resource after water, used for concrete, electronics, glass, aeronautics, etc., is expected to be in high demand given the expanding urbanisation in the developing world. The most used building material worldwide is concrete, of which 70–80% is aggregates such as sand, gravel, and rock. Demographic pressure in Asia made expansion towards the sea an obvious option. Singapore holds the world record in shifting sands, having grown by 20% since the 1960s. Artificial islands and airports on artificial land expand in South China Sea. Some parts of the world already suffer from sand shortages. In India, illegal sand mining is widespread, while in North Africa beaches and dunes disappear. Ironically, desert sand, which is abundant, is not suitable for building, which requires sand extracted from quarries and waterbeds and transported to the building sites. Research and innovation for artificial sand and alternative building materials are crucial, given the estimations for annual growth of sand demand of 5–6% by 2030 (EP 2017).

2.3.2 Preventing Food Waste, a Major Polluter and GHG Emitter

FAO estimated that each year, approximately one third of global food produced for human consumption is lost or wasted. The environmental footprint of food wastage, including both food loss and food waste, along the food supply chain, impacts much climate, water, land, and biodiversity. A model has been developed to assess the magnitude of food wastage impacts on the environment and the main sources of these impacts, in terms of regions, commodities, and phases of the food supply chain involved (FAO 2013).

The figures speak by themselves and tell inconvenient truths. The global volume of food wastage is estimated to be 1.6 Gtonnes of 'primary product equivalents' and the total wastage for the edible part of food is 1.3 Gtonnes. The total agricultural production for food and non-food uses is about 6 Gtonnes. Without accounting for GHG emissions from land use change, the carbon footprint of food produced and not eaten is estimated to be 3.3 Gtonnes of CO_2 equivalent. Globally, the blue water footprint, that is, the consumption of surface and groundwater resources, of food wastage is about 250 km^3, equivalent to three times the volume of Lake Geneva. Finally, produced but uneaten food vainly occupies almost 1.4 billion hectares of land, close to 30% of the world's agricultural land area. With such figures, it seems clear that a reduction of food wastage at global, regional, and national scales would have a substantial positive effect on natural and societal resources. Food wastage reduction would not only avoid pressure on scarce natural resources but also decrease the need to increase food production by 60% to meet the 2050 population demand (New Economics Foundation 2014, NRC 2012b).

Researchers at the Johns Hopkins Centre for a Livable Future in Baltimore shed light on the wasted U.S. seafood supply and examined all stages of the seafood supply chain. The study looked at seafood waste by focusing on the amount of seafood lost annually at each stage of the food supply chain and at the consumer level, during the years 2009–2013, and concluded with recommendations, also in the light of the 2010 U.S. Dietary Guidelines. The main result is that 40–47% of the edible U.S. seafood supply went uneaten during 2009–2013. This is a too high proportion, given the prospects for future availability of seafood, threatened by overfishing, unsustainable seafood farming practices, ocean pollution, and acidification.

Fish is wasted as it moves through the supply chain from hook or net to plate. The greatest part of this loss (51–63%) is attributed to consumption and occurs at the levels of individual consumers, in and out of the home. It seems that individual consumers throw out 1.3 million pounds of edible seafood each year. The second largest portion comes from bycatch, the incidental capture of nontarget species discarded by commercial fishers (16–32%), and is followed by losses during distribution and retail operations (13–16%). According to conservative estimates, this waste represents 208 billion g protein, 1.8 trillion mg omega 3 acids, and 1.1 trillion kilocalories (Love et al. 2015).

The 2010 U.S. Dietary Guidelines encourage citizens to improve the health of their diets and recommends consuming a variety of seafood in place of some meat and poultry to reach eight ounces per person per week. Yet achieving those levels would require doubling the U.S. seafood supply or eliminating nearly all seafood waste. The seafood that is lost could fill 36% of the gap between current and recommended consumption and the U.S. Waste reduction has the potential to support increased seafood consumption without further stressing aquatic resources. It is essential to quantify waste levels to effectively target and design waste reduction interventions. As another way of understanding the magnitude of loss, this lost seafood could provide the total yearly target quantity of protein for 10.1 million men or 12.4 million women. Waste prevention strategies should engage governments, businesses, and consumers to reduce seafood loss and create a more efficient and sustainable seafood system.

The study concludes that much of the wasted seafood could be prevented by changing consumer behaviour. For waste to be effectively reduced, measures have to be taken along every step of the way, from fishing, to production processing, packaging, selling, and consuming. The researchers recognise that loss cannot be completely avoidable, but they do hope that the significance of their estimates will stimulate a response. This is absolutely necessary at a time that seafood sources are under pressure and consumers are being urged to eat more healthy food with less impact on the environment.

The U.S. administration announced in 2014 the first-ever goals for cutting food waste nationwide. The Environmental Protection Agency and the Department of Agriculture announced an ambitious goal to reduce food waste in the United States by 50% by 2030. This is fully reflected on the SDG 2030. On the business side, the Consumer Goods Forum, which represents more than 400 companies across 70 countries, recently adopted a resolution to reduce food waste among members by half by 2025.

The European Commission suggests that food loss and waste in industrialised countries are as high as in developing countries, but their distribution differs. In developing countries, more than 40% of food losses happen after harvest and during processing, whereas in industrialised countries, more than 40% occurs at the retail and consumer level. Factors contributing to food waste include insufficient shopping and meal planning, stock management issues for manufacturers and retailers, inadequate storage/transport at all stages of the food chain, and overproduction or lack of demand for certain products at certain times of the year. Underlying all these problems is an overall lack of awareness, by many actors, of the sheer scale of the problem and the benefits that come from reducing food waste. School are a great place for raising awareness on food saving and good nutrition.

2.3.3 *Seafood for Urban Food and Nutrition Security*

During the industrial revolutions, increases in agricultural productivity have served as a driving force for urbanisation. As farms became more profitable, excess labour from agriculture moved to industrial jobs in cities. However, although past urbanisation promoted growth and prosperity, the current urban explosion in the emerging world is occurring faster than the pace that allows cities to integrate newcomers without major shocks and risks and to provide them opportunities for a better life. Poverty, food insecurity, and malnutrition are shifting rapidly from rural areas to urban centres. Furthermore, political instability in many emerging countries make cities inadequately equipped in infrastructures, and vulnerable to natural disasters, urban poverty, food insecurity, and malnutrition, and the least able to sustainably address these challenges.

Coastal regions and cities are normally more open to seafood. The Economist Intelligence Unit Global Food Security Index 2017 highlighted that landlocked countries, especially the wealthier ones, consume on average less seafood than coastal ones Fish consumption equals about a third (30.2%) of total fish and meat supply in coastal countries compared with less than a fifth (17.8%) in landlocked ones (EIU 2017).

The link between proper nutrition and normal physical and cognitive growth is a hardwired system. Access to sufficient, nutritious, and affordable food is not only a basic human right but provides the foundation for citizens' ability to grow, learn, prosper, and reach their full potential. Emerging cities have to ensure food for all slum populations, especially the youths less than 18 years old facing a high malnutrition risk and expected to form 60% of the urban slum population in 2030.

Food from the ocean could be decisive for the food security of coastal but also inland cities, which often are highly dependent on imported foods. In many cities in the developing world, 30–50% of food staples and vegetable oil found in the markets are imported and most often cost more than urban garden products or street vendor foods. Urban dwellers, and especially the urban poor, are more affected by international food prices than small farmers. Food security in urban areas tends to be tied to consumption patterns and affordability, whereas food security in rural areas is related to the availability of food.

Fisheries provided humankind for a long time with good quality food until overfishing became a major global problem. Other major problems for fisheries include eutrophication, as algal blooms can produce toxins that infect seafood. Cities could cooperate with states in eliminating harmful subsidies and close their harbours to illegally harvested fishes. And they can cooperate with educational, public health, and cultural institutions and organisations and support the health of global fisheries by ensuring that products of artisanal fisheries have their place in the seafood markets (WRI 2016).

Fisheries and aquaculture are a vital source of nutritious food, economic opportunities, and jobs (World Bank et al. 2014). They already account for 16.6% of all animal protein consumed globally and 6.5% of all protein for human consumption. They are a precious food for the full spectrum of populations, rich and poor, urban and rural. The contribution of fish to food is likely to increase as consumers become more conscious of the nutritional value of seafood. Stronger sustainability performance of the value chains could also have an important impact in terms of jobs and growth, in particular in coastal areas (FAO 2016).

One of two citizens of the developing world takes their animal protein from the sea: 97% of the world fishermen are from the developing world. Urban fish and wet coastal markets concentrating the catch of the day are gradually replaced by supermarket chains with aquaculture products. The global development of aquaculture has been spectacular. Since 2014, aquaculture contributes more than capture fisheries to seafood at the world level (OECD-FAO 2017).

Fish production could well change because of global climate change. Warming oceans could result in distributional changes, with fish moving polewards or into deeper and warmer waters. Substantial changes in the dominant species are possible in the world's most productive marine ecosystems. In addition, ocean acidification is likely to change the composition of the plankton, with unpredictable effects on fish populations.

As the world capture fisheries production has limited potential for further growth, future demand will mainly rely on a substantial increase in aquaculture production. Global aquaculture has grown at an impressive rate during the past decades and could provide two thirds of world fish production by 2030. Further substantial expansion of this production is possible, and even essential, if world supplies of fish products are to keep pace with human population growth. China, the biggest world producer, importer, and exporter of fisheries products, and other countries are increasing their investments in aquaculture to help meet this mounting demand. Vietnam is expected to become the second world export overcoming Norway. European aquaculture, subject to strict environmental and animal health rules and very high standards of safety and sustainability, is speeding up its efforts (FAO 2016; OECD-FAO 2017).

Sustainable fisheries can contribute much to the food and nutrition security of cities. Food security issues have been among the most forgotten urban issues, of paramount importance for eradicating extreme hunger and poverty, instability, and uncertainty. The international community should acknowledge that poverty and food insecurity and malnutrition issues are serious urban problems. A new framework for addressing food and nutrition security in both rural and urban areas is critical. The Chicago Council proposed an Urban Food Model designed for policy makers and city leaders as an analytical tool to evaluate a city's food security situation and examine policy options (Chicago Council on Global Affairs 2013).

The importance of addressing urban food and nutrition security is increasingly highlighted by cities. Especially in the developing world, rapid demographic growth and urbanisation, inadequate infrastructure and green spaces, dietary consumption changes, and more frequent and extreme weather phenomena linked to climate change bring enormous challenges to cities, many of which already face the overwhelming burden of providing basic health and education services to citizens. Discontent over poverty and food insecurity can be politically explosive in cities (FAO 2009).

Seafood seems well placed to help close the food gap to 2050, providing high-quality protein with lower impact on the environment. It would also help fight malnutrition, which progresses in all countries in the world and is a leading global driver of disease, as highlighted by the Global Nutrition Report 2016. The 2014 Rome Declaration on Nutrition achieved consensus on the multiple challenges of malnutrition, a vision for the future, and commitments to specific actions. The launch of the United Nations Decade of Action on Nutrition, in 2016, has been a rallying cry involving the whole of society.

On World Food Day, 16 October 2015, at the Milan Expo "Feeding the Planet, Energy for Life," more than 100 cities signed the Milan Urban Food Policy Pact, the first pact engaging political leaders representing together more than 400 million citizens to develop sustainable food systems to provide healthy and accessible food to all, to protect biodiversity, and to fight against food waste. The Pact is an initiative of the Mayor of Milan who decided to bring together urban areas from the north and the south of the world to map out a common path towards new food policies. The Pact aimed to create an international network of cities committed to developing and implementing sustainable food policies. It involves the joint commitment to policies that promote fair, sustainable, and resilient food systems and a framework of action that focuses on food waste, distribution, production, social and economic justice, sustainable diets, and governance.

The experiences by the Milan Urban Food Policy Pact signatory cities and EUROCITIES have been analysed in the framework of an EU-supported study, which highlighted the emerging role of cities in the development of sustainable food and nutrition systems. For a long time, food production was considered beyond the sphere of competence of cities, mainly because food is usually produced outside the urban limits. Cities manage their food-related activities in various ways, and some develop comprehensive strategies and long-term plans in cooperation with local actors and by considering the impact across policy areas and municipality departments. Research results suggest that food strategies are either the result of the cities' own initiatives or stem from collaboration among cities. The main policy tools include citizen involvement and social innovation, public procurement, and multi-stakeholder governance. The domains in which cities are most active bring forward inclusive governance, sustainable diet and nutrition, food production, supply and distribution, and food waste (EUROCITIES et al. 2017).

Sustainable aquaculture promotes consumption and production patterns that preserve the natural resource. Main challenges include adequate space, feed, and breeding technology. More space for marine aquaculture can be freed up by the development of extensive, offshore production systems that can use remote locations. The development of offshore wind farms could create new areas suitable for aquaculture. Integrated multi-trophic aquaculture, that is, the combined farming of algae or shellfish and finfish, can ensure the optimal use of available space. Progress in feed technology can help marine fish to be produced more efficiently and with a higher proportion of feed of plant origin. Improvement of breeding technology could make new species available for sustainable aquaculture. Marine shellfish farming requires limited inputs and provides important environmental services by removing nutrients from the water. This type of farming is particularly sensitive to water quality, and research is focusing on selective breeding of the most resistant shellfish families. Well-designed integration of activities such as angling and tourism with aquaculture may add environmental and landscape management services.

Aquaculture needs appropriate sites and can be in competition for space with other coastal economic activities. Strategic planning tries to enhance synergies among activities which are often seen as antagonistic such as shipping, tourism, recreational fishing, and aquaculture. Possible adverse environment impacts of aquaculture also must be prevented and addressed, including localised eutrophication events, changes in benthic communities, contamination from antibiotics, and introduction of nonindigenous species. The development of genetically modified marine organisms tailored for high aquaculture productivity may offer benefits but also raise ethical issues for marine environmental protection and for social acceptance (EEA 2016).

In the European Union, the High Level Group of the EC Scientific Advice Mechanism has published an independent opinion on Food from the Oceans in 2017. Drawing on the best available scientific and technical evidence, the group identified the potential for a large increase in food obtained from the ocean, up to three- or fourfold, in particular from mariculture and the cultivation of marine organisms such as algae and shellfish. The policy recommendations for realising this potential in a sustainable way suggested the need for mainstreaming a responsible culture and capture approach to ocean food into policy agendas, and enabling the full potential of mariculture, including greater attention to ecosystem-based spatial planning and sustainable fisheries partnership agreements between the EU and southern partner countries (EC 2017b).

The City of Bergen, UNESCO Creative City of Gastronomy, hosted on 18 June 2017 the Sustainable Gastronomy Day with a wide range of activities related to food from the oceans. The Academia Europaea Bergen Hub and the Science Advice for Policy by European Academies discussed themes such as the common use of algae in meals, fisheries, possible use of bycatch fish and products, and potential for additional food and biomass. A marketplace on the quayside welcomed fishing boats and their harvests and proposed stands from local restaurants for tastings of bycatch and seaweed.

Cities can offer the testing grounds and support the development of alternative fish production systems, investing, for example, in demonstration projects of urban aquaponics, combining aquaculture with hydroponics, growing plants and vegetables in mineral nutrient solutions in a symbiotic environment. Aquaponics has been included in the ten technologies that could change our lives (STOA 2015). In an aquaponic system, water from an aquaculture system is fed to a hydroponic system and the by-products are used by the plants as nutrients. The water is then recirculated back to the aquaculture system, increasing eco-efficiency and ecological value.

2.4 Ecological Cities and Protected Areas Under the Water

2.4.1 Urban Eco-Cells, Eco-Organs, and Eco-Quarters

Each city is a living organism with infinite possibilities of regeneration. Cities are not simply concentrations of people and activities, but hives of intense relationships and synergies, in which the whole can greatly exceed the sum of the parts and enhance the conditions for regeneration. The balance among co-evolving policy objectives is dynamic and impacts ecological processes and the distribution of wealth. The opportunities for renewal depend on the health of its cells and organs, its places and living conditions (Orsenna and Gilsoul 2018).

Buildings can be seen as the eco-cells of eco-responsible cities and influential public buildings and spaces as the city's eco-organs. Schools can be the most instrumental eco-organs of a city and great places for promoting social interaction, environmental awareness, and behavioural change. The international award programme 'Eco-Schools' created a movement of schools embedding sustainability principles into school life and action. School children lead the project in all its aspects and help carry out audits to assess and improve the environmental performance of their school. They also learn principles and practices to apply in other spaces, for example, waste prevention and recycling in their families. Many "eating local, seasonal, organic" or "cleaning the beach" action programmes in cities have been initiated by local schools, influencing public values.

Opening schools to the surrounding communities and sharing resources with all is critical. In New York, through the Schoolyards to Playgrounds programme, part of the OneNYC plan, Parks and the Department of Education have partnered to expand access to community space, opening school play space during non-school hours. The project 'Schoolyards to Playgrounds' introduced an innovative way to create more parkland for residents and families. Many of the schoolyards received capital improvements, including playground equipment, painted sports surfaces, trees, and benches to accommodate public needs. In 2016, the number of sites participating in the Schoolyards to Playgrounds programme exceeded 250, and this helped NYC Parks bring 85% of New Yorkers within walking distance of a park by 2030. More than 98,000 public school buildings are open to their neighbourhoods across the US and host community and civic events, elections, recreation leagues, and other activities. Also, local communities are beginning to use one most important school asset, the kitchens, which are typically overlooked. New research by the "Kids' Safe and Healthful Foods" project offers recommendations on legal, financial, and other issues surrounding public use of school kitchens based on the early successful local initiatives (New York City Global Partners 2013; NYC 2015a,b, 2017).

Many cities have introduced transformational places and functions to achieve the overall sustainability transition for their fabric and their heritage. In Malmö, Västra Hamnen, the Western Harbour, provides a fine example of a previously industrial site transformed into an ecological coastal area. Its regeneration began with residential developments, mostly linked to energy self-sufficient districts. Among the recent towers, the twisting 190-m-tall 'Turning Torso' quickly became the city's skyscraper landmark. The long boardwalk to the Western Harbour has become a favourite summer leisure environment for citizens, especially students, and visitors. The Western Harbour is often cited as the EU first carbon-neutral district.

Environmental, social, and cultural sustainability are given high attention, and much effort is invested in the quality of the natural and built environment, through new plantations for biodiversity and the construction of eco-innovative green buildings. In Hanover, the design and construction of the Kronsberg district, created in the context of Expo 2000, incorporated state-of-the-art ecological materials and techniques, providing a turn-of-the century best practice. Construction waste was reduced by 80% through reuse and recycling measures, but also social and educational models such as the waste-free breakfasts for children are featured (Mega 2010).

In Stockholm, Hammarby Sjöstad is an innovative district on the former waterfront industrial land set aside for the ultimately unsuccessful bid for the 2004 Olympics. The concept includes bioclimatic design, sound surroundings, and organisation of local life according to sustainability principles. The eco-cycle model of Hammarby aims to make the district autonomous throughout its life cycle. The Royal Seaport, the city's latest flagship sustainability project, set on land previously partly industrial and royal hunting grounds, aims at creating an exemplary organic part of the city. In addition to housing and offices, the area intends to be fossil fuel free and host urban parks, cultural and recreation spaces, and a harbour for cruise ships (City of Stockholm 2014).

Since 2008, SymbioCity, the Swedish trademark for sustainable urban development, has promoted improved, mutually reinforcing, urban realities. The essence of the concept is the generation of environmental and economic benefits through synergies between urban systems. The SymbioCity approach has been used as a platform to support sustainable urban development around the globe. A network of Swedish environmental technology companies and organisations promote the SymbioCity model in their international exchanges. SymbioCity is scalable and adaptable and has key components in common with the circular economy model. For example, excessive heat from an industry can warm a household or the waste from an industry can provide useful material for a service. The concept evolved towards a new era integrating a stronger focus on poverty reduction and gender equality. City projects are designed based on experience from previous pilot projects in Eastern Europe, Asia, and Africa.

Coastal and port cities can offer a fertile environment for eco-innovation to thrive in the articulated land and sea domains and functions and can serve as a magnet for talent, financial capital, and entrepreneurship. These cities can provide a competitive environment to maritime companies and ancillary services and serve as an open blue–green technology platform that gives partners an unmatched opportunity to develop, test, and validate technologies at a large scale and under real-world conditions. Green urban procurement and voluntary practices such as eco-labels can accelerate and amplify change (EC 2016a).

The Hanseatic triangle of the major harbours of Antwerp, Rotterdam, and Hamburg provides interesting examples. Hamburg led a thorough urban regeneration and created the emblematic HafenCity (Harbour City) on abandoned port industrial space. Hamburg's bustling harbour, at the edge of the city centre, once hosted the infrastructures and warehouses of the former flourishing shipping industry. But the advent of bigger ships required the creation of a new harbour, leaving behind a vast empty space which provided the fundamental resource for HafenCity. It aspires to become a breathing part of the city, a place where citizens want to both work and enjoy leisure time. Early promising achievements include a groundbreaking Ecumenical Forum designed according to the strict sustainability standards of the HafenCity eco-label. In 2011, the district welcomed 'Osaka 9,' the HafenCity sustainability pavilion with an exhibition space on sustainable urban development. By the time HafenCity is finished in 2025, it will stretch 1.5 km between the city centre and the Elbe River (HafenCity Hamburg 2013).

The jewel in the crown is the Elbphilharmonie, one of the largest and most acoustically advanced concert halls in the world. Inaugurated in early 2017, the popularly nicknamed Elphi is already the dominating feature of the Hamburg horizon. Designed by the architecture firm Herzog & de Meuron and in construction for 10 years, the prestigious project tested the city's patience and expectation. The crystal building immediately became a landmark. With its curved windows and white-tiled crested roof, it resembles a hoisted sail or wave, in dynamic contrast with its utilitarian brick foundation, a former cocoa warehouse built on 1963 near the historical Speicherstadt district. The freely accessible plaza offers citizens and visitors breathtaking views of the Hamburg skyline, the Elbe River, and the container port bristling with cranes.

2.4.2 From Urban Sprawl to Compact Cities

Soil, air, and water are the fundamental resources that support human settlements. Urbanisation increases pressure on all ecosystems. Soil is a living organism, a multifunctional and largely nonrenewable resource that has to be sustainably managed to transform waste into nutrients. Urban soil sealing, covering ground converted into urban land by an impermeable material, is one of the main causes of soil degradation. The expansion of grey infrastructure and urban engineering projects has disastrous consequences for biodiversity. Soil sealing often affects fertile agricultural land and fragile shores and results in the loss of important ecosystem services, such as food production, flood protection, and biological diversity.

Everywhere in the world, land consumption has grown at more than twice the rate of population growth. In the European Union, an area five times that of Greater London has been consigned to urban sprawl during the first decade of the twentieth century. Europe loses an area the size of Berlin through urban and infrastructure expansion every year. Unplanned urban growth undermines other drivers and determinants of sustainable development, especially dependency on private cars. For example, for every 10% increase in sprawl, there is a 5.7% increase in per capita CO_2 emissions and a 9.6% increase in per capita hazardous pollution (EC 2016b; EEA 2006).

Sustainability asks for density and intense use of space in cities. Key sustainability policy strategies encourage dense and contiguous development at urban fringes, the retrofit of existing built-up areas, and enhance diversity and quality of life in urban centres. At the other end of the spectrum, urban sprawl can disproportionately increase the ecological footprint of cities and aesthetically deform their peripheries, the "lost cities," as these are called in Latin America. The threats are very high for cities that are surrounded by trademark landscapes or very fragile water and marine ecosystems. The benefits of coastal spatial planning include generation and enhancement of synergies between different activities, encouragement of investment by instilling transparency and predictability, and reduced conflicts and better ownership by the local stakeholders (OECD 2012b).

Urban sprawl and the surge of suburban satellite settlements brought many unsustainable developments to cities and increased the fragmentation and degradation of critical ecosystems, including sensitive coastal ecosystems. The extension of urban areas into previously green areas allows some citizens to enjoy more living space in single-family houses, surrounded by gardens and with direct access to nature. But it can also create serious negative environmental, social, and economic problems for society, in particular if low-density and scattered urban residential developments are far from essential places such as work, home, school, and services, and highly dependent on private cars. Important ecosystem services, such as the removal of pollutants from air and water, mitigation of floods and drought, protection of shores from erosion, and providing open space and wildlife habitat for recreation, are threatened to be disrupted or impaired (Beach 2002; Elmqvist et al. 2013).

Compact settlements imply a clear boundary between urban and rural spaces and encourage functional diversification of land uses at the neighbourhood level and high-quality public transport. Compact cities favour urban mix for "a real neighbourhood gene, an organic part of the city everywhere in the city." The reintegration of urban functions should reinforce identity but also improve communication and openness. The Dutch compact city policy aimed at enhancing scarce space as efficiently as possible, creating multiple residential environments and curbing the overall ecological burden. Diversity and mixed land uses are linked to enhancing the city's unrivalled character in fragile equilibrium with the sea.

Compact cities have the potential to manage resources at remarkably lower levels of material and energy consumption, and lesser generation of waste and pollution, comparing to diffuse settlements and dispersed populations. The decentralised concentration of Danish cities highlights the importance of all these components. The renaissance of Copenhagen, described as the largest ongoing recycling project in Denmark, is founded on principles of quality and equality and aims at ensuring that sustainable development enhances all natural, marine, and human resources. As early as 1989, the regional plan for Greater Copenhagen tried to promote a "better city instead of a larger city." The Copenhagen "Five Finger Plan" has been a prime example of strategic planning directing the future expansion of the city into the countryside surrounding the coasts, preserving the green wedges and the marine environment, and promoting highly performing public transport (Mega 2012).

Copenhagen North Harbour welcomed recently the UN City, an eight-arm star hosting the UN Headquarters, larger UN hub in Scandinavia. The iconic white building on Marble pier highlights openness to the world and exemplifies UN efforts for sustainable development. The building, certified platinum by LEED, has 1400 solar panels on its roof, features innovative shades on its façades to absorb sunlight, and uses cold seawater in its cooling system. It offers a citizen-friendly environment for eight organisations to work for development, prosperity, and peace in the world, in harmony with the blue planet. UNICEF has constructed a state-of-the-art, fully automated warehouse, from which humanitarian assistance is shipped to world destinations every day.

2.4.3 *Marine Protected Areas: Living Laboratories for a Breathing Ocean*

The marine and coastal protected areas have been proposed, designed, and developed as a tool for conservation and sustainable management to preserve ecosystems, habitats, and endangered species as well as natural resources. In 2016, the percentage of terrestrial protected areas was 14.7%, the percentage of marine protected areas within national jurisdiction 10.1%, and the percentage of marine protected areas in high seas 4.12% (UN 2017).

Science-based and effectively managed marine protected areas (MPA) are currently increasing towards the 10% target, but only limited parts are truly preserved reserves, implying a complete ban of extractive or polluting activities. MPAs provide ocean life and their habitats refuge from human impacts and also allow depleted marine resources to recover. Many MPAs also serve as living laboratories, critical to scientific research and development. Effectively managed MPAs support the blue economy by helping to sustain marine resources. The International Union for Conservation of Nature has adopted scientists' recommendations for preserving at least 30% of the ocean by 2030 to maintain biodiversity, boost fisheries, and safeguard the ocean's economic, cultural, and life-supporting benefits.

Marine protected areas are critical to address key pressures on marine ecosystems including overfishing, pollution, habitat destruction, climate change, and invasive species. MPAs create space for the conservation of species and ecosystems, and preserve the ocean's capacity as a carbon sink. The design and management of marine protected areas can be well described by the 2017 OECD study, which offered good practice insights for effectively managing marine protected areas. Although global coverage of MPAs has been increasing during the past two decades, further efforts are required to meet the SDG14 target and to ensure that MPAs are implemented with other policy instruments, to more comprehensively and effectively manage the multiple pressures on marine ecosystems. It is important to evaluate accurately the involved costs and rising benefits, to design strategically located MPAs, with monitoring and financing tools, and embed these in wider policy portfolios (OECD 2017a).

Some initial estimations of the total ecosystem service benefits of achieving 10% coverage of MPAs, according to the SDG14 target, suggest $622–923 billion for the period 2015–2050. But designation does not automatically mean effective protection. Of the 3.41% global MPA coverage in 2014, only 0.59% was established as no-take MPAs, that is, areas that prohibit extractive practices such as fishing and mining (OECD 2017a).

Marine protected areas must be strategically located sanctuaries to maximise synergies and benefits. In the US, the MPA Centre uses science to assess the nature and use of MPAs to sustain healthy marine ecosystems. It aims at enhancing stewardship by strengthening capacity for planning, management, and evaluation. There are many different types of MPAs including national marine sanctuaries and national parks, national wildlife refuges, and state areas for the protection of habitat, fish, and wildlife. National systems of Marine Protected Areas help weave all these areas together into an effective network that can protect species which move through various habitats during different life stages.

Marine Protected Areas in European territorial waters are progressing. Germany, France, Norway, the Netherlands, and Belgium have the highest share of protected territorial waters. However, the percentage of territorial waters that are protected is still small, and marine protection is not in any case a panacea for rebuilding the ecological capacity of the sea. Good practices include Belgium's marine spatial planning process, through which political leadership, public participation, science, and effective communication have allowed the country to create a vision for managing competing uses of its small, but very busy, part of the North Sea (EUI 2017).

The Mediterranean Sea, one of the richest seas in biodiversity, is considered to be one of the world's priority eco-regions. Although it represents less than 1% of the global ocean surface, it hosts almost 20% of global marine biodiversity and has a high level of endemism, nearly 50% for some groups. It is an important breeding area for several key pelagic species, some of which are threatened. There is intense human activity, urbanisation, tourism, shipping traffic, overfishing, and pollution. Of the Mediterranean waters, 7.14% is under some designated protection status in 1231 sites, including 186 MPAs covering 1.6% of the Mediterranean surface, 898 marine Natura 2000, fisheries restricted areas, and the Pelagos sanctuary for marine mammals. These sites present a great variety of conservation designations, but little is known about the measures taken and if they are effective at maintaining or restoring the biodiversity under protection. Of these nationally designated MPAs, 78% is more than 10 years old, which is considered the average minimum age for an MPA to reach a certain maturity; 46 sites are more than 20 years old. The majority of the Mediterranean protected areas are near the European coasts. The National Marine Park of Alonissos and Northern Sporades, the first marine park established in Greece, is currently the largest marine protected area in Europe (MedPAN 2012, 2016).

Chile has committed to protecting one third of its territorial waters. In 2017, Chile declared the Rapa Easter Island MPA, aiming to safeguard the waters surrounding the volcanic island, a World Heritage site 4000 km west of mainland Chile that is home to at least 142 unique endemic species, including 27 that are threatened or endangered. In addition, Easter Island's waters are important spawning grounds for many migratory species, such as tuna, marlin, and swordfish. This MPA, one of the largest in the world, will protect an area roughly the size of Chile's land area from industrial fishing, mining, and other extractive activities, within the exclusive economic zone that surrounds the island. Public consultation resulted in a positive vote, limiting extractive techniques to those that are traditional to the Rapa Nui people. Such efforts highlight the need for large-scale interventions that both benefit ocean health and impose limited costs on local populations.

In 2016, 24 countries and the EU agreed to the creation of the Ross Sea Marine Protected Area in the Antarctic, enforced as of December 2017. Covering 1.55 million km^2, it is the outcome of skilful negotiation highlighting the long-term benefits of protecting an area that could produce up to three quarters of the nutrients that sustain life in the rest of the global ocean. Cities could lead ocean governance and protection of the high seas and adopt sanctuaries and preserved spaces. Landlocked countries and cities could become active in coalitions of the committed for the global ocean and demonstrate openness to and engagement with the world.

References

(The) Chicago Council on Global Affairs (2013) Feeding an urban world: a call to action. Emerging leaders perspectives. Chicago Council on Global Affairs, Chicago

Arup and Sydney Water (2014) The future of urban water: scenarios for urban water utilities in 2040. Arup and Sydney Water, Sydney

Baabou W et al (2017) The ecological footprint of Mediterranean cities: awareness creation and policy implications. Environ Sci Pol 69:94–104

Beach D (2002) Coastal sprawl: the effects of urban design on aquatic ecosystems in the United States. Pew Oceans Commission. Washington DC

City of Hamburg (2013) HafenCity. Hamburg

City of New York (NYC) (The) (2017) Progress report: OneNYC. New York

City of New York (NYC) (The) (2015a) OneNYC. The plan for a strong and just city. New York

City of New York (NYC) (The) (2015b) Building a smart + equitable city. New York

City of Stockholm (2014) Stockholm Royal Seaport. Stockholm

Economist Intelligence Unit (The) (EIU) (2017) Global food security index. London

Ellen MacArthur Foundation (2017) Urban biocycles. Geneva

Elmqvist T et al (2013) Urbanization, biodiversity and ecosystem services: challenges and opportunities. a global assessment, New York, Springer

Eriksen M et al (2014) Plastic pollution in the world's oceans: more than 5 trillion plastic pieces weighing over 250,000 tonnes afloat at sea. PloS One 9(12):e111913

EUROCITIES, City of Milan and Cardiff University (2017) Food in cities: study on innovation for a sustainable and healthy production, delivery, and consumption of food in cities. Brussels

European Commission (EC) (2016a) Opportunity now! Europe's mission to innovate. Brussels

EC (2016b) Sustainability now! A European vision for sustainability. Brussels

EC (2017a) Report on the Blue Growth Strategy. Towards more sustainable growth and jobs in the blue economy. Brussels

EC (2017b) Food from the oceans by the Scientific Advice Mechanism. Brussels

European Environment Agency (EEA) (2006) Urban sprawl in Europe. The ignored challenge. Copenhagen

EEA (2013) Signals – building the future we want. Copenhagen

EEA (2014) Marine messages: our seas, our future. Copenhagen

EEA (2016) Seafood in Europe – a food system approach for sustainability. Copenhagen

European Parliament (2017) Global trendometer. Brussels

Food and Agriculture Organisation (FAO) (2009) Food for cities. Rome

FAO (2013) Food wastage footprint: impacts on natural resources. FAO, Rome

FAO (2016) The state of world fisheries and aquaculture. FAO, Rome

Friends of Europe (2013) Global outlook on urban air quality. Brussels

Gilsoul N (2011) La ville fertile, Paris

GIZ and ICLEI (2014) Operationalizing the Urban NEXUS. Towards resource-efficient and integrated cities and metropolitan regions. Freiburg

Global Ocean Commission (GOC) (2014) From decline to recovery. a rescue package for the global ocean. Oxford

GOC (2016) The future of our ocean. Next steps and priorities. Oxford

Gore A (2013) The future: six drivers of global change. Random House, London

HafenCity Hamburg (2013) Essentials. Quartier projects. Hamburg

ICLEI (2012) Green urban economy: new drivers for a sustainable urban development. Bonn
International Energy Agency (IEA) (2016) World Energy Outlook special report. Energy and air pollution. Paris
International Panel for Biodiversity and Ecosystem Services (IPBES) (2018) IPBES assessment report, on global land degradation and restoration. Medellin
International Union for Conservation of Nature (2017) National marine plastic litter policies in EU Member States: an overview. Brussels
Jambeck et al (2015) Plastic waste inputs from land into oceans. In Science 347:768–771
London School of Economics Cities (LSE) and ICLEI (2012) Going Green. How cities are leading the next economy. London-Bonn
Love DC et al (2015) Wasted seafood in the United States: quantifying loss from production to consumption and moving toward solutions. Glob Environ Chang 35:116–124
McDonald R (ed) (2014) World's water stressed cities. In: Global environmental change. The Nature Conservancy, Arlington
MedPan (2012) The status of marine protected areas in the mediterranean sea. MedPAN & RAC/SPA. Ed, MedPAN Collection
MedPAN (2016) The status of marine protected areas in the Mediterranean Sea 2016. Marseilles
Mega V (2010) Sustainable cities for the third millennium. The odyssey of urban excellence. Springer, New York
Mega V (2012) Bio-diver-cités. Editions Universitaires Européennes, Saarbrucken
Mega V (2013) Quintessential cities, accountable to the future. Innovation, sustainability and leadership. Springer, New York
Mega V (2015) Conscious coastal cities. Sustainability, blue green growth and the politics of imagination. Springer, Cham
National Research Council (NRC) (2012a) Water reuse: potential for expanding the nation's water supply through reuse of municipal wastewater. Washington, DC: NAP
NRC (2012b) A sustainability challenge: food security for all: report of two workshops. NAP, Washington, DC
New Economics Foundation (2014) Urgent recall. Our food system under review. London
New York City Global Partners (2013) Best practice: converting schoolyards to community playgrounds. New York
OECD-FAO (2017) OECD-FAO agriculture outlook 2017–2026. OECD-FAO, Paris
Organisation for Economic Co-operation and Development (OECD) (2012a) OECD environmental outlook to 2050: the consequences to inaction. OECD, Paris
OECD (2012b) Compact city policies. A comparative assessment. OECD, Paris
OECD (2015) Green growth in fisheries and aquaculture. OECD, Paris
OECD (2016) The ocean economy 2030. OECD, Paris
OECD (2017a) Marine protected areas: economics, management and effective policy mixes. OECD, Paris
OECD (2017b) Review of fisheries: policies and summary statistics 2017. OECD, Paris
Orsenna E, Gilsoul N (2018) Désr de villes. Robert Laffont, Paris
Sesini M (2011) The garbage patch in the oceans: the problem and possible solutions. New York: Columbia University
STOA (2015) Ten technologies which could change our lives. European Parliament Research Service, Brussels
UN Convention of Biological Diversity (CBD) (2010) Strategic plan for biodiversity 2011–2020, including Aichi Biodiversity Targets. Nagoya
UN Convention of Biological Diversity (CBD), Stockholm Resilience Centre and ICLEI (2012) Cities and biodiversity outlook. Action and policy. UNEP, Paris
UN Environmental Programme (UNEP) (2016a) Global environmental outlook, 6th edn. Paris
UNEP (2016b) Marine plastic debris and microplastics – global lessons and research to inspire action and guide policy change. UNEP, Paris
UNESCO/IOC (2017) Global Ocean Science Report. Paris
UNEP/GRID-Arendal (2016) Marine litter vital graphics. Nairobi

UNEP/International Solid Waste Association (ISWA) (2015) Global waste management outlook. United Nations Environment Programme, Nairobi
United Nations (UN) (2016) First ocean assessment. New York
United Nations (UN) (2017) Progress towards the sustainable development goals 2017. Report to the UN Secretary-General, New York
West FM (2016) Megachange: economic disruption, political upheaval, and social strife in the 21st century. Brookings Institution Press, Washington, DC
World Bank (2012) Inclusive green growth. The pathway to sustainable development, Washington, DC
World Bank et al (2014) Fish to 2030. Prospects for fisheries and aquaculture. The World Bank, Washington, DC
World Bank (2017) The sunken billions revisited: progress and challenges in global marine fisheries. World Bank Group, Washington, DC
World Business Council on Sustainable Development (WBCSD) (2017) The business case for reducing ocean waste. Geneva
World Health Organisation (2017) World health statistics 2017: monitoring health for the SDGs, Sustainable Development Goals. World Health Organization, Geneva
World Resources Institute (WRI) (2012) Impacts of hurricane sandy and the climate change connection. Washington DC
WRI (2016) Shifting diets for a sustainable food future. Washington, DC
World Wide Fund for Nature (WWF) (2015a) Living blue planet report. Species, habitats and human well-being 2015. Headquarters in Gland, Switzerland
WWF (2015b) Reviving the ocean economy: the case for action. Headquarters in Gland, Switzerland
WWF (2016) Living planet report 2016. Headquarters in Gland, Switzerland

Selected Websites and Social Media

#breakfreefromplastic
#CleanSeas
#EarthDay
#EUGreenWeek 2018
#worldoceansummit
@RethinkPlastic
www.cbd.int/doc
www.cityofboston.gov/climate/bostonsplan/
www.climatesummitformayors.dk
www.closedlooppartners.com/ocean/
www.dublinked.ie
https://ec.europa.eu/food/safety/food_waste_en
www.eea.eu.int
www.environnement.brussels
www.eolss.net
www.europa.eu.int/comm/environment/urban
www.footprintnetwork.org
www.hamburg.de
www.iclei.org
www.icriforum.org/about-icri
www.ipcc.org
www.kk.dk

www.marinedebris.noaa.gov
www.medtrends.org
www.myclimate.org
www.oceannetworks.ca/
www.oneplanetliving.net
www.ourocean2017.org
www.panda.org/livingplanet
www.pewtrusts.org/en/research-and-analysis/collections/2017/05/the-ocean-conference
www.plasticchange.org
www.rethinkplasticsalliance.eu
www.sdg.iisd.org/
www.stockholm.se
www.stockholmresilience.org
www.surfrider.org
www.symbiocity.se/
www.thebluecarboninitiative.org
www.un.org/esa/sustdev
www.unep.org
www.unfccc.int
www.unhabitat.org
www.wbcsd.org
www.worldwatchinstitute.org
www.wpci.nl
www.wri.org

Watercolour 5
Paris, EU Innovation Capital 2017 and the Inspiration from the Sea (Tribute to Frank Gerry)

Chapter 3
The Future of Climate Will Be Decided in Cities and Oceans

Abstract This chapter argues that the battle for climate change will be won or lost in cities and the seas. Oceans and cities are the ecosystems that most affect, and are most affected by, climate. The oceans have a vital role in regulating the global climate system, generating oxygen and absorbing carbon dioxide: cities constitute the 121 largest emitters and potentially most-efficient energy users, but also the highest concentrated human resolve to change the status quo. Sea rise, ocean warming, and acidification may have disastrous effects on the global ocean and the future of the planet. Cities impact climate not only because of their sheer size but also because of their democratic traditions and their global networks, which can have a great role in climate justice, and their power to disseminate influential patterns to the world. Climate change, subsidence, and sea level rise are important risks, especially for low-elevation, densely populated urban areas, but eco-responsible city developments can prevent disaster. Riverside, coastal, and deltaic cities must bear the worst effects of the extreme events linked to climate change. Low-lying developing megacities are extremely vulnerable, but developed cities and assets are also at risk and should share knowledge and technologies and disseminate the best approaches.

3.1 Cities, Seas, and Climate: Highly Interlinked in Ways Yet Unexplored

3.1.1 Systemic Climate Action: A Race Against Time and Short-Sightedness

Climate change is already perturbing the planet and greatly impacting cities and oceans. Numerous scientists, philosophers, theologians, and communities share this concern. In 2015, the Pope's 'Laudato Si' referred to the ethical causes and the need to bring the whole human family together to seek sustainable and integral development. But after 3 consecutive years of record-high temperatures, CO_2 emissions rose for the first time in 4 years, in 2017, which is also among the 3 hottest years on record and the hottest year ever without an El Niño event. As a consequence, 2017

V. P. Mega, *Eco-Responsible Cities and the Global Ocean*,
https://doi.org/10.1007/978-3-319-93680-2_3

was marked by major wildfires in the US, Chile, and Portugal, leading to significant economic costs and losses. Some cities are particularly hit by extreme weather events including sea storms and sea level rise: the melting of the ice caps in Greenland and Antarctica is raising global sea levels, and the melting of the Arctic cap is opening up a new and fragile world. Urgent action to reduce greenhouse gas (GHG) emissions and stabilise the climate is an overarching imperative for oceans and cities (Pope Francis 2015).

The Paris Agreement on climate change is the first major multilateral pact of the twenty-first century. It legally entered into force with unprecedented speed, ahead of COP22 in Marrakesh, exactly 11 months after the Oceans Day in COP21 in Paris, a month after the ratification by countries representing 55% of the world's emissions. COP22 in Marrakech provided the next opportunity for all parties to debate implementation paths.

The day before the Paris Agreement entered into force, UNEP's 2016 emissions gap assessment revealed that the emission reductions corresponding to pledges made by countries in the framework of the Paris Agreement still leave the world on a trajectory to 3.4 °C by 2100, well above the 2 °C level. That degree of climate change would almost certainly be catastrophic for the planet (UNEP 2016).

The Paris Agreement differs fundamentally from previous climate agreements. It is a universal bond that aims at limiting the rise in global temperatures to well below 2 °C with an aspirational goal of 1.5 °C, fostering resilience and the ability to adapt, and making finance flows compatible with a low-emissions, climate-resilient development pathway. Contributions to emission reductions beyond 2020 are set out in the Nationally Determined Contributions (NDC).

In October 2016, the Kigali Amendment to the Montreal Protocol provided a timetable and mandate for countries to reduce, over time, the production and usage of hydrofluorocarbons, a more ozone-friendly substitute for earlier chemicals used in refrigerators and air conditioning, but which are also powerful greenhouse gases. The G20 Presidency in 2016 advanced on critical green finance issues and some countries, such as France, have made climate disclosure mandatory. In June 2017 at the G20 Leaders' Summit, 19 of the 20 world leaders reaffirmed their commitment to implementing the Paris Agreement.

In autumn 2017, the eighth edition of UNEP's 2017 emissions gap report, published ahead of COP23, suggests that national pledges only bring one third of the reduction in emissions required by 2030 to meet climate targets, with private sector and subnational action not increasing at a rate that would help close this worrying gap (UNEP 2017). At the same time, the World Meteorological Organisation (WMO) announced that atmospheric carbon dioxide concentrations continue to rise rapidly and have reached 145% of pre-industrial (before 1750) levels, the highest in 800,000 years. And the Lancet Countdown's 2017 report concludes that the human symptoms of climate change are unequivocal and potentially irreversible, and the delayed response to climate change over the past 25 years has jeopardised human life and livelihoods. However, the past 5 years have seen an accelerated response, and momentum is building (Lancet Countdown 2017).

The Facilitative Dialogue to be held in 2018 will be the first opportunity to take stock of countries' collective mitigation efforts. It will also inform the preparation of subsequent NDCs. As transparency and trust build, there is hope that countries will be more ambitious in the emissions reduction commitments taken in each successive round of engagement. Every 5 years thereafter, global stocktakes will provide further assessments of collective progress towards achieving the Paris Agreement, covering the full spectrum of mitigation and adaptation but also issues such as finance, technology, and capacity building.

The decarbonisation of world economies is inexorable and offers tremendous new opportunities for a future-proof world. The EU proved that low-carbon growth is possible. It has increased its economy by 46% since 1990 while lowering emissions by 23%. And the world entered a new phase with flat energy emissions for 3 consecutive years of growth in 2014–2016. The OECD suggested that "Investing in Climate is Investing in Growth" as economic conditions in many countries provide an opportunity to boost growth and investment that will drive the transition to a post-carbon future (EC 2016a, b; IEA 2017; OECD 2017a).

The Intergovernmental Panel on Climate Change has had an important role in providing authoritative scientific assessment of the urgency of the challenge and the drivers and likely impacts of change. Although many pinpointed the knowledge gaps, there was an overwhelming suggestion that humanity knows enough to act. Public authorities have a responsibility to respond to the evolving scientific evidence and implement the best possible portfolio of policies (IPCC 2014a, b, c).

3.1.2 Cities and Oceans on the Climate Scene: Need for Accelerating and Amplifying

The UN climate negotiations progressively incorporated the ocean and urban dimensions. The UN Framework Convention on Climate Change (UNFCCC), adopted in 1992, had already explicitly referenced the Ocean in Articles 4–1 (d and e) suggesting that the Parties "... promote and cooperate in the conservation and enhancement, as appropriate, of sinks and reservoirs of all greenhouse gases not controlled by the Montreal Protocol, including biomass, forests and oceans as well as other terrestrial, coastal and marine ecosystems" and "… cooperate in preparing for adaptation to the impacts of climate change; develop and elaborate appropriate and integrated plans for coastal zone management…".

Since 1995 and the beginning of the Conferences of the Parties (COP), discussions have focused mainly on issues related to the reduction of greenhouse gas emissions, adaptation, and funding for developing countries. In 2015, COP21 promised a new age of commitment on climate and stakeholders advocated for urban and ocean issues to be among the priority issues. The oceans are vital in regulating the global climate system, generating oxygen and absorbing carbon dioxide from the atmosphere. Cities, hosting more than half of the population and producing more than two thirds of the emissions, would be an indispensable partner: limiting global warming to less than 2 °C is essential to maintain the regulation capacity of the oceans.

The Paris Agreement underlines "...the importance of ensuring the integrity of all ecosystems, including oceans, and the protection of biodiversity, recognised by some cultures as Mother Earth, and noting the importance for some of the concept of "climate justice," when taking action to address climate change. It also states the importance of non-state actors, among which cities are prominent.

Cities represent a leading force for global action on climate change, as both drivers of climate change and influential places of concentrated climate responses. Economic growth, mainly taking place in cities, has been the main cause of the continuous increase of GHG emissions. Climate change is partially the cumulative result of millions of decisions made by the world's citizens within their immediate environments. In 2007, China overtook the US as the largest emitter in the world because of higher levels of income and unsustainably increasing consumption patterns. Urgent action is needed in both the developed and the emerging world, in anticipation of a new investment cycle of capital renewal and expansion, mainly for urban infrastructures.

Cities have a responsibility to be frontrunners in the double front of mitigation and adaptation to climate change. Acting both locally and in network, they can have a meaningful global impact. Although each city is unique in its morphology, culture, infrastructure, municipal services, and potential impacts to and from climate change, the underlying drivers of emissions in all cities largely reflect biodiversity decline, unsustainable resource and waste management, flawed economic models and behaviours, and traffic congestion. The IPCC Fifth Assessment Report (AR5) was the first IPCC assessment report with a dedicated chapter on urban areas, which recognised the urgency for climate change mitigation as well as the adaptation of cities. It pinpointed that many emerging climate change risks are concentrated in urban areas and impact an increasing number of people and assets exposed to climate change threats. The greatest potential for mitigating GHG lies in rapidly developing megacities which have also the greatest needs to protect the most vulnerable world citizens from adverse climate effects (IPCC 2014a, b, c).

COP21 brought a historic climate-oriented gathering of mayors at the Climate Summit for Local Leaders, organised by the Mayor of Paris and the United Nations Secretary-General's Special Envoy for Cities and Climate Change, the former New York City Mayor. This event provided a forum for a dynamic discussion of climate targets and links to urban action. On the ocean front, the Ocean Climate Platform highlighted the role of marine and coastal ecosystems as natural carbon wells, forming the blue carbon, the importance of healthy and functional ecosystems to face climate change, and the need of a consistent and resilient network of marine protected areas. It underlined the importance of developing marine renewable energy, while preserving marine biodiversity and guiding maritime transport sustainability transition. The platform proposed priority adaptation measures for the most vulnerable regions, including coastal zones in developing countries, territories, and SIDS.

The majority of world cities seem convinced of the effects of climate change and 70% are taking actions. Despite, or probably because of, the decision of the U.S. President to withdraw from the Paris Agreement in June 2017, 13 states representing almost 40% of the U.S. economy, along with the mayors of nearly 200 cities and more than 900 business leaders, have signed different pledges to continue reducing greenhouse gas emissions. The United States Climate Alliance, which was established by the governors of California, New York, and Washington on 1 June 2017, in response to the presidential announcement, was quickly joined by Connecticut, Delaware, Hawaii, Massachusetts, Minnesota, Oregon, Puerto Rico, Rhode Island, Vermont, and Virginia. The bipartisan coalition is committed to achieve the US contributions made under the Paris Agreement, aiming at reducing emissions by 26% to 28% below 2005 levels by 2025. Hawaii has even already passed a law committing to the Paris Agreement. Further, 125 cities signed the "We are still in" declaration, including the mayors of Chicago, Los Angeles, Miami Beach, New York City, Pittsburgh, Salt Lake City, San Francisco, and Washington, DC. The declaration features a dedicated website for further engagement and exchanges.

International city-engaging networks are essential for advancing faster and further. The Cities Climate Leadership Group (C40) provides a shining example. Created and led by cities, C40 connects 91 of the world's greatest cities, representing one quarter of the global economy and 1 of 12 world citizens. The coalition is committed to tackling climate change and driving urban action that reduces greenhouse gas emissions and climate risks, while increasing the health, well-being, and economic opportunities of citizens (C40 and Arup 2016).

Active networking and advocacy for the ocean intensify and expand. The interlinkages of the oceans with climate are multifold and the subject of ongoing research. Climate change greatly affects marine and coastal ecosystems and their dependant populations. Changes to the climate, caused by increasing levels of greenhouse gases in the atmosphere, which thus lead to changes in the oceans, put marine ecosystems and coastal communities at risk. The AR5 synthesised research regarding the impacts of climate change on physical and biological processes in the oceans, at both global and regional levels, and expressed high confidence for observed and projected impacts of climate change, including ocean warming, acidification, and sea level rise. Ocean deoxygenation is increasing because warmer waters hold less oxygen (IUCN 2016).

As on land, many balances of the sea began seeing disturbances in the 1970s. Between 1971 and 2010, the oceans have absorbed approximately 93% of the excess heat caused by global warming, leading to several major changes such as the increase in stratification, limitation in the circulation of nutrients from deep waters to the surface, and sea level rise. In addition, the oceans have absorbed 26% of anthropogenic CO_2 emitted since the beginning of the Industrial Revolution, which has resulted in ocean acidification (WWF 2015a, b).

Together, these processes strongly affect the abundance, geographic distribution, and migration patterns of marine and coastal species. As a consequence of these complex changes, marine and coastal human activities, such as fisheries, aquaculture, aquatic tourism, and offshore wind and marine energy, are in turn at risk. Impacts to aquaculture are expected to be generally negative, depending on location, species, and method. Coastal tourism could be affected to a lesser extent, through, for example, a polewards redistribution of tourism flow because of increased temperatures and the loss of beaches through coastal erosion and sea level rise. Impacts to human health primarily concern the increased frequency of disease transmission and expansion, deaths associated with storm surges and flooding, increased harmful algal blooms, and reduced food and water security.

The IPCC AR5 global estimates of observed or projected interactions from multiple stressors were limited, with studies rarely addressing adaptation to multiple stressors over a longer time, which is essential for both cities and seas. This limitation suggests a stronger need for improved understanding of ecosystems and species sensitivity to various sources and forces of change. Improved knowledge is also needed with respect to the drivers of human adaptation in cities. Climate change narratives require explicit connections between the biophysical and human impacts, as well as options for climate-resilient pathways.

Cities and the seas have many ethical implications for climate change, related to the local impacts of a global process and the experienced and imposed effects of climate change. Climate justice relates to the allocation of responsibility for anthropogenic climate change. Mobilisation in world cities has been intense, and many suggest that climate change is an urban sea change that involves special responsibility for world developing megacities and small developing island states (City of New York (NYC) (The) 2015a, b).

Think tanks, NGOs, and knowledge organisations were very active in co-organising the Oceans Day at COP21 on 4 December 2015. The Institute for Sustainable Development and International Relations (IDDRI) and Pierre et Marie Curie University coordinated "The Oceans 2015 Initiative," which further discussed the impacts and the prospects, and their degrees of confidence. The IDDRI's "Oceans 2015 Initiative," Part II, provided an updated synthesis of the essential science about climate change implications and risks for marine and coastal socioeconomic activities that form the blue economy. The synthesis is based upon the AR5 and scientific articles and grey literature more recently published. The IDDRI advocated with the "Because the Ocean" Initiative, together with the Global Ocean Commission, the governments of Chile and France, the Prince Albert II of Monaco Foundation, and the organisation Tara Expeditions.

The Consortium for Ocean Leadership, a non-profit organisation that represents more than 100 leading public and private ocean research and education institutions, aquaria, and industry with the mission to advance research, education, and ocean policy, launched a statement with the European Marine Board, in advance of COP21, highlighting the critical role of science as part of the societal response to climate change. This Ocean Climate Nexus Consensus Statement proposed key priorities for ocean research and called on funders to invest in critical and long-term ocean–climate research and observation programmes at a level that matches the urgency and seriousness of the impacts of the changing ocean and climate.

The ocean–climate nexus brings the challenge to enhance human understanding of fundamental processes and to develop observational and modelling frameworks that will allow integrated assessment of the interconnected ocean–climate–human system. Combined with a deeper understanding of the social and economic drivers, such a knowledge base can underpin effective ocean management at local, regional, and global scales.

Climate change has the potential to disrupt ocean currents and thus the oceanic redistribution of heat, which may impact regional climates and the attractiveness of locales. A warmer ocean also has consequences for the formation, extent, and persistence of sea ice in polar and continental margin regions. Variability of sea surface temperatures has been linked to changes in regional and local climate. The uptake and storage of heat and carbon by the ocean, and its exchange with the atmosphere, varies among regions and over annual to multi-decadal timescales.

The marine environment is essential to climate, for example through the biological pump that transfers carbon from the surface ocean to the deep ocean. The strength and impact of this pump depend upon many complex processes such as ocean circulation, supply of nutrients, and marine food web processes. Changes in ocean physics and chemistry, along with pollution, are impacting marine ecosystem productivity and services through effects on the physiology, life cycles, and distribution of marine organisms. However, the impacts of these changes on biodiversity and the functioning of marine food webs are highly complex, and knowledge gaps abound. Richer, more diverse ecosystems are better able to cope with disturbances such as extreme events and the emergence of diseases. Biodiversity is vital for maintaining stable and productive marine ecosystems, and enhances the ability of ecosystems to respond to change by allowing greater scope for development of alternative structures with changing environmental conditions. Conversely, loss of biodiversity may reduce the resilience of marine ecosystems and their capacity to adjust to both gradual changes and rapid fluctuations (IPBES 2018; UN 2017).

One year after the Paris Agreement was entered into force, COP23, chaired by Fiji, asked for swifter and more ambitious action. It launched the Ocean Pathway embodying the nexus between the Ocean and Climate Change and advocating for the nexus to be firmly anchored in the UNFCCC process. The pathway indicates the possibility of a UNFCCC agenda item and a work programme by 2019, ensures the insertion of the ocean into the Nationally Determined Contributions, and enhances the opportunity to support ocean health and maintain critical ecosystems. It supports existing priorities that affect and are impacted by ocean and climate and reinforces mobilisation and cooperation for the conservation and enhancement of the resilience of the global ocean. This pathway links to displacement and migration, coastal infrastructure, marine ecosystem services, cities, ocean food security, and ocean energy.

3.2 Ocean Warming and Acidification: Grief for the Reefs and Not Only Reefs

3.2.1 Ocean Warming and Complex Interacting Effects of Climate Change

Science highlighted, with an extremely high certainty, that the upper ocean had warmed during the past 40 years and predicted, with very high confidence, that this trend would continue over the coming century. It was considered very likely that this increase in temperature had contributed to significant global mean sea level rise. High agreement among data provided evidence that stratification caused by increasing sea temperatures has caused declining oxygen concentrations in seawater. The IPCC report expressed high confidence that the oceans are absorbing anthropogenic carbon and that the resulting chemical reactions cause ocean acidification (IPCC 2014c).

Increasing sea temperatures affect the geographic distributions of organisms, with observations lending evidence to polewards migrations of species. Temperature was also shown to be affecting the timing of major life events such as reproduction and migrations. Net primary productivity was projected to decrease moderately by 2100 in the open ocean under both low- and high-emission scenarios, paralleled by an increase at high latitudes and a decrease in the tropics. Despite the lack of field observations, ocean acidification was predicted to have a significant effect on many aspects of the physiology, behaviour, and population dynamics of organisms.

The IPCC suggested with high confidence that the various environmental drivers act simultaneously on organisms, causing complex, interacting effects. AR5 highlighted a need for a better understanding regarding the extent of warming in deep water masses (below 700 m), the likelihood of climate-induced changes to major systems, and the ways in which climate-induced changes in the physiology and biogeography of individual species may alter ecosystem structures, species interactions, and food webs.

In 2016, the International Union for Conservation of Nature and Natural Resources (IUCN) provided an ultimate report on ocean warming for a renewed interest among the political, social, and scientific communities. It summarises the main insights in the conclusion that increasing human activities have injected vast quantities of heat into the ocean and shield humanity on land from the worst effects of climate change. This regulating function, however, happened at the cost of profound alterations to the ocean's physics and chemistry, that led especially to ocean warming and acidification and consequently to sea level rise (IUCN 2016).

Sea surface temperature, ocean heat content, sea level rise, melting of glaciers and ice sheets, CO_2 emissions, and atmospheric concentrations are increasing at an accelerating rate, with significant consequences for humanity and for ocean species and ecosystems. There is likely to be an increase in mean global ocean temperature of 1–4 °C by 2100. As the consequence of many regional unbalances, the greatest ocean warming overall is occurring in the Southern Hemisphere and is contributing to the subsurface melting of Antarctic ice shelves. Since the 1990s, the atmosphere in the Polar Regions has been warming at about twice the average rate of global warming. Arctic warming and ice loss could possibly lead to the essential disappearance of the summer Arctic sea ice within the next few decades.

Political decisions involve value judgments, but scientific knowledge has a key role in, for example, analysing the potential effects of climate change, and their likelihood, identifying and evaluating the potential consequences of various policy responses, expanding the portfolio of possible options, and improving the effectiveness of policies. Robust and long-term climate observations are crucial for reliable foresights and evaluations to illuminate and orient political action. A call for a new era of climate change science and policy highlighted the importance of 'citizen-inspired' research, which not only improves understanding of the causes and consequences of climate change, but assists decision makers at the local, regional, national, and international levels in mitigation and adaptation actions (NRC 2010).

The declaration by the UN of the decade of ocean science as 2021–2030 reminded humanity that only 10% of what happens under the waters is known, including the most fascinating forests and dramatic valleys. Major biological manifestations of the impacts from ocean warming and other stressors have resulted in changes in biogeographic, biodiversity, community size, and species abundance as well as ecological shifts. Such shifts often interfere, or are predicted to interfere, with the benefits expected from the ocean. Whilst rising CO_2 levels and increasing warming can occasionally have positive effects, the overwhelming evidence and predictions are for a cocktail of negative effects, which are only now starting to be understood. Ocean warming and climate change are ultimately contributing to global homogenisation of biodiversity, as vulnerable species become extinct and 'non-native' species from different biogeographic regions spread and are established across the world's oceans (IUCN 2016). The introduction of invasive aquatic species to new aquatic environments is a major threat to ocean biodiversity and can result in serious impacts on infrastructure and livelihoods. The spread of invasive species is mainly associated with shipping, aquaculture, offshore structures, and marine debris.

3.2.2 *The Health of Coral Reefs, the Red Gold, as a Litmus Test*

Coral reefs, the most fascinating ecosystems, are facing severe threats that jeopardise their survival worldwide. Coral reefs make up about 1–2% of the ocean, yet harbour almost one fourth to two fifths of its biodiversity. They are among the most extraordinary and vulnerable ecosystems, because of their complexity and the intricate relationships among their vital species and aspects. The deterioration of these incredible ecosystems is a systemic risk, because they protect many vulnerable coasts from storms, are nurseries for invaluable fish, and provide food and livelihood for more than 1 billion people. Turning the tide is desperately needed to ensure that reefs continue to exist and recover their structure and functions. Scientists and policy makers across the globe are developing and implementing measures to protect and conserve these ecosystems against a variety of local and global stressors. The OECD estimated that coral ecosystems provide an average value of $172 billion to the world economy (OECD 2017b).

In 2011, a groundbreaking report by the World Resources Institute shed light on the extent to which reefs are at risk and revealed some inconvenient truths. Mass coral bleaching has occurred in every region of the world. Approximately 75% of coral reefs worldwide are threatened by a combination of local and global stressors: this percentage reaches 95% of coral reefs in Southeast Asia. More than 75% of the coral reefs in the Atlantic, 65% of the coral reefs in the Indian Ocean, and nearly 50% of coral reefs in the Pacific are threatened. Climate change is a major but not the only stressor. Coral reefs are experiencing higher ocean temperatures and acidity than in the past 400,000 years and more than 60% of the reefs are directly experiencing one or more local stresses.

Unsustainable fishing practices, such as overfishing, are considered the most significant non-climate-related threat affecting more than 55% of all world reefs. The percentage of threatened coral reefs increased by 30% in the first 10 years of the millennium, and fishing-related threats increased by 80%. Without urgent action taken to minimise local stressors, the part of coral reefs at risk could rise to 90% by 2030 and be close to 100% by 2050. Because of ocean acidification, it is projected that by 2050 only about 15% of coral reefs will be in areas where aragonite levels are adequate for coral growth. Twenty-seven countries and territories are identified as highly vulnerable to reef loss across the world's reef regions, and of these 19 are small island states (WRI 2011).

Coral reefs hold a vital symbolic and bequest value. Approximately 14% of Australia's coral reefs are threatened, although this ocean nation is ranked as the least threatened coral reef region. Significant parts of the emblematic Great Barrier Reef off Australia's eastern coast have been severely degraded or lost during past years, both a tragedy and a very serious wake-up call. The Great Barrier Reef is the world's largest coral reef system, composed of more than 2900 individual reefs and 900 islands stretching for more than 2300 km in the Coral Sea, off the coast of Queensland. The Reef supports an amazing diversity of life and is the world's largest single structure made by living organisms that can be seen from outer space. The Great Barrier Reef has long been known to and used by the Aboriginal Australian and Torres Strait Islander communities and is an important part of local cultures and heritage. It is also a very attractive destination for tourists.

A large part of the reef is protected by the Great Barrier Reef Marine Park, which helps limit the impact of human activity, especially fishing and tourism. It seems that climate change took a heavy toll and caused mass coral bleaching of the reef, which also suffered from population outbreaks of the crown-of-thorns starfish. A study published in 2012 by the National Academy of Sciences reported that the reef has lost more than half its coral cover since 1985. Storm damage accounted for 48% of the decline, while the crown-of-horns starfish contributed 42%. Coral bleaching, caused by warmer water, accounted for 10% of coral loss. Further reports stated that coral bleaching was more widespread than previously believed, seriously affecting the northern parts of the reef as a result of warming ocean temperature.

The World Future Council (WFC), created to give a voice to future generations, decided to acknowledge in 2012 some exemplary coastal and ocean policies. In partnership with the UN Secretariat of the Convention on Biological Diversity (CBD), the Global Environment Facility, and FAO, WFC examined outstanding policies, including marine protected area programmes and integrated coastal zone management plans, policies regulating fisheries, trade in marine products, marine litter, and land–sea interactions. Inspiring best practice often comes from the developing world and SIDS.

The Palauan government has been the first to ban commercial fishing of shark species in its territorial waters. Palau's Shark Haven Act of 2009 aims to protect more than 100 species of open water and reef sharks in national waters. Up to 73 million sharks are killed every year, and the ecological effects of diminishing shark populations are being felt in the collapse of marine food webs and the increase in jellyfish blooms. Any sharks caught in other nets must be released unharmed, with substantial fines for violators. Palau confirmed its ambition to be a leader in shark conservation, by pushing for international regulation banning shark finning and trade of shark products. A number of other countries including Honduras, the Maldives and the Bahamas have since enacted similar laws and banned shark fishing in their territorial waters (WFC 2012).

More than 130 million people rely on reef ecosystems in the Coral Triangle, at the epicentre of the world's richest marine biodiversity. The Tubbataha Reefs, a World Heritage Site and hotspot of coral reef biodiversity, are located in the Sulu Sea at the heart of the Coral Triangle. The Philippines' 2010 Tubbataha Reefs natural park act ensures the effective management of the reefs, by strengthening the legislative mandate of its managing bodies. It provides the legal arsenal to combat illegal fishing and poaching by foreign fleets and helps to safeguard a long-term food source for local communities. Tubbataha has demonstrated that with sensitively negotiated stakeholder agreements, local communities need not bear the burden of closed protected areas, but rather can be their main beneficiaries. Benefits such as user fees from divers are distributed amongst stakeholders as well as being used for monitoring and management. As an important nursery site for fish and invertebrates, the reef supports local artisanal fisheries and operations. Fishers report higher catches since the protection of the area.

Namibia's Marine Resources Act adopted in 2000 to stop harmful subsidies to the fishing industry is another recognised case. When Namibia gained independence in 1990, it inherited severely overexploited and unregulated fisheries. The Namibian Constitution set a new course, with the sustainable use of ecological resources for the benefit of current and future generations as a guiding principle. During the past two decades, the country has largely reversed the trajectory of its fish stocks, through regulations addressing the key drivers of degradation of marine capture fisheries: bycatch, illegal fishing, overcapacity from subsidies, and harmful fishing gear. Namibia's Marine Resources Act seems exemplary for instituting an ecologically and economically viable fishing industry, founded on scientific evidence and a rights-based management system. Notably, Namibia has achieved this by avoiding government subsidies to support the fishing industry. Most fisheries show signs of stability, and the sector provides employment for approximately 14,000 Namibians. Access to the fisheries is fully controlled, contributing to an ecologically and economically sustainable fishing industry (WFC 2012).

Caribbean coral reefs face significant stress. Across the Caribbean, many marine protected areas are merely "paper parks:" these are recognised and protected by law, but there is no enforcement because of the lack of resources. As the result, 75% of coral reefs are threatened, with 30% ranking in the 'high' or 'very high' threat category. But there are cases that can inspire. In Cuba, the "Jardines de la Reina" (Queens' Gardens), a coral reef and mangrove complex 60 miles off the country's southern coast, offers lessons and hope for coral reef ecosystems throughout the world. Although overfishing presents the most pervasive challenge, the region's reefs also suffer from pollution and coastal development. The Gardens comprise a rare, healthy coral reef ecosystem, with top predators and large fish still present. The coral reef has been spared much of the pressure to which most Caribbean reefs are subject, partly because of its location, far from shore and fishing pressure. It also enjoys many mangroves, which help to filter pollution.

In 1996, a large no-fishing sanctuary was created in the Gardens through a partnership between the Cuban government and an entrepreneur with an interest in establishing an eco-tourist hotel and a dive operation in the Gardens. In exchange for exclusive rights as the dive operator in this area, fishing regulations were enforced. In addition, the number of visitors is strictly limited to 1000 divers per year. This unique collaborative arrangement between the government and a tourism operator, with a vested interest in maintaining a healthy coral reef, has kept human pressure under control, and allowed the coral reef and mangrove ecosystems to flourish. The Gardens are struggling with invasive lionfish, a prolific, voracious species than can wipe out native fish populations. But the region is already taking action to address this problem with lionfish hunting patrols run by the snorkelling resort, catching and killing the invasive lionfish and feeding them to the native sharks and groupers that can control invasive lionfish on pristine reefs.

At the 2018 Economist World Ocean Summit, the Global Mangrove Alliance, bringing together three of the world's largest conservation groups—Conservation International, The Nature Conservancy, and WWF—pledged restoring 20% of the world's mangrove habitats by 2030. Bloomberg Philanthropies would develop a list of the 50 most critical coral reefs to protect. The Summit also brought forward the first insured coral reef after a trust fund was created by governments, academia and NGOs for insuring the Mexican part of the Mesoamerican coral reef from hurricanes. The purpose is to raise awareness on the coastal protection offered by coral reefs against hurricanes and sea level rise and save funds for restoring coral reefs after been hit by hurricanes.

3.3 Cities, Climate and Intertwined Extreme Events

3.3.1 Far-Sighted Climate-Conscious Cities Design Climate Action

The impacts of climate change pose significant social, environmental and economic threats and risks to the urban, national and global communities. More frequent and extreme weather events, especially storms and floods, heat waves and droughts, increasing temperatures and rising sea levels, seriously affect livelihoods, food and energy supply, infrastructure, ecosystems, society and the economy as a whole. Concerted responsible action, unshakable commitments, and performance monitoring and reporting should be key elements of all urban strategies. Climate-conscious cities involve citizens and local stakeholders in climate commitments to promote action and provide guidelines for implementing policies to reduce greenhouse gas emissions (Mega 2012, 2013, 2015).

Cities are the very places where most of the decisions impacting climate change are being taken, and most solutions are being co-shaped and can be most efficiently and effectively implemented and co-assessed. High urban population densities and short distances yield many opportunities for low-carbon lifestyles, such as the use of cycling paths or public transport. In most countries, energy use per capita of urban residents is lower than the national average. In emerging world cities, climate impacts may aggravate poverty conditions and social inequalities. Climate change could also exacerbate other existing knotty environmental problems such as air pollution and inadequate water supply. More and more experts warn that poor cities and communities are the most vulnerable to climate change. As the coastal cities of Africa and Asia expand, many of their poorest residents are being pushed to the edges of liveable land and into informal settlements, often clustering on low-lying river shores with inadequate infrastructures and services (World Bank 2013).

Climate change happens everywhere but it impacts various cities very differently. Water scarcity in some areas and overabundance of water in others are the hallmarks of climate change in many world regions. Inconsistences in the monsoon season and unusual heat extremes may have extremely serious consequences. Dhaka, Kolkata, and Mumbai may be confronted with increased flooding, intense cyclones, sea level rise, and warming temperatures. In Southeast Asia, coastal cities may have to face intense stress. A sea level rise of 30 cm, possible by 2040 if inaction continues, would cause massive flooding in cities and inundate low-lying cropland with saltwater corrosive to crops. The Mekong Delta, in Vietnam, a global rice producer particularly vulnerable to sea level rise, could lose much of its production (World Bank 2013).

Most cities are engaged with climate mitigation, established strategic objectives with their stakeholders, and introduced measures to achieve them, often in a noble emulation with their partners. Pioneer European cities, such as Amsterdam, Copenhagen, and London, acted early and formulated climate action plans with strategic objectives and targets, conceived mitigation and adaptation policies, and created dedicated agencies to monitor progress. To confront the challenge of climate change as an opportunity rather than a threat, cities set up climate-proof programmes with the aim of resilience. Many more cities see adaptation as an opportunity for better urban planning and policy to develop the adequate infrastructures, improve quality of life, and create new innovation trajectories and employment possibilities (Amsterdam Climate Office 2008).

Copenhagen pledged to become CO_2 neutral by 2025 and to provide the world with a gold standard for sustainable cities. Copenhagen's 2009 Climate Plan was a conscious milestone and the first in Scandinavia. To achieve its ambitious goal, the city has established precise targets including energy efficiency, renewable energy sources, and green building standards. All new buildings have to be carbon neutral by 2020, and sustainable transport is the preference of most citizens demonstrating a true desire for cycling. The city developed a smart bike equipped with sensors to provide real-time information not only to riders but also to administrators for open data aggregation on issues of air quality and traffic congestion.

Copenhagen has a symbiotic relationship with the sea, strengthened and demonstrated in times of heavy thunderstorms and exceptional rain levels. A network of temporary reservoirs was built in the 1990s to store excess rainfall and wastewater, preventing the overflow of sewage systems and the risk of flooding. The works also improved the surrounding environment and the quality of the water in the harbour, clean enough for swimming. More frequent intense rain would require the extension of this infrastructure to maintain the achieved high standards, along with other measures to increase the city's resilience to climate change. A torrential rain in 2010, a once in 100 years event, stressed the vulnerability of the city and put enormous pressure on its drainage infrastructure.

The Copenhagen Climate Plan gave emphasis on green roofs expanded in Europe and the world. In Rotterdam, more than 185,000 m^2 of green roofs were installed in 2014 alone; in Recife, in Brazil, the Green Roof Law approved by the Municipal Council in 2014 requires buildings with more than four floors to have their roofs covered with native vegetation. The Torre Charles Darwin, a 35-story building, served as the first example of a green roof building. The building features a cover crop of 2.8 million square feet, along with a rain-harvesting tank powering the air conditioning system. As much as 46% of the city's total area is green, 60% of which is protected under conservation laws. A plan for enhancing bike lanes and the creation of bus corridors to facilitate commuters and reduce private cars are among the initiatives undertaken by the city government to reduce its impact on the environment and greenhouse gas emissions. The Apibaribe River Navigability Project is focussed on ensuring that the six rivers and 66 canals of the city are used as alternative sustainable routes.

In Tokyo, the top world city, a climate change strategy introduced in 2007 aimed at a Carbon-Minus Tokyo. One of the five major initiatives was the first urban cap and trade programme to reduce its GHG emissions and reach its target of 25% less emissions by 2020 compares to its 2000 emissions. The scheme comprised 1400 installations, including 1100 business facilities and 300 factories, which are large

CO_2 emitters. Although these account for only around 0.2% of some 700,000 industrial and commercial facilities, their carbon dioxide emissions in 2007 stood at about 20% of total metropolitan emissions. The aim for the first compliance phase (2010–2014) had set the cap at a level of 6–8% below 2000 emissions and the second compliance period (2015–2019) at a level of 15–17%. The higher compliance factors (8% and 17%) apply to office buildings, and district and cooling plant facilities, excluding facilities that use large amounts of district heating and cooling. Carbon revenues had to be truly and transparently invested in sustainability projects. Campaigns invested in the good communication of carbon taxes as necessary elements of a structural reform towards sustainability (Tokyo Metropolitan Government 2007).

The results exceeded expectations. It seems that the system contributed to more than a 20% reduction in emissions, but there were also other drivers, including the energy savings triggered by the earthquake in East Japan in 2011. The introduction of high-efficiency heat sources and light fittings have been key activities in reducing emissions. Reductions have continued alongside increases to gross floor space, indicating a decrease in emissions intensity of the building sector in Tokyo. Monitoring is permanent, and compliance assessment is integrated in each phase. Mandatory reporting of emissions is an important prerequisite and asks for the cooperation of all, so that consensual action is effectively implemented. Annual reports insist on a well-designed consultation process federating all stakeholders (ICAP 2018).

Singapore, a low-lying island state, has pledged to cut its emissions intensity by 36% below 2005 levels by 2030 and stabilise emissions peaking around 2030. Improving energy efficiency is essential for reducing emissions across the industry, transport, buildings, residential, resource, and water sectors. Increased awareness, reinforced regulations, capacity building, and government support go hand in hand. By publicly disclosing progress on an annual basis, the city-state has empowered citizens to assess achievements, while initiating a city-wide shift towards a post-carbon future (National Climate Change Secretariat 2016).

In the US, Portland's 2015 Climate Action Plan aims at an integrated set of strategies by 2020 to keep the city on a path to reduce GHG emissions 40% by 2030 and 80% by 2050. The proportion of citizens travelling primarily by public transport, cycling, or walking is expected to rise to 50%, and the number of electric vehicles is set to increase fourfold to 8000. The plan aims to reduce energy use in existing buildings by 1.7% annually, resulting in an annual GHG emissions reduction of 280,000 metric tonnes in 2020. "Sharrows," bike lane markings with double arrows, connect low-traffic neighbourhood greenway streets throughout the city, providing cyclists with safer options.

Boston, labelled the 'city of science and ideas,' has been designated among the cities of the first cohort of the U.S. Climate Action Champions (White House 2014). The Boston Climate Action Plan aimed at reducing greenhouse gas emissions 25% by 2020 and 80% by 2050, through formal planning processes, engaging communities, and developing innovative businesses and skills to take advantage of climate action opportunities. A Climate of Progress, launched on Earth Day 2011, encompasses the recommendations from the Climate Action Leadership Committee and the Community Advisory Committee. Boston's update climate action plan, launched in 2013, focussed on climate preparedness and community engagement. The city is the first in the region to adopt Green Building Zoning and to include climate resilience in the new construction review process. Additionally, a Green Ribbon Commission brings together businesses, non-profit organisations, and community leaders from an array of sectors to develop shared strategies for fighting climate change in coordination with the city's Climate Action Plan, which is being updated every 3 years (Boston Foundation 2011).

Major cities create independent bodies to advise them on climate risks and resilience and assist them achieving their targets of reducing GHG emissions. The New York City Panel on Climate Change (NPCC) assists achieving the most ambitious target of reducing GHG emissions by 80% by 2050. The NPCC climate projections up to 2100 include coastal flood risk maps to the end of the century for the current 100-year and 500-year coastal flood events, enhanced dynamic modelling of future coastal flooding that includes the effects of sea level rise, an overview of climate-linked health risks for citizens, and a process for enhancing a New York City climate indicators and monitoring system. Mean annual precipitation has increased at a rate of approximately 1.2 inches per decade since 1900, nearly twice the observed average global rate of 0.5–0.7 inches. These trends are expected to be aggravated in the future, as the mean annual temperatures are projected to increase and the mean annual precipitation to grow from 4% to 11% by the 2050s. The frequency of heat waves is projected to increase from two to around six per year by the 2080s. Projections for sea level rise in New York City, comparing to the period 2000–2004, suggest an increase between 11 inches and 21 inches by the 2050s and between 22 inches and 50 inches by 2100, with a worst case projection of as much as 6 feet by 2100 (NPCC 2015).

Alliances among cities working together for climate protection are extremely important for promoting awareness, emulation, and excellence. The C40 coalition created new partnerships with the World Bank and ICLEI to accelerate climate action in cities through streamlined financing, emission accounting, and uniform reporting. Both for cities and companies, measuring and disclosing the amount of emissions are the first steps for preventing and mitigating global warming and also creating a new culture for adaptation. The Carbon Disclosure Project (CDP), a non-governmental international endeavour, launched a transformative global system for cities and businesses to measure, disclose, manage, and share climate change and water information. The CDP brought together thousands of organisations willing to report their greenhouse gas emissions and share their assessed climate change risks and opportunities, to set reduction targets and to improve performance. Some leading companies have moved to become carbon neutral, whereas other organisations managed to radically reduce greenhouse gas emissions (CDP 2014).

Cities have to reduce the climate risks faced by citizens and businesses through investment in well-performing infrastructure and services and by developing impactful policies and incentives. The benefits that business brings to cities, including jobs, tax revenue, and services, are among key drivers for cities to improve their climate resilience. Similarly, businesses are reliant on public infrastructure and policies to support and guide their operations. Public and private actors can benefit from a greater understanding of shared climate change risks, and companies can help embedding local adaptation needs within their operations (WBCSD 2010a, b, 2014).

The CDP Public Procurement Programme is designed to enable local and national governments to create new climate-friendly markets. This plan is an effective way for local and national governments to ask their suppliers about energy use, greenhouse gas emissions, and global warming implications. This information allows governments to better understand the climate change risks, which in turn can help work towards building a low-carbon government supply chain.

Network emulation dynamics are powerful drivers of change at the fronts of both mitigation and adaptation. Transparent disclosures, including to investors and citizens, enables local and regional authorities to underline commitment, reduce perceived risks and associated insurance costs, and demonstrate their visions and values in a competitive globalised world. Cooperation between governments and business can improve climate resilience. Companies worldwide seem already ahead of their governments in planning for climate change risks, costs, and opportunities. They are calling for clear pricing and regulatory certainty to better plan their activities and their climate-related investments, and more secure, interconnected carbon markets (CDP 2014).

Climate change risk assessment is increasingly being adopted by cities. In search of quantitative evidence of the costs and benefits of mitigation and adaptation in cities, the EC-supported RAMSES project presented urban climate impacts and estimates of adaptation using a common currency. The aim is to empower policy makers, businesses, and the civic society to make direct comparisons across cities and activities. The project offers a generalised approach on key infrastructures and characteristics and their relationship to climate mitigation and adaptation; this is complemented with selected case studies and simulations of the effects of climate change. The analysis of the institutional and political context can help establishing the transition strategies for specific cities including Antwerp, London, New York, and Rio de Janeiro.

3.3.2 The Adaptation Journey of Cities

Mitigation and adaptation are communicating vases. In contrast to mitigation, adaptation is a recent concern for many cities. A 2014 survey of 200 European cities across 11 countries highlighted that 72% of them had no adaptation plan in place and 35% had no mitigation plan. Cities increasingly introduced adaptation actions in their climate plans to reinforce their capacity to withstand threats. Resilience fortifies the immunity of cities against irreversible capital losses. Resilient cities and communities are those that have citizens, businesses, institutions, organisations, assets, and flows which can withstand economic, social, environmental, and political threats without serious interruptions or irreversible damage. Cities reaching turning points in many regions of the world should invest as early as possible to prevent disasters. European cities have started providing models for the design and organisation of urban transformation processes and the mainstreaming of adaptation measures into spatial planning and policy (UNISDR 2012).

In the European Union, the "EU Cities Adapt" initiative was launched in 2012 and the European Climate Adaptation Platform provided information and policy advice. The "Mayors Adapt – the EU Covenant of Mayors Initiative on Climate Change Adaptation" was promoted to engage cities in taking decisive action to adapt to climate change. Signatory cities engaged with the EU Adaptation Strategy, through a comprehensive local adaptation strategy or the integration of adaptation measures into relevant existing policies (EEA 2016).

The 2013 Bonn declaration of Mayors, during the adaptation forum, expressed concern that the historical threshold of 400 ppm CO_2, passed in May 2013, underlined the interlinkage between climate change action and the global development agenda and confirmed commitment to globally coordinated local climate action. The declaration welcomed the Durban Adaptation Charter as a powerful mechanism to enhance urban adaptation action by building networks of communication and cooperation among world local governments, creating a global platform for the dissemination of best practice.

In 2014, Metropolis, the World Association of major metropolises, a leading international organisation that brings together cities and metropolitan regions with more than a million inhabitants, initiated a dialogue for cities to adapt to climate change. The Metropolis Initiative "Integrated Urban Governance – Successful Policy Transfer," led by the city of Berlin, concluded the "No Regrets Charter," a declaration of key principles for climate change adaptation in cities. A 'no regrets' or precautionary approach, holistic, integrated, and participative, has been judged as the best model for disaster risk reduction, water and flood management, spatial planning, and the natural and built environment (Metropolis 2014).

Initiatives on urban adaptation in Europe can lead to attractive, climate-resilient cities. Cities and their stakeholders, investors, and citizens must be prepared for climate change so long as there still is leeway for action. This plan allows working together to develop specific concepts, whether for existing assets or for future projects such as renewable energy and water supply concepts. A review and subsequent adjustment of urban adaptation to climate change by local governments and by supporting regional, national, and European institutions, researchers, and other relevant stakeholders offers many lessons (EEA 2016).

Copenhagen's Cloudburst Management Plan introduced a range of interlinked initiatives and invested in 300 projects to increase the level of flood protection across the entire city. The plan promotes an integrated system of green areas dotted throughout the city to help slow rainfall runoff and reduce the risk of flooding, including green streets and pocket parks that will function as water retention areas and water basins. The projects are tightly interlinked with other key plans for sustainable, social, and economic development in the Danish capital. Climate adaptation is an opportunity to create green growth and to improve the quality of life and leisure (City of Copenhagen 2012).

Green and blue areas could help the city to cool down during summer months and cope with the expected higher temperatures. Seas and vegetation areas have an immediate refreshing effect as they hold moisture and release it into the air. Conversely, artificial surfaces tend to absorb and retain heat. Furthermore, beaches and parks also increase recreation spaces for residents and visitors and improve conditions for biodiversity and human well-being. Clean urban beaches, often after 'dirty waters,' as in Boston, provide clean bathing waters.

The Global Movement of 100 Resilient Cities (100RC), pioneered by The Rockefeller Foundation on its Centennial in 2013, started working with an initial group of 32 cities in December 2013. In 2014, 35 cities were selected, while the third 100 Resilient Cities Challenge closed in November of 2015 and the final group of cities was constituted in May 2016. Cities in the 100RC network are provided with the resources to develop a roadmap to resilience along four main pathways: financial and logistical guidance for establishing an innovative new position in city government, a Chief Resilience Officer, leading the city's efforts in developing a robust Resilience Strategy, and actively participating in a global network of member cities who can learn from and help each other (Rockefeller Foundation 2017).

The City of New Orleans adopted in 2015 its Resilience strategy, which resulted from a joint effort between the city and the Rockefeller Foundation's 100 Resilient Cities initiative, which linked New Orleans to public and private partnerships as well as funds to employ a chief resilience officer. The City's first Chief Resilience Officer is responsible for working across silos to implement the resilience strategy, advise the mayor, promote resilience thinking, and act as a global thought leader, coordinating efforts across government and multi-sector stakeholders (City of New Orleans 2015). The city adopted its Climate Action Strategy in 2017 as a roadmap for New Orleans to combat climate change. Climate Action for a Resilient New Orleans proposes 11 strategies and 25 actions to reduce greenhouse gas emissions 50% by 2030 (City of New Orleans 2017).

Future Melbourne provided the impetus for a comprehensive climate change risk assessment, culminating in an integrated City Adaptation Strategy and supporting Action Plan and the ongoing implementation of action enhancing the resilience and diversity of urban forests, and the design of cool roofs to reduce the Urban Heat Island effect. The 2008 "Future Melbourne – City of Melbourne" plan acknowledged that climate change presents key strategic risks for the community and must be collectively addressed. By 2030, Melbourne is likely to be significantly affected by warmer temperatures, drought and heat waves, lower rainfall, intense storm events, sea level rise, and flash flooding. The plan was developed through open collaborative public engagement with key stakeholder groups including professional networks, other city councils, the Victorian and Australian Governments, universities, and non-government organisations (C40 Cities and Siemens 2014).

The Durban Adaptation Charter, adopted in 2011, recognises that the majority of climate change impacts and the proactive and reactive responses will occur at the local level. It marks a clear milestone in inviting local governments to adopt a holistic ecosystems-based approach for developing secure city–region food systems, work for urban poverty eradication, protect and enhance local biodiversity, and strengthen urban resilience. The charter stressed that the costs of adaptation and loss and damage resulting from climate change should be covered from local, regional, national, and global sources. It called for new modalities of interaction for local and subnational governments to fulfil their role as critical stakeholders and decisively contribute to national and global effort for a better future for all.

The Rotterdam Climate Initiative aimed at improving the climate for the benefit of people, the environment, and the economy in a complex multiple deltaic environments. Rotterdam's adaptation embraced a tailored 'inner-dyke/outer-dyke' approach consisting of the giant sea gate Maeslantkering, a flexible storm surge barrier, permanent sand dunes along the coast, and dykes along the rivers. The inner-dyke city, mostly below sea level, is protected by smaller secondary dykes. The outer-dyke city, at 3–5.5 m above the sea level, where about 40,000 citizens live, is forging its resilience against rising sea levels. It is being adapted through use of innovative technologies, such as floating buildings, and more traditional approaches such as insulation of building facades. Rotterdam promoted green roofs and is experimenting with climate proofing to address fiercer rainfall threats. It has built tidal parks and water storage spaces, including the Museumpark car park underground water storage with capacity of 10,000 m^3, and is integrating Blue-Green Corridors, watercourses, and ponding areas into the urban landscape (City of Rotterdam 2013).

The comprehensive Paris Adaptation Strategy is transforming the French capital into a climate-resilient city. Paris has to address a wide range of climate-related challenges, including heat waves, drought, flooding, fires, and food and water security. The 2015 Paris Adaptation Strategy, unanimously approved by the Paris City Council, sets out clear objectives for how the city can respond and adapt to the evolving challenges while improving the resilience of its fabric and functions and the well-being of its citizens. The strategy sets out four priority areas: protecting Parisians against extreme climate events; ensuring water, food, and energy supply; living with climate change; and fostering new lifestyles and boosting solidarity. Greening actions will help preserve and encourage more biodiversity in the city. Actions have been also designed to help Parisians to cool down during extreme heat, ensuring that all residents live within a 7-min walk of a cool place, and asking that city parks remain open 24 hours a day during hot spells (C40 and Arup 2016).

Building a culture of resilience is a shared responsibility among governments, the private sector, and citizens. In the US, the White House adopted, in 2014, measures to help cities adapt. A task force of governors, mayors, and other leaders focused on recommendations for policies to incorporate climate change and provide tools to help local communities to better design their adaptation measures. A Web-based climate resilience toolkit has been designed to help local leaders adopt measures to prepare cities for facing extreme events of rising sea levels, droughts, diseases, and other climate impacts. Sixteen local and tribal communities were selected as Climate Action Champions for their decisive action to cut carbon pollution and build resilience. The designated communities benefit from facilitated peer-to-peer learning and mentorship (White House 2014).

Reinforced resilience and advanced decarbonisation have to go hand in hand, and most cities introduce adaptation measures together with mitigation aiming at 80% GHG emission reduction by 2050. The overarching purpose of climate protection is often reflected in the goals of the smart cities movement. In Yokohama, the Smart City Project strives to more effectively manage energy use and mitigate climate change through energy management systems. The city's action plan sets the target of an 80% CO_2 emission reduction by 2050. The project is designed to engage citizens and stakeholders as a key factor of successful implementation.

3.3.3 Sea Rise and Storm Surge Threaten the Symbiosis of Cities with Seas

Coastal cities are the frontiers among cities and seas. The future of the world's coastal cities and the threats originating from the sea are of particular concern to humankind (West 2016; WEF 2018). Ocean warming dominates the increase in energy stored in the climate system, accounting for more than 90% of the energy accumulated between 1971 and 2010 (high confidence), with only about 1% stored in the atmosphere. On a global scale, ocean warming is greatest near the surface, and the upper 75 m warmed by 0.11% per decade over the period 1971–2010. It is virtually certain that the upper ocean (0–700 m) warmed from 1971 to 2010, and it likely warmed during the period 1870–1971. Advancing climate change holds more threats for coastal cities as predictions suggest a global rise by 52–98 cm by the year 2100 in the case of high emissions, which would threaten the survival of coastal communities and entire island nations. And even with aggressive GHG emissions reductions, sea rise could reach 28–61 cm and impacts on many coastal areas could still be serious, especially in relationship to coastal erosion and flooding. The risk of increasing floods creates a major political and institutional challenge for coastal cities and can undermine their symbiosis with the sea. Ambitious and proactive actions at the local level are needed to help prevent large-scale flood disasters (IPCC 2014c).

At the end of the century, densely populated coastal areas around the world could be underwater as the result of land loss from erosion and sea level rise. The International Panel of Climate Change suggests that, over the period 1901–2010, global mean sea level rose by 0.19 m. The average rate of sea level rise since the mid-nineteenth century has been larger than the rate during the previous two millennia. Effects could reduce public access and recreational opportunities and harm natural habitats and the species that support and depend on them. Mitigating shore erosion requires coastal planning and control that help preserve the natural features of coastal shorelines (IPCC 2014b).

Vulnerable cities need a Noah's Ark or a full protection system, which represents a major investment. In times of scarce resources, the mobilisation of such resources could be a real political and institutional challenge for local authorities, especially in the developing world. Present urbanisation and flood defence patterns have been designed for past conditions, and even a moderate change in sea level is sufficient to make them inadequate, thus magnifying flood losses to catastrophic levels. Local subsidence amplifies the effect of global sea level rise on local water levels. Subsidence is accelerated by human activities such as water extraction.

Climate protection measures have been designed by many cities. Venice plans ingenious gates, the Experimental Electromechanical Module, a flood barrier system, to seal the lagoon and tame the destructive tides. The city is notoriously vulnerable to flooding and 'aqua alta,' the Adriatic's high tide water, which partly caused its sinking by 23 cm during the twentieth century and became more frequent during the past years. The rise in sea level could have great impact, because it is estimated that, in the absence of drastic measures, a 30-cm rise in sea level would flood St. Mark's Square 360 times a year. The completed project is expected to include 78 mobile barriers blocking the three inlets to the Venice lagoon. The barriers will be housed in enormous tanks anchored to the seafloor and will rise whenever high water threatens to flood the city. Once the threat is over, the released air will allow the barriers to sink down again. The completed system would only be mobilised to address the most extreme risks.

Climate change, subsidence, and excessive population growth and urbanisation are the main threats for coastal cities. In 2007, an OECD study on urban vulnerability revealed a first estimate of the exposure of the world's large port cities to coastal flooding from sea rise, storm surge, and high winds. The study also investigated the impact of climate change on the exposure to coastal flooding by the 2070s, alongside subsidence and urbanisation, focussing on the 136 port cities around the world with more than 1 million inhabitants. Most of these largest port cities are located in Asia (38%), and many of them (27%) are in deltaic sites. The latter tend to present higher coastal flood risk as a result of their lower location and experience significant subsidence (Hallegatte et al. 2013).

Across all cities, about 40 million urban residents, almost 0.6% of the global population, were exposed to a 'one in 100 years' coastal flood event. The exposure of population and assets to such events highly depends on their particular topographical and morphological conditions. The exposure is concentrated in a small number of cities. The10 cities with the highest population exposure account for almost half the total exposure and the top 30 cities for about 80% of the global exposure. Of these 30 cities, 19 are located in deltas.

Populations in vulnerable cities need to be prepared and protected. In terms of exposed population, the list of the proposed most vulnerable cities included Mumbai, Guangzhou, Shanghai, Miami, Ho Chi Minh City, Kolkata, Greater New York, Osaka-Kobe, Alexandria, and New Orleans, on the basis of 2005 data. The ten cities with highest population exposure are almost equally split between developed and emerging countries. Alexandria is the only most vulnerable city outside Asia and Northern America. This taxonomy is, however, highly variable depending on the size of cities, the magnitude and nature of the extreme phenomena, and the adopted policies.

Urban economic assets also need to be strengthened and protected. In terms of exposed assets, the list of the most vulnerable cities includes Miami, Greater New York, and New Orleans, followed by Osaka-Kobe, Tokyo, Amsterdam,

Rotterdam, Nagoya, Tampa-St. Petersburg, and Virginia Beach. These cities contain 60% of the total exposed assets, but belong to only three countries: the US, Japan, and the Netherlands. The total value of assets exposed in 2005 was estimated to equal around 5% of the global GDP. The situation is very dynamic and one can expect changes to have already occurred, especially with the rise of many ports on the Chinese shores.

By the 2070s, the total exposed population could grow more than threefold to about 150 million people from the combined effects of sea level rise and increased storm surge, subsidence, population growth, urbanisation, and socioeconomic change. The total asset exposure could grow even more dramatically, more than ten times the current levels, and rise to roughly 9% of projected annual GDP by 2070.

The symbiosis with the sea can be challenging and has to be carefully planned and monitored. Risk management policies, including upgraded defence engineering, could mitigate the impacts of rising sea levels. Coastal defensive structures should make the best possible use of the rich protection already generously offered by nature. Marshes, seagrass beds, coastal forests, and coral reefs provide natural buffers that absorb the energy from waves and storms, making it easier to protect cities. In addition, artificial constructions could also be required to provide full protection, such as barriers to manage safe access to ports.

Cities can keep their flood risk under control by better managing urban development. Smart coastal protection infrastructure has to be integrated in land use planning and policy. Emerging countries have a real window of opportunity, as urbanisation is ongoing at very fast rates and the infrastructure has not yet trapped the future into unsustainable schemes. Smart urban policies, based on disaster risk management and governance, could drive new developments towards safer places and prevent the urbanisation of the most vulnerable and costliest to protect areas. Weak institutions and institutions, lack of enforcement of building norms, and extension of informal settlements are strong obstacles.

Better models could help citizens understand the frequency and strength of natural disasters. The human component has an ever-increasing role in reducing the risk from natural disasters, in terms of both risk management and addressing the anthropogenic causes of climate change. Cloud-based information systems for disaster are being proposed, offering preparedness mobile and web applications. Such public resources could prepare families for earthquakes, floods, and typhoons and provide real-time platforms for responsible and reliable communication among local governments, academies, the private sector, civil society organisations, NGOs, and affected communities.

Public awareness is extremely important, as vulnerable cities have to be prepared for a broad range of measures from individual protection to large-scale evacuation, especially in the most exposed places. They also need concerted contingency plans to manage crisis and emergency services and recovery and reconstruction schemes, well-regulated insurance systems, and post-disaster measures. Forging awareness, well before extreme events happen, is primordial for economies and cities to rebound, and the participation of all stakeholders is of the highest importance (Pelling and Blackburn 2013).

Preventing urban disasters from the sea and taking appropriate measures is vital. In sheer economic terms, the stakes of sea level rise in urban areas are particularly high. An added 0.5 m of ocean water by the year 2050 would put $8 trillion in assets at risk in the world's 136 major port cities. On the northeast coast of the US, the expected maximum rise in sea level would threaten Baltimore, Boston, New York, Philadelphia, and Providence. Without climate action, sea level rise on the East Coast and the shift to an arid climate in California are likely to have a high cost (WWF and Allianz 2009).

Coastal and sea defence planning and flood risk governance are critical options. In the past, engineering provided the preferred protection solutions, including dams, dikes, and seawalls, to the prospect of encroaching seas. The Thames Barrier, a 570-yard-wide floating span of rotating gates, is a good example. It allows navigation on the river while protecting the city against storm surges and high tides. However, responses to sea level rise increasingly encompass soft engineering solutions such as green infrastructure and wetland restoration. Such could even include measured human retreat from coastal areas, allowing the rising seas to follow their way (Wilson 2013).

The Tokyo Metropolitan Government has a long record in facing extreme events, especially storm surges and floods, with an array of control measures. Transport and telecommunications infrastructures, notably a most modern teleport, made the global metropolis highly liveable for a huge number of residents and businesses. The global megacity had to work very hard to strengthen its resilience in the face of severe disasters and earthquakes and more frequent typhoons and floods. As streams flow through highly urbanised areas, special situational measures are often needed to quickly ensure safety against flooding. Such measures include regulating reservoirs to store part of the flood water or diversion channels to let down part of the water. The Kanda River provides a good example with its Loop Road No.7 Underground Regulating Reservoir, which prevented important flood damage when Typhoon Melor struck Tokyo in October 2009.

The eastern lowlands of the Tokyo Metropolitan Area are composed of fragile ground, and most regions are below sea level, the result of land subsidence caused by post-Meiji period groundwater pumping. These areas are especially susceptible to damage caused by natural disasters such as storm surges and earthquakes. They are being bolstered against flood damage through river improvement projects, the representative example being the Super Levee Project. A super levee is an especially wide embankment built in cooperation with riverside urban redevelopment projects. It assists in effective land use, the strengthening of earthquake countermeasures, the development and maintenance of a healthy river, and the improvement of the urban environment.

Extreme phenomena such as exceptional storms, earthquakes, and tsunami further challenge the vulnerability of maritime cities, ports, and nuclear structures to coastal floods. The disaster of 11 March 2011 caused by the earthquake in East Japan, and subsequently the tsunami and damage to the Fukushima nuclear plant, also had an effect on Tokyo, 400 km from the epicentre. Improvements brought over time to the seismic performance of buildings and risk management policies resulted in very few fatalities and limited damage. The city prevented fires, thanks to 'my-com-meter' system, introduced post-1995, shutting down residential gas supply at the first sign of an earthquake. Transport infrastructures also resisted well after considerable reinforcement during the past years.

A few months after the Fukushima disaster, the International Union of Architects (IUA) held its XXIV Congress in Tokyo in September 2011 to debate the visions for urban design for 2050. The IUA Tokyo declaration expressed a commitment to learn from overcoming disasters, rebound around the globe and progress towards a balanced economic, social, and environmental sustainability. The declaration is also an invitation to exchange the many diverse global and local initiatives for equity and solidarity and to promote responsibility beyond national, cultural, and mental limits (IUA 2011).

Extreme weather events resulting in hazards such as heat waves, floods, and droughts are expected to happen more frequently and affect quality of life. Urbanisation, population ageing, and resource scarcity interact with climate change and can compromise public health, reduce productivity, and constrain the functionality of infrastructures. The replacement of natural vegetation with artificial surfaces and buildings impacts temperature, moisture, wind, and rainfall patterns. Excessive amounts of rainwater cannot drain into the ground, especially in sealed urban areas, and this can generate or worsen floods. Artificial surfaces store heat and cause higher temperatures in cities compared to their surroundings. An ageing population in OECD cities increases the share of citizens vulnerable to heat waves and asks for additional urban services. Awareness about the effects of climate change before extreme events occur is necessary for drastic changes in urban and regional management. Innovations bringing multi-dividend responses are the most powerful. Strong ecosystem-based spatial planning, especially for risk-prone areas, can be an effective instrument.

Present and future buildings and urban infrastructure need huge investments, alongside those for climate proofing, to continue supporting urban services even under extreme phenomena. Urban planning and architecture should incorporate lessons from humanitarian emergencies and the creation of temporary shelters for refugees. Social innovations can also unlock multiple opportunities. Climate change adaptation concerns have to be incorporated into building design and standards and eco-retrofitting activities, to ensure that systems can respond under heavier conditions. Adaptation is strongly reinforced through green and blue infrastructure such as parks, wetlands, green walls and roofs, and ocean gardens, spaces that offer protection and fulfilment.

3.4 Strengthening Resilience in Riverside, Coastal, and Deltaic Cities

3.4.1 Struggling to Tame Water: Gold Standards Against Threats from the Seas

The Netherlands offer the world's best practice in fighting with elevating sea levels. For Dutch cities, striving or thriving at or below sea levels has had been an existential question. Between 1950 and 1997, the low-lying country constructed vast delta works, including 250 miles of dams and other barriers. Over the years, cities developed integrated, flexible strategies to manage the permanent symbiosis with the sea. The country rediscovered the compatibility of hard and soft solutions. In 2007, a new Delta Committee, the second in five decades, recommended raising the flood protection levels of coastal dikes by a factor of 10. The committee advocated two new strategies that rely on sustainability principles instead of engineering capabilities. The first, "Building with nature," involves beach nourishment as well as restored natural estuaries and tides to prevent flooding. The second, "A room for the river," asks for strategies requiring adequate administrative, legal, and financial frameworks.

Designing policies to tame waters has been vital for Rotterdam, largely below sea level, the first European port city that has built up centuries of expertise in this domain. An historic review of the port of Rotterdam encapsulates some strengths and weaknesses of many ports. Strategically located at the Rhine-Meuse-Scheldt delta on the North Sea and at the heart of a dense rail, air, road, and inland waterways network, Rotterdam is a true gateway to Europe and the largest port of the world from 1962 to 2004. It stretches over a distance of 40 km and includes the historic city harbour. It has continuously expanded, as the city of Rotterdam developed from a small town into a major harbour city. Since the nineteenth century, links between Rotterdam and the North Sea have been poor, with a large estuary/delta area and many small waterways. In 1872, the 'New Waterway' was constructed to connect the Rhine and Meuse Rivers to the North Sea. Over the years, the port was further developed seawards through docks and harbour-basins and enlarged by the construction of the Europoort, its Gate to Europe complex along the mouth of the New Waterway, including the industrial park Maasvlakte. The extension of the port to the west continued, after years of discussions, protests, and planning. The infrastructure has been laid on reclaimed land and industry, shipping, sea defences, and transport locations have been created as part of the Main Port Development Rotterdam.

Protection and accessibility of the Rotterdam region form the dual precondition for the port city to strengthen its position on the global chessboard. The port and its surrounding area have to be equipped against possible storm surge from the North Sea, a risk largely underestimated until the flood of 1953, which killed almost 2000 citizens and caused widespread damage. The flood acted as a clarion call, and a Delta committee was assigned with drawing up a plan to protect the areas from future flooding and to provide clean drinking water. In 1959, the Delta Law was approved for the implementation of the Plan and the construction of the Delta Works, a chain of flood protection structures within and around the Rhine-Meuse-Scheldt delta. The aim was to shorten the coastline and convert it into freshwater lakes. The project helped draining the low areas that flood frequently and regulated the flow of saltwater from the sea, providing both drinkable water and freshwater for irrigation.

The Delta Works flood protection plan considered various options for the floodgate before a unique design, the Maeslantkering, was constructed in 1997 as the city's huge first line of defence. Its volume has been compared to the size of two tubular Eiffel towers toppled over. This flood barrier consists of two huge doors that normally rest in a dry dock beside the New Waterway. When a flood of 3 m above sea level is predicted, the gates are floated into positions, and sunk. When the water level recedes enough to open the gates, the latter are floated back into their docks. Since its construction, the monumental Maeslantkering kept open the immense waterway that serves the entire port of Rotterdam. It did not actually served to prevent a flood, but it is regularly tested in preparing for the unexpected. The next challenge is cyber-security as the cyber-resilience of the whole system has to be beyond any suspicion.

Nature can help much in reconciling cities with the sea and protecting them from global risks. It offers protection which usually is less costly, intrusive, and perturbing than engineering works and can effectively and efficiently shield cities and the surrounding ecosystems. The use of green infrastructure and the enhancement of natural processes should become an essential systematic part of flood planning. Many cities could benefit for creating urban retention areas for flood waters and transforming unused assets into protective systems. Transport infrastructure, abandoned or underutilised, can also complement green infrastructure for flood control.

"Live with the water rather than struggle to defeat it" has been a prevailing principle for Rotterdam, a city that was levelled by bombs in World War II and recovered from decades of hardship, investing in environmental ingenuity, breathtaking architecture, and innovative design. With its harp-like Erasmus bridge as its emblem, the city has reinvented itself and pioneered the creation of multi-function facilities, such as garages and rowing courses which act as retention ponds and collect floodwater in emergencies. On the other side of the Atlantic, Boston, a great part of which is built on landfill, prepared for apocalyptic storms. The city has a long past of design and engineering flood defence projects that recapture the nineteenth-century vision of an emerald necklace surrounding the city. Boston's location, with three rivers flowing into a sheltered harbour well suited for waterborne trade, helped it grow into a major city. But history bears witness to the city's vulnerability. In 1978, Boston's then-new Charles River dam, built 23 years after Hurricane Diane flooded New England in 1955, was barely in place when the city was hit by a 3-day blizzard. The resulting tide rise came within 0.4 m of overtopping the dam, designed to protect against a storm surge of 3.8 m above average sea level. The accuracy of predictions matters. In the absence of the Charles River dam, the dramatic floods of 2010, which inundated parts of Rhode Island, could also have been disastrous for Boston.

Climate Ready Boston is a vision to develop resilient solutions to prepare the city for future climate and storm conditions. Boston residents are already affected by extreme heat, rain, snow, and flooding, and these trends are expected to continue. The City launched Climate Ready Boston as an ongoing planning activity for the future impacts of climate change. A comprehensive study was launched in 2016 to engage with the community and work with stakeholders towards realising the vision for a Climate Ready Boston. Various events were organised in various parts of the metropolis to mobilise local stakeholders and engage with them in more detailed place-oriented responses. Climate Ready Boston is coordinated with "Imagine Boston 2030," the first citywide plan in 50 years, which aims to make Boston a more affordable, equitable, connected, and resilient city (City of Boston 2016, 2017).

On 29 October 2012, Hurricane Sandy, the largest hurricane to ever form in the Atlantic, hit New York and the most densely populated US region, with tragic devastating effects. Wave heights reached a record of 32.5 feet in New York Harbour. Water levels at Battery Park rose to a record 13.88 feet, surpassing the previous record of 10.02 feet more than 50 years ago. At least 159 people were killed, as either a direct or indirect result of the hurricane, and hundreds of thousands of homes and businesses were damaged. The Hurricane Sandy Rebuilding Task Force designed a rebuilding strategy to help the affected region and serve as a model for communities facing huge risks from extreme events. The strategy proposed 69 policy recommendations, to help residents stay in and repair their homes, conduct small businesses, and revitalise local economies. The purpose was to ensure that entire communities are more capable to withstand and recover from future storms (Hurricane Sandy Rebuilding Task Force 2013).

Hours after the storm, the National Oceanic and Atmospheric Administration (NOAA) began surveying obstructions in the waterways that lead to critical petroleum facilities within the Port of New York/New Jersey, helping restore the flow of emergency fuel supplies. Over a span of 5 days, maritime commerce was restored in all major ports. This step ensured that citizens in the region had access to food, fuel, and supplies, preventing wide disruption to the economy.

The main lessons learnt from New York included aligning funding with local vision and cutting red tape to help communities reconstruct their critical infrastructure. Re-design and co-piloting of innovative resilience strategies has to ensure a regionally coordinated, robust approach to infrastructure investment. Resilient rebuilding has to be founded on current and future risk assessment, provide safe, affordable, and secure housing options, support to small businesses and local economies, and strengthening public policy for long-term resilience (WRI 2012).

3.4.2 Vulnerable Megacities with Fast-Growing Dense Populations

Emerging megacities face far greater challenges. Rising seas might trigger major disasters that go far beyond ecology and engineering, especially on intrinsically vulnerable short-term equipped urbanised coastlines that are inadequately prepared for widespread inundations. Vulnerability to climate change depends on natural and anthropogenic factors and the adopted policies to protect the concentration of people and assets. Dhaka and Phnom Penh, both capital cities of least developed countries, top the rankings as having the lowest adaptive capacity to climate change.

Dhaka is an extremely vulnerable megacity with an estimated population of more than 15 million, setting at sea level in the broad and flat heart of the Bengal delta. One of the world's most densely populated cities, which doubled its population in just more than a decade, is sinking from ground subsidence and sea level rise. Dhaka, as all large cities that are low lying and subject to violent storms, needs to refrain from building on wetlands and take better care of its water and ecological resources (World Wide Fund 2009).

The IPCC suggested that saltwater intrusion could reach as far as 100 km from the coast and that could influence the availability and quality of freshwater and rice production and affect the natural ecosystems, such as mangroves, that currently safeguard the coast from storm surges. Salt-tolerant mangroves, ranging from shrubs to towering trees 45 m tall, with extensive roots above and below the water, provide home to many marine species and precious ecosystems and act as important carbon sinks. The estimated 10,000 km^2 of mangroves within the Ganges–Brahmaputra Delta still limit the effects of sea level rise, but massive storm surges have already had catastrophic consequences. In the past, tropical cyclone-induced 6-m storm surges have led to 500,000 deaths in 1970 and 150,000 deaths in 1991. Future losses could be even more serious without mangrove forests to buffer the effects.

Kolkata has a high overall vulnerability and has to strengthen its resilience, moving in tune with the rhythm of the Ganges Delta and its rising and falling dynamics. Being situated on the banks of the Hooghly River, and within the Ganges Delta, Kolkata is only metres above current sea levels. Urban explosion reclaimed significant amounts of surrounding wetland. The metropolitan area is also within a considerable seismic zone and thus prone to earthquakes. Sea level rises accompanied by ground subsidence seem the most severe threats. The combined effects of sea level rise and subsidence could allow saltwater to intrude into the Ganges Delta 100 km from the coast, seriously impacting groundwater supplies. For Kolkata, better water management is crucial to minimise the effects of salinity and overextraction of groundwater. The megacity needs increased protection, enforcement, and restoration of mangroves to limit the impact of saltwater intrusion and fiercer storm surges (World Wide Fund 2009).

Vulnerable cities have to plan early investing in protection versus relocation. The retreat, as happened in New Orleans after Katrina, is an extreme case and policy option, and a powerful inducement to cities to study, discuss with stakeholders, appraise, and plan alternative options. Learning from nature and bringing together the most advanced technologies would be most appropriate for many cities. A balanced blend of hard and soft engineering responses helps reinforce protections. The creation of climate and ocean parks and retention areas, and the early training of citizens to best respond to sea storm surge, are the first steps.

In Lagos, the megacity of slums, the largest slum quarter on stilts in Africa risks sinking under the mounting saltwater of Benin River. Makoko, the growing waterworld slum, features floating constructions, stinking waters, and a labyrinth of canals. Two of three Lagos residents live in a slum with no reliable access to clean drinking water, electricity, or waste disposal. Nigeria's oldest slum is home to a population of around 100,000 residents, banded together to create an informal city-on-stilts at the edge of the lagoon. Many of the original residents of Makoko are fishermen attracted from across the region to hopes of a better life in Nigeria, West Africa's oil-rich economic powerhouse. The crime-ridden Makoko, a seething mass of people crammed into the steamy lagoons of southwest Nigeria, had its landmark. The Makoko Floating School offered free education to children living in nearby huts on stilts and became a vital meeting point for this unique community soon after completion in 2013, rising like a beacon from the murky lagoon. An inspiring triangular timber structure, designed by Nigerian architect K. Adeyemi, three stories high, was buoyed on floating barrels and locally sourced timber.

The aim of the school was to generate skills and resilient sustainable building for Africa's coastal regions. Unskilled local workers were hired to build the structure, with the objective that they would then build their own homes with the same materials and principles. Inexpensive and elementary to assemble, the floating structure adapts to the tidal changes and varying water levels of the lagoon, making it invulnerable to flooding and storm surges. Another impact of the school was the powerful sense of ownership that Makoko inhabitants derived from it, the only public space for the community. The floating school had made an important statement for the citizens living there.

Unfortunately, the floating school has collapsed only 7 months after its official opening in 2015. Heavy rains brought down the pyramid-shaped wooden school. None of its nearly 50 pupils were in the building when it collapsed, but the damaged civic pride will be difficult to restore. Classes had already been moved to another location after heavy downpours at the start of the rainy season began to affect classes. Many houses surrounding the floating school also collapsed. The natural barriers of the city become more and more fragile; the lagoon is disappearing, exposing the large African port to the ocean.

Some years after the devastating flooding of 2011, Bangkok unveiled its Resilience strategy, assisted by the 100RC. With this strategy the Bangkok Metropolitan Administration is embedding a lasting capacity for the city to build resilience. Resilient Bangkok was developed in consultation with major stakeholders and more than 50 grassroots district organisations. The strategy, with its comprehensive approach to resilience building, aims at enabling the city to grow not only bigger but stronger and smarter. It outlines a vision for a more liveable and competitive Bangkok and takes into account the needs of all citizens, starting with the most vulnerable, ensuring that every citizen has a place and voice in the city.

The "Resilient Bangkok" strategy, developed with the assistance of 100RC, focuses on three interconnected areas for action, each with specific goals, initiatives, and projects to strengthen the city and its communities: 'Increasing Quality of Life: Bangkok' will be a city for all, making it a more pleasant and accessible place to live and work. Key actions to support this vision include screening for diabetes and high blood pressure in adults, improving the working and living conditions of migrant workers, and improved public transport; reducing risk and increasing adaptation will prepare its citizens to better adapt to natural and human hazards and the increasing risks posed by climate change. Key actions to support this vision include waste collection from canal communities, improved drainage across the city, and youth education for disaster safety; last but not least, driving 'a strong and competitive economy, through economic diversification, both in terms of industries and in employment opportunities for all residents which involves promoting and supporting urban agriculture, skills development programs for tourism operators, and financial and social support for vulnerable communities (Bangkok Metropolitan Government 2017).

References

(The) UN Office for Disaster Risk Reduction (UNISDR) (2012) How to make cities more resilient. In: A handbook for local government leaders, Geneva
Amsterdam Climate Office (2008) New Amsterdam climate – summary of plans and on-going projects, Amsterdam
Bangkok Metropolitan Government (2017) Resilient Bangkok, Bangkok
Boston Foundation (2011) The measure of poverty. A Boston indicators special report, Boston
C40 and Arup (2016) Deadline 2020: how cities will get the job done, outlining a vision to reduce emissions by 2020 and help meet the Paris agreement targets, Paris
C40 Cities and Siemens (2014) City climate leadership awards, New York
CDP (2014) Global corporate use of carbon pricing. Disclosures to investors, London
City of Boston (2016) Climate ready Boston, Boston
City of Boston (2017) Coastal resilience solutions for East Boston and Charlestown, Boston
City of Copenhagen (The) (2012) Cloudburst management plan 2012, Copenhagen
City of New Orleans (2015) Resilient New Orleans. Strategic actions to shape our future city, New Orleans
City of New Orleans (2017) Climate action for a resilient New Orleans, New Orleans
City of New York (NYC) (The) (2015a) OneNYC. The plan for a strong and just city, New York
City of New York (NYC) (The) (2015b) Building a smart + equitable city, New York
City of Rotterdam (2013) Rotterdam climate change adaptation strategy, Rotterdam
European Commission (EC) (2016a) Opportunity now! Europe's mission to innovate, Brussels
EC (2016b) Sustainability now! A European vision for sustainability, Brussels
European Environment Agency (EEA) (2016) Urban adaptation to climate change in Europe 2016 – transforming cities in a changing climate, Copenhagen
Hallegatte S, Green C, Nicholls R, Corfee-Morlot J (2013) Future flood losses in major coastal cities. Nat Clim Change 3:802–806
Hurricane Sandy Rebuilding Task Force (2013) Hurricane sandy rebuilding strategy, Washington, DC
International Carbon Action Partnership (ICAP) (2018) Japan: Tokyo cap-and-trade program, Berlin
International Energy Agency (IEA) (2017) World energy outlook, Paris
International Union for Conservation of Nature and Natural Resources (IUCN) (2016) Explaining ocean warming: causes, scale, effects and consequences, London
International Union of Architects (IUA) (2011) Tokyo declaration. Beyond disasters, Tokyo
International Panel of Climate Change (IPCC) (2014a) Climate change 2014: impacts, adaptation and vulnerability. Bonn
IPCC (2014b) Climate mitigation, Bonn
IPCC (2014c) Climate change fifth assessment report, Bonn
Intergovernmental Platform on Biodiversity and Ecosystem Services (IPBES) (2018) Regional assessment report on biodiversity and ecosystem services for Europe and Central Asia, Summary for policymakers
Lancet Countdown (The) The (2017) report, London
Mega V (2012) Bio-diver-cités. Editions Universitaires Européennes, Saarbrucken
Mega V (2013) Quintessential cities, accountable to the future. Innovation, sustainability and leadership. Springer, New York
Mega V (2015) Conscious coastal cities. Sustainability, blue green growth and the politics of imagination. Springer
Metropolis (2014) No regrets charter. Principles for climate change adaptation in cities, Berlin
National Climate Change Secretariat (2016) Singapore's climate action plan: take action today, for a carbon-efficient Singapore, Singapore
NRC (2010) America's climate choices. NAP, Washington, DC

New York City Panel on Climate Change (NPCC) (2015) Building the knowledge base for climate resiliency, New York City
Organisation for Economic Co-operation and Development (OECD) (2017a) Investing in climate, investing in growth. OECD, Paris
OECD (2017b) Marine protected areas: economics, management and effective policy mixes. OECD, Paris
Pelling M, Blackburn S (2013) Megacities and the coast. Risk, resilience and transformation. Routledge, London
Pope Francis (2015) LAUDATO SI': Encyclical letter on care for our common home. Holy See, Vatican
Rockefeller Foundation (2017) Cities taking action. How the 100RC network is building urban resilience, New York
Tokyo Metropolitan Government (2007) Climate change strategy. A basic plan for "10-year project for a carbon-minus Tokyo", Tokyo
United Nations (UN) (2017) Progress towards the sustainable development goals. Report of the Secretary-General. New York
UN Environmental Programme (UNEP) (2016) The emissions gap report 2016, Paris
UNEP (2017) The emissions gap report 2017. A UN environment synthesis report, Paris
Wilson (2013) Shifting ground: redefining Boston's landscape infrastructure for a changing climate. Thesis at the Harvard Graduate School of Design, Cambridge, MA
World Business Council on Sustainable Development (WBCSD) (2010a) Vision 2050, Geneva
WBCSD (2010b) A world of sustainable cities, Geneva
WBCSD (2014) Action 2020, Geneva
World Economic Forum (WEF) (2018) The global risks report (13th edition), Geneva
West D (2016) Megachange: economic disruption, political upheaval, and social strife in the 21st century. Brookings Institution Press, Washington, DC
White House (2014) U.S. communities recognized as climate action champions for leadership on climate change, Washington, DC, p 16
World Bank (2013) Turn down the heat: climate extremes, regional impacts, and the case for resilience, Washington, DC
World Future Council (WFC) (2012) How can we save the world's oceans and coasts? The 2012 future policy award, Hamburg
World Resources Institute (WRI) (2011) Reefs at risk revisited. Washington, DC
WRI (2012) Impacts of Hurricane Sandy and the climate change connection, Washington, DC
World Wide Fund for Nature (WWF) (2009) Mega-stress for mega-cities: a climate vulnerability ranking of major coastal cities in Asia. Headquarters in Gland, Switzerland
WWF (2015a) Living blue planet report. Species, habitats and human well-being
WWF (2015b) Reviving the ocean economy: the case for action. Headquarters in Gland, Switzerland
WWF and Allianz (2009) Major tipping points in the Earth's climate system and consequences for the insurance sector, Berlin

Selected Websites and Social Media

#CitiesIPCC
#movethedate
@EMarineBoard
www.100resilientcities.org
www.bestclimatepractices.org/
www.C40.org
www.climate-adapt.eea.europa.eu

http://en.klimatilpasning.dk/media/665626/cph_-_cloudburst_management_plan.pdf
www.earthobservatory/nasa.gov
www.europa.eu./climate
www.icap.org
www.imo.org
www.ipcc.org
www.ises.org
www.iucn.org
www.lancetcountdown.org/
www.metropolis.org
www.nyc.gov/plany
www.oecd.org
www.oneplanetsummit.org
www.reefresilience.org/coral-reefs/
www.resilientnola.org/wpcontent/uploads/2015/08/Resilient_New_Orleans_Strategy.pdf
www.sdg.iisd.org/
www.unep.org
www.unfccc.org
www.usmayors.org/climateprotection/agreement.htm
www.wbcsd.org
www.wearestillin.com
www.whitehouse.gov/share/climate-action-plan
www.who.int
www.wri.org/tags/coral-reefs
www.wwf.org

Watercolour 6
The Gold of the Ocean: Fire Corals in Peril

Chapter 4
The Paths to Decarbonisation Through Cities and Seas

Abstract Eco-responsible cities have to ensure access for all, at all times, to clean, secure, and affordable energy. Investments in energy efficiency and sustainably managed renewable energy resources can only have beneficial effects. Renewables, the central pillar of the energy transition, have conquered most cities of the world. In 2015, renewables globally surpassed coal to represent the largest source of installed power capacity: they then continuously broke records. Still, this is not sufficient to cap temperature increase to 43°, less than 2 °C. Global emissions from energy reached their peak also in 2015. The world has inexorably entered an era of lower energy intensity. Offshore wind energy generation and ocean blue–green energy, including tidal and wave energy, can far extend possibilities on the scene of the future. Sustainable mobility and accessibility are fundamental for an equitable society and for linking the first and the last mile of all urban journeys to terrestrial and maritime networks and the broader world through integrated, smarter, and cleaner transport. This chapter offers an insight into evolving energy production and consumption patterns, and the concerted efforts to raise the bar of sustainability excellence and overcome national and supranational energy targets. Intelligent mobility and connected and automated driving will be essential for the shift towards post-carbon cities, making the most of digital technologies and speeding up the deployment of zero-emission alternatives. Automated eco-ships may also be a reality in the near future.

4.1 Renewables and Efficiency as Cornerstones of the Energy Transition

4.1.1 Towards Clean, Secure, Competitive, and Inclusive Energy

Sustainable cities must ensure continuous access for all citizens to secure, economically affordable, socially inclusive, and environmentally sound energy. Sustainable energy has been described by the United Nations (UN) as the golden thread

V. P. Mega, *Eco-Responsible Cities and the Global Ocean*,
https://doi.org/10.1007/978-3-319-93680-2_4

which connects economic growth, social equality, and a healthy environment. Transformative change in the energy sector, the source of at least two thirds of greenhouse gas emissions, is essential to reach the shared objectives. The aspiration of less than 2°C and the need for improving air quality in cities ask for the energy systems to become carbon neutral by 2100 (IEA 2016a, b, 2017a).

The world is at a turning point in decoupling economic growth and energy emissions. Growth in energy-related CO_2 emissions flattened for the 3 years 2014-2015-2016, mainly because of gains in energy efficiency, the cleaner and cheaper way to energy transition. Globally, energy efficiency attained its highest annual level (IEA 2017a). The expanded use of cleaner energy sources, mostly renewables, has also contributed, at a time when investment in upstream oil and gas has fallen sharply. The IEA reminds us that, in 2016, subsidies for fossil fuels in the world were more than double the subsidies for renewables and that the potential of clean energy technologies remains underutilised (IEA 2016b, 2017b).

Energy is crucial for achieving all the Sustainable Development Goals 2030. It has an important role in the eradication of poverty, improvement of education, food, water supply and industrialisation, cities, and climate change. The proportion of the global population, especially urban, with access to electricity has increased steadily. Recent global progress has largely been driven by Asia, where access is expanding at more than twice the pace of demographic growth. Policy attention to air quality is rising, and global emissions of all the major pollutants fall in projections, but their health impacts remain severe. However, overall progress in all fronts of sustainable energy falls short of what is needed to achieve energy access for all and to meet targets for renewable energy and energy efficiency (UN 2017).

Significant improvements will require higher levels of policy commitments and investments. The UN informs us that, globally, 85.3% of the population had access to electricity in 2014, an increase of only 0.3% since 2012, which means that 1.06 billion world citizens, predominantly rural dwellers, still lack electricity. Half of those persons live in Sub-Saharan Africa. More than 3 billion people, the majority of them in Asia and Sub-Saharan Africa, are still cooking without clean fuels and more efficient technologies (UN 2017).

The world is very diverse in its ways of energy intensity, efficiency trajectories, and the use of renewable sources. The longer-term objective of decarbonisation by 2050 demands investment in energy efficiency and renewable energy sources in all world regions. Irrespective of the technological choices, higher energy efficiency and shares of renewables are the best options. Improving energy efficiency and incorporating renewable energy infrastructure, including offshore installations, can provide multiple local benefits. The equipment of emerging world cities with sustainable energy infrastructures and the replacement of ageing infrastructures in developed countries represent unique opportunities for the decarbonisation of cities and the world in the most efficient way.

The European Union is the most carbon-efficient major economy of the world and the one that most strongly boosted the rise of renewable energy sources. The strategy Europe 2020 is aimed at a reduction of greenhouse gas emissions of at least 20% below 1990 levels, 20% of EU energy consumption from renewable resources and a

20% reduction in primary energy use through improved energy efficiency. The Union is confident that the objectives will be met by the 2020 horizon (EC 2016b).

Science, technology, and innovation have essential roles in improving energy efficiency and in exploring and capitalising on the potential of renewable energy options, speeding up developments and to guarantee higher and better penetration of renewables in the energy market. The European Commission's Energy Union strategy, adopted in February 2015, built on the ambition to achieve a fundamental transformation of Europe's energy system in a cost-effective way. The transition should be accelerated by moving to smarter, more flexible, more decentralised and integrated, sustainable, secure, and competitive ways of providing energy to consumers. The ambitions for 2030 are much higher: included are a binding EU target of at least 40% reduction of greenhouse gas emissions, a binding target of at least 27% of renewable energy used at EU level, an energy efficiency increase of at least 27%, to be reviewed by 2020 having in mind an EU level of 30% for 2030, and the completion of the internal energy market by reaching an electricity interconnection target of 15% between Members States (EC 2015a).

The 2015 Integrated SET-Plan provided the overall framework for promoting strengthened cooperation in research and innovation between the EU, Member States, and stakeholders, to step up the efforts to bring new, efficient, and cost-competitive low-carbon technologies to the market faster and thus support the energy transition. The Plan has been devised following the assessment of the energy system prospects, their importance for energy transition, and the potential to create growth and jobs in the EU. It addressed the whole innovation chain, from fundamental research to market uptake, financing, and regulatory framework (EC 2015b).

Cities can be a testing ground for proof-of-concept urban interventions entering quickly into the market through emerging technologies and business models. Sustainable energy services should satisfy citizens' demands and respect both the planet and the rights of future generations. Cities and local governments should knowledgeably lead communities into the post-carbon age with investments in energy efficiency and renewable energy sources as major drivers. A smarter energy system asks for engaged communities to manage energy consumption and actively participate in the production of energy. Increased use of information and communication technology for the management, amongst others, of large volumes of data will be required to optimise and manage energy flows. The renewable electricity market has witnessed an unprecedented acceleration in recent years, and it broke another annual deployment record in 2016, with solar photovoltaics as a main driver. A powerful vision would be for cities to become net producers instead of net consumers of energy. Urban buildings and districts could be converted into active energy generators, and the seas could also provide green energy (Mega 2015; IEA 2016b).

Renewable technologies increasingly move to the heart of the energy system. Driven by climate and energy targets, the share of renewable energy in the EU has almost doubled during the past 10 years, which has enabled Europe to lead the global development and large-scale installation of renewable energy technologies. The European electricity grid continues to offer a high-quality supply. Battery costs are dropping rapidly. The US, Japan, and Korea are leaders in this sector, and they are expanding their manufacturing capacity.

Energy efficiency in the EU is seen as a market opportunity, especially for the construction value chains. A responsible architecture is a powerful force that can bring solutions both in the medium term and the longer term, from design and renovation, and, throughout the life cycle of buildings, and contribute to reducing energy and resource consumption while fostering the resilience of the built environment.

From the EU to the world, Electricité de France (EDF) is committed to energy and electricity research and innovation (R&I) in Asia, especially China and Singapore. EDF's Singapore Centre of Excellence is at the cutting edge of sustainable city developments in Asia. Created in 2014, this EDF Lab, the Asian Sustainable Cities Centre of Excellence, has been devoted to energy research in urban planning and policy across sectors and frontiers. It cooperates with local authorities in Singapore, as well as other cities in the region, and uses futures methodology and scenarios, 3D modelling, and energy narratives to improve sustainable environments in world cities.

4.1.2 The World Energy Landscape and Outlook

Energy options depend on the macroeconomic environment, geopolitics, technological developments, international markets, trade and investments, citizen preferences and environmental concerns, and global governance, especially in relationship to climate change. Energy dependency can have serious consequences, such as supply uncertainty, higher and unpredictable energy prices, and exposure to the political instability of exporting world regions.

2015 marked a turning point for the world: renewables have surpassed coal to become the largest source of installed power capacity, thanks to strong policy support in key countries and sharp cost reductions. Led by wind and solar, renewables represented more than half the new power capacity around the world, reaching a record 153 gigawatts (GW), 15% more than the previous year. Most of these gains were driven by record-level wind additions of 66 GW and solar photovoltaic (PV) additions of 49 GW. About half a million solar panels were installed every day around the world in 2015. China, where for the first time in 2013 new renewable power capacity surpassed new fossil fuel and nuclear capacity, accounted for about half the wind additions and 40% of all renewable capacity increases in 2015, equal to two wind turbines installed every hour. Expectations have increased but remain modest compared with the huge untapped potential of renewables (IEA 2016b).

According to the IEA 14th edition world energy outlook, four megachanges seem to shape the global energy landscape in 2017: the rapid deployment and falling costs of clean energy technologies, the growing electrification of the world, the shift to a more services-oriented economy and a cleaner energy mix in China, and the resilience of shale gas and oil in the US, the first world producer. The key forces underpinning these megachanges include a global economy growing at an average rate of 3.4% per year, a population expected to increase to more than 9 billion in 2040, and a process of urbanisation that adds a city the size of Shanghai to the world's urban population every 4 months (IEA 2017a).

More emerging countries, led by India, move towards centre stage. The largest contribution to energy demand growth, almost one third, comes from India, whose share of global energy use rises to 11% by 2040, which is still well below its 18% share in the anticipated global population. Southeast Asia, another heavyweight in global energy, sees its energy demand growing at twice the pace of China. Overall, developing countries in Asia account for two thirds of global energy growth, with the rest coming mainly from the Middle East, Africa, and Latin America.

Improvements in efficiency have a huge role, as the projected rise in final energy use would more than double without them. Renewable sources of energy meet 40% of the increase in primary demand, and their explosive growth in the power sector marks the end of the boom years for coal. In India, the share of coal in the power mix is expected to drop from three quarters in 2016 to less than half in 2040 (IEA 2017a).

Renewables are set to capture two-thirds of global investment in power plants until 2040 as they become, for many countries, a most competitive source of new electricity generation. Rapid deployment and declining costs of solar photovoltaics, led by China and India, helps solar become the cheapest and largest source of low-carbon capacity by 2040. The IEA suggested a two-speed world for renewable electricity over the medium term. Although Asia takes the lead in renewable growth, this only covers a part of the continent with a high rise in electricity demand. China alone is responsible for 40% of global renewable power growth, but this represents only half the country's growth in electricity demand, in sharp contrast with the European Union, Japan, and the US, where additional renewable generation is outpacing electricity demand growth.

Electricity is the rising force among worldwide final energy uses, making up 40% of the rise in final consumption to 2040. Renewables are expected to cover more than 60% of the increase in world electricity generation. However, policy uncertainty persists in many countries, slowing down the pace of investments. Rapid progress in variable renewables such as wind and solar PV is also exacerbating system integration issues in a number of markets, and the cost of financing remains a barrier in many developing countries.

The direct use of renewables to provide heat and mobility worldwide could double. Advancing industry initiatives and policy support suggest projection for the global electric car fleet up to 280 million by 2040, from 2 million in 2016. In Brazil, the share of direct and indirect renewable use in final energy consumption would rise from 39% to 45%, compared with a global progression from 9% to 16% up to 2040.

However, the era of fossil fuels is not over. Oil demand continues to grow to 2040, but at a steadily decreasing pace. In 2040, the world energy supply mix looks almost equally divided into oil, gas, coal, and renewable sources. Each part faces particular challenges. Oil supply until 2025 is projected to come mainly from non-OPEC countries, the US, Canada, and Brazil, but this seems to decline by the mid-2020s. Natural gas is the fastest growing fossil fuel, especially in liquefied form, and offers some protection against supply disruptions. Growth of coal demand is constrained by air pollution and climate policies, mainly in the US and China. The accidental release of devastating hydrocarbons into the oceans during transportation has serious consequences to marine environments. Oil spills are particularly dangerous and often unfold into major disasters. The investigation of the Macondo Well Blowout Disaster in 2010, following the explosion of the Deepwater Horizon, which caused a seafloor gusher and spilled oil into the Gulf of Mexico, provided many world lessons about the fragility of marine resources but also the challenges for global governance to urgently end such calamities.

Natural gas clearly seems to be the transition, and its use is expected to rise by 45% to 2040. It became a polyvalent source of energy and gained ground in all domains, including power generation, industry, heat, and transport. Global gas markets went under profound change, notably because of developments with shale gas in North America. With liquefied natural gas, trade has become increasingly global since transport has become more independent from pipelines. Unconventional gas is seen as a promising new source of energy supply and a bridge fuel from coal to renewables (Gore 2013).

Carbon capture and storage (CCS) reduces greenhouse gas emissions from the energy sector and carbon-intensive industries. Currently, four large-scale demonstration projects in the EU are expected to be operational around 2020, but none has taken the final investment decision as the business case for CCS in Europe has not yet been established. The first CCS power plant in the world started operation in Canada.

Nuclear energy faces many uncertainties despite technological improvements, notably for combining safety, security, and efficiency. China will account for almost half the expected growth, overtaking the US by 2030 to become the largest producer of nuclear-based electricity. India, Korea, and Russia will collectively account for a further 30%. In response to the 2011 Fukushima nuclear accident, the EU further strengthened its already very high standards for nuclear safety. The EU leads in the adaptability of reactors to variation in demand and the world's first geological repositories for high-level and heat-generating long-lived radioactive waste.

The ocean has been bountiful in fossil fuels. One third of the global oil and a quarter of the natural gas come from underwater fields. Offshore oil and gas drilling took off in the 1970s and it gradually moved further and deeper offshore. It is growing in the Arctic, which possesses 13% of oil and 30% of gas, respectively, the Mediterranean, and East Africa. This trend represents multiple threats, especially for the extremely fragile Arctic environment. Marine ecosystems can be seriously disturbed by seismic surveys, and spills are hard to address in deep and remote waters. Binding safety and environmental standards are essential for the offshore energy industry, including risk management and capacity (GOC 2016).

4.1.3 The EU Bore the Torch of Renewables to the World

Through renewables, the European Union (EU) sent a green wave to the world. Solar, wind, biomass and biofuels, geothermal, and ocean and tidal energy are fundamental vectors towards a sustainable energy future. Starting from a rather marginal role in the previous century, renewable energy dramatically increased its contribution. Enthusiasm about renewable energy is escalating, barriers are being removed, and infrastructure is expanding. Climate change mitigation is a powerful driver for renewables, followed by the need to address reducing air pollution and diversifying energy supplies to improve energy security.

In the European Union, renewables account for 80% of new capacity, and wind power could become the leading source of electricity soon after 2030 because of strong growth both onshore and offshore. Policies continue to support renewable electricity worldwide, and the transformation of the power sector is amplified by millions of households, communities, and businesses investing directly in distributed solar PV.

A rising numbers of cities, regions, and states will decide to transition to renewable energies. Many cities have a renewable energy target, including Vancouver and San Francisco, which have ambitious targets to be 100% powered by renewable energy. Munich is advancing towards supplying the entire city with 100% green electricity by 2030, in cooperation with the municipal utility. A first milestone has been reached in 2015, when the utility-operating plants started producing enough green electricity to power all private households in the city, and also the underground and the public tram system (EPRS 2017).

Renewable energy technologies have entered the policy and market mainstream. When they become mature, investment costs decline. Solar, wind, and biomass are the technologies progressing most rapidly. Solar and wind develop for electricity generation. Biomass, solar, and geothermal energy have huge potential for the heating and cooling sector. Renewable energy technologies have been gradually integrated into the market with reduced or no support, and should over time contribute to the stability and security of the grid on a level with other electricity generators (EC 2012a).

Solar energy has the highest theoretical potential for energy production. The sun is the universal primary source of energy. Solar energy systems can harness solar rays and produce electricity and heat. Solar energy represents a real chance for distributed energy in sunny cities, also in the emerging world, where micro-power is often cheaper than extending the grid. Photovoltaics use solar cells to convert light directly into electricity. The produced energy goes directly to the grid or is stored in batteries. Many trajectories of renewable energy growth progressed more than initially expected, and promises for the future abound. In the EU, installed solar capacity reached 100 GW in September 2016, already four times more than the installed capacity in the US, almost half of the world total (IEA 2016b).

Wind power deployment has more than doubled during the years 2008–2013, approaching 300 GW of cumulative installed capacity, led by China, the US, and Germany. Wind power provides 2.5% of global electricity demand and up to 30% in Denmark, 20% in Portugal, and 18% in Spain. Policy support has been instrumental in stimulating this growth. Conflicts with neighbouring communities led to a lower allocation of local permits and the search for new spaces. Terrestrial wind power generation is already competitive where wind resources are strong and financing conditions are favourable, but it still requires support in most countries.

The best wind resources are offshore and are largely undeveloped. Offshore wind technology costs levelled off after a decade-long increase but are still higher than terrestrial costs. Offshore wind farms seem to cause more limited scenic and acoustic impact and less tension with local communities. The precautionary principle must, however, be cautiously adopted and the impacts on the marine ecosystems have to be carefully examined. The IEA suggested that recent developments in Europe for large-scale offshore wind projects indicate possible 40–50% cost reductions for new plants by 2021. Offshore wind energy grew by 25% in 2017, mostly in the UK and Germany, bringing the entire capacity to 15.8 GW. Eleven offshore wind parks currently under construction are expected to add a further 2.9 GW. Offshore wind capacity is forecast to triple over the period at the global scale with China's capacity scaling up fast. France and Korea have progressed in ocean energy for many reasons, including technology improvements, increased competition, and enhanced policy support in key markets.

The geographic pattern of deployment of wind energy is rapidly changing. Although OECD countries led early wind development, since 2010, non-OECD countries are installing more wind turbines. After 2030, emerging countries will have more than 50% of global installed capacity. A policy roadmap developed by the International Energy Agency assumes that the cost of energy from wind will decrease by as much as 25% for land-based and 45% for offshore turbines by 2050 on the condition of strong research and technology development to improve design, materials, and reliability, optimise performance, and reduce uncertainties.

Bioenergy sources are plentiful, including residential, organic, agricultural, forest residue, and algae. Biomass is versatile and can generate electricity, heat, and/or transport fuel. Furthermore, it can transform waste into an energy asset. Producing energy from waste, including marine waste, is the most efficient potential energy generation. Second-generation biofuels, made from agricultural, aquatic, and forest waste, offer a promising avenue for creating assets from liabilities. Third-generation biofuels from algae can further expand opportunities. Food security crises have reminded the world that biofuels should not be in conflict with food safety.

Second- and third-generation biofuels are expected to provide multiple benefits to the economy, society, and the environment. Many cities use biogas for their operations. For some years Amsterdam has already been using biomass from municipal waste to generate green electricity and heat. The metro and trams operate on green electricity. The Waste and Energy Company is the larger producer of sustainable energy. More than half the incinerated waste of non-fossil origin is used as biomass (Amsterdam Climate Office 2008).

Many cities invest in fuel cells producing energy from hydrogen and oxygen in an intrinsically clean and efficient way. In contrast to batteries, fuel cells do not store energy, but support a continuous flow process. Fuel cells may replace, in the medium term, a large part of the current combustion systems in industry, buildings, and road transport. In the long term, fuel cells and hydrogen are expected to form an integral part of renewable energy supply and lead to a significant international market for fuel cells in transport and industry.

Hydrogen is a vital energy carrier for a future sustainable energy economy. It is abundant and perfectly clean. It provides a unique pathway for gradually increasing the contribution of renewable energy sources. Hydrogen can be used in fuel cells for all final applications, stationary and mobile. Cost-effective transport, distribution, and storage of hydrogen are major issues, together with the creation of an appropriate infrastructure. Amsterdam is one of the leaders of hydrogen-fuelled public transport, including buses and boats. In cooperation with other urban regions, the city provides a test site for large-scale sustainable transport operations (Amsterdam Climate Office 2008).

Global geothermal resources can contribute more than 2% to global electricity production. Ocean energy, wave and tidal, represents some of the most plentiful sources, and the EU invests much to harness their power. Ocean cities can be pioneers in engaging in projects for harnessing the power from the waves that touch their shores; they can be true strongholds of blue–green energy and interfaces between local generation and national grids.

4.1.4 Decentralised Energy Production and Consumption: Prosuming

Local communities are crucial in energy transition, and the digital transformation gives them many opportunities for empowerment and action. As buildings, renovation projects, public spaces, and entire districts and mobility networks must progressively decarbonise, cities should discuss with citizens and stakeholders all dimensions of their participation in this shared future.

Behavioural change will be a most critical force in the transition. A real change in consumer and society lifestyles is still to be accomplished even in the most advanced cities and countries. Eco-responsible organisations support public and private stakeholders in the transition. Advanced technological insights, decision-making tools, assessment methods, and new business models are tested in living lab areas all around the world.

Centralised electricity generation is inherently inefficient. An essential change in electricity supply is the transition towards networks of smaller decentralised power plants nearer the consumers. Decentralised micro-generation from renewable energy sources is advancing rapidly despite an overall respite linked to the economic crisis. Micro-generation implies that every home and district can become a mini-power station. Decentralised energy can be much more efficient, as it allows the energy losses and financial costs associated with long-distance national transmission to be radically reduced (Greenpeace 2005; UNEP/ICLEI 2015).

Cities are impactful partners in moving towards decentralised energy production for households, industry, offices, shops, public and private activities, and transport. The integration of decentralised energy resources and renewable energy into the main grid is expected to change the energy paradigm of urban societies, with electricity generated in large power plants and delivered to the consumers through a passive distribution infrastructure. The involvement of all stakeholders, including utilities, independent producers, central and local governments, regulators, industry, port authorities, and consumers, is a major precondition for achieving the transition.

The energy demandpatterns of cities present key differences but also many common points. They depend on socioeconomic conditions, climatic and cultural features, population density, and urban form and functions, but also political decisions at all levels. Northern cities with colder climates and low population density can be huge energy consumers but may also develop a culture of high energy efficiency. The severity of winters is mirrored in energy consumption peaks, whereas warmer summer conditions are reflected on a growing number of cooling systems. However, many advanced Nordic cities have limited their energy consumption. Extreme events linked to climate change increase uncertainty and demand greater participation of all actors.

Electricity has been the symbol of industrial societies and economies. Industrial energy consumption is being stabilised, mainly as a result of the transition to a digital, knowledge-based, and service-oriented economy. Investment in technology and innovation has enabled industry to become more eco-efficient. Heavy industry, such as the automobile sector, has demonstrated significant reductions in energy consumption during past years. Environmental regulation and enforcement have been key driving forces for the development of eco-innovations. Energy consumption by the transport sector, depending almost entirely upon oil, has increased steeply, mainly because of the continuing growth of road transport for passengers and freight, and the explosion of air transport, fuelled by the rise of low-cost leisure trips.

Emboldening consumers as micro-producers is crucial for removing barriers to renewable energy growth. Micro-power generation is expected to continue emerging gradually alongside the grids and increasingly incorporate renewable energies in smarter systems. Solar, wind, biomass and geothermal power, and combined heat and power systems can significantly reduce the need for power from the grid, for households, commercial and industrial spaces, and public buildings. As consumers become "procumers"(producers-consumers), they also gain a stronger sense of ownership and control over their energy use. Empowered smart micro-prosumers could help increase public awareness on the amount and the price of the electricity services and encourage behavioural change at the building, neighbourhood, and city levels.

Heat production is predominantly decentralised, including individual heating systems and dedicated heat stations with their associated local networks. Needs range from household and office heating and cooling to steam production for industrial port uses. Solar district heating systems are used in all continents, including Antarctica. The Danish city of Silkebord had the largest solar thermal plant in operation in Europe at the end of 2016, using 157,000 m^2 of solar thermal collectors, enough to cover 22 football fields and 20% of the annual heating demand.

Combined heat and power generation can substantially increase energy efficiency. The cogeneration process uses waste energy from electricity production for heating, and, at the same time, prevents impacts from additional heat generation. Cogeneration can use renewable energy sources and reach very high efficiencies from the inherent characteristics of the process. The produced heat can be used locally for residential district heating. Tri-generation, combined heat and power with additional production of cooling, holds potential for ever higher efficiencies.

The Copenhagen district heating system is supplying 97% of the city with clean, reliable, and affordable heating that uses waste heat from refuse incineration plants and combined heat and power plants. The process saves energy and substantially reduces CO_2 emissions and pollutants. Waste heat, usually sent into the sea as a by-product from the incineration plants and combined heat and power plants, is pumped through a 1300-km network of pipes straight into homes. The system maintains water temperature providing homes with heat from waste. In Gothenburg, the district heating system supplying heating to more than 90% of apartment blocks and around 9000 family homes uses an energy mix of recovered heat, 18% renewable energy, and 8% fossil fuels (EPRS 2017).

Local authorities can boost green electricity through innovative public procurement and high standards, invigorating partnerships with citizens and associations, access to adequate infrastructure, and energy audits. They can remove constraints in the value chain and foster accessibility to renewable energy sources. All municipal buildings should be lighthouses of the energy transition and demonstrate state-of-the-art renewable energy technologies and energy-efficient spaces. Urban governments will have to drastically reduce energy consumption related to their spaces and functions by carrying out the required renovation works and the necessary audits to continuously monitor and improve energy performance. They also have to weave the links among all levels of governance, stakeholders, and networks with other cities.

It is also crucial to introduce schemes for the effective recovery of heat losses from electricity and industrial production and to enhance cogeneration of heat and power. Port authorities can also contribute much. The "WaterNet" Waste Water Treatment Company at the Port of Amsterdam is operating a successful monitoring system for electricity and combined heat and power. This approach leads to the significant reduction of greenhouse gas (GHG) emissions, higher energy efficiency, and reduction of costs.

4.1.5 *Energy from the Ocean Can Far Exceed All Present and Future Needs*

Energy from the sea is essential for cities and for maritime transport. The global ocean energy resource far exceeds all present and projected future energy needs. Exploiting this endogenous resource would help countries and cities to address dependence on fossil fuels and enhance energy security. This concept may be particularly important for island nations and regions, which, totally dependent on the ocean, can contribute to energy self-sufficiency and replace expensive imported energy (EC 2012c, d, 2016a, c).

Offshore wind energy is progressing rapidly and is expected to increase. The EU is the global leader in offshore wind energy with more than 90% of the world's installed capacity (EWEA 2014). Europe counts two offshore floating wind demonstration projects, in Portugal and Norway. Floating platforms anchored to the seabed can lead to more cost-effective technological and policy options. Nearby cities should be involved in the decision-making process, ensure that impacts on marine ecosystems have been properly addressed and that synergies with other activities are being carefully explored.

Ocean energy can be instrumental for fuelling coastal urban economies and societies. Its growth might be followed by tidal energy and maybe in a later stage by wave energy. Most of the technologies are still in the demonstration phase. The UK Carbon Trust estimated that the global wave and tidal energy market could be worth up to €535 billion by 2050. Ocean energy has also the potential to create new, high-quality jobs in technology development, manufacturing and operations (EC 2014a).

Ocean energy could help to balance out other renewable energy sources to ensure a steady aggregate supply of clean energy to the grid. Ocean energy devices, entirely or partially submerged, have a low visual impact and do not disturb local communities. As for offshore wind energy, the potential impact on marine ecosystems has to be carefully studied. Supporting policies could draw from the early measures to encourage offshore wind power development in the 1980s and 1990s. Since then, the wind sector, including offshore wind, has grown exponentially, having benefitted from targeted policy support. Ocean thermal energy conversion technology has a strong potential in tropical regions, displaying the highest difference of temperature between surface and deep waters. Local deployment can provide for the needs of the coastal communities and contribute to their independence.

Second- and third-generation biofuels, from marine waste and algae, respectively, are a major energy avenue for transport in coastal cities. Fuels from fish residues can transform a liability into an asset. Algae have been already used as food, feed, and fertilisers for centuries. Their size ranges from micrometres of unicellular microalgae to macroalgal seaweeds of tens of metres. Algae contain several high-value molecules, such as lipids, proteins, and carbohydrates, and present a growing interest for an array of possible uses. Approximately 200 species are used worldwide in different sectors, from pharmaceuticals to the cosmetics industry and the production of ethanol or biodiesel, and research is ongoing on genetic engineering of microalgae (EC 2014b).

The global governance architecture is decisive for the decarbonisation of the world. Governments, at all levels, have to lead by example and invest in post-carbon, energy-efficient systems, goods, and services. Urban governments will have to drastically reduce the energy consumption related to their spaces and functions by conducting exemplary smart renovation works and audits to continuously monitor and improve energy performance. They also have to weave the links among all levels of governance, stakeholders, and the networks with other cities, and ensure the decarbonisation of the commons.

4.2 Urban and Local Excellence for the Post-carbon World

4.2.1 *Cities and Citizens Have to Act Consciously and Urgently*

Cities, as the most concentrated energy demand centres, must participate in all decisions on energy transition and act at the earliest possible stage for the effective decarbonisation of energy systems. Energy infrastructures are long lived, and energy investments take time to bring results. Decisions made at present already impact the energy reality of 2050. The Covenant of Mayors' Climate Summit of Local and Regional Leaders, during COP23 in Bonn, demonstrated that non-state actors from across the world can come together to share best practice and promote the transition to low-emission and climate-resilient economies. Cities are increasingly becoming game changers when it comes to addressing climate change and energy issues, leading the transition to a post-carbon future. By addressing unsustainable energy and climate change patterns at the local level, cities can bypass national political discord and create synergies and opportunities that accelerate action and amplify mitigation and resilience (Friends of Europe 2017).

Zero-emission cities multiply, and the Birmingham City Council and the World Business Council for Sustainable Development Zero Emission Cities (ZEC) initiative have identified the Smithfield development as the focus for a ZEC engagement. The Smithfield visioning document outlines the requirement for the development to meet high standards of sustainable design and construction that will be essential in creating an adaptable environment which will stand the test of time. This vision is accompanied by a proposed action-oriented sustainability framework. The Framework draws together a range of sustainability standards and examples of best and aspirational practice that, if embedded, will make the vision come true (WBCSD 2017).

The residential, transport, and tertiary sectors hold the highest potential for improvements and deserve enhanced attention. Cities can have a major role in increasing awareness through appropriate regulation, public procurement policies, and economic signals. Increased action on efficiency can serve as a unifying energy policy foundation that brings multiple benefits, including substantial energy security and financial profits. Decarbonisation will require a large quantity of biomass for heat, electricity, and transport. In transport, a mix of several alternative fuels will be needed to replace oil. Biofuels will probably be a main option for aviation, maritime transport, long-distance road transport, and rail.

Cities are generally willing to go for renewables given the significant local benefits for air quality and public health and for the integrity of ecosystems. Decarbonisation can be an advantage for champion cities early engaging with the energy transition, contributing to reduce import dependency and exposure to the volatility of fossil fuel prices. Digitalisation can support centralised large-scale systems, such as nuclear and gas power plants, and decentralised renewable energy systems to work together. The magnitude of the transformation and the minimisation of investments in carbon-intensive assets can best be orchestrated at the local level.

Local energy production from renewable sources asks for a more intelligent grid to handle the variable generation from many distributed sources. With more decentralised generation, from terrestrial and offshore sources, smart grids, new network users, such as electric vehicles and ships running on biofuels, there is a greater need for a more integrated approach on transmission, distribution, and storage. Cities should be among the key stakeholders to organise the dialogue, invest in evaluating the policy options and fully involving informed citizens in the technological choices and the pricing mechanisms that must remain transparent. Technological choices have to take account the local environment and infrastructures, including also the offshore possibilities. Vulnerable customers could need specific support in the transitional period. Institutional investors could become greater players in the financing of promising costly offshore energy developments.

4.2.2 *Energy Efficiency of Buildings: The First and Best Option*

Cities are seas of buildings. The energy efficiency of the built environment is crucial for all urban governments. Higher energy efficiency in new and existing buildings is vital and feasible. Nearly zero energy buildings should become the norm. Buildings, including homes, could produce more energy than they use. Materials, equipment, and appliances gradually have to fulfil the highest energy efficiency standards. In transport, efficient vehicles and incentives for behavioural change are required. Citizens will gain with better information on technological issues and policy options. With smart technologies, such as home automation, consumers will be better equipped to monitor and transform their consumption patterns.

Urban planning can contribute to saving energy in the medium and long term, with all land and marine uses having a complementary contribution. Local authorities should train advisers to help citizens optimise energy conditions, such as insulating buildings, reducing heating and cooling, and review the necessary infrastructure and networks to systematically use the waste heat of electricity generation in combined heat and power plants.

The great majority of the housing stock of European cities is old and its energy performance has high potential for improvement. In Amsterdam, almost half the housing units were built before the Second World War. The Amsterdam Climate Office collaborated with stakeholders, private landlords, owners, and residents to promote insulation, double glazing, efficient heating installations, and the use of renewable energy. The much better energy performance of new houses has however been counteracted by an increase in the number of housing units and thus of electricity consumption.

Housing associations, owning more than 50% of the housing units in Amsterdam, are important players along with tenant organisations. They have invested for many years in better insulation and higher efficiency. Since 2007, the municipality of Amsterdam has headed the alliance of leaders in which the housing associations participate in concrete initiatives to achieve the best performance. Model houses to reduce emissions are the joint results of common efforts. If all housing associations undertake serious renovation by 2025, a CO_2 reduction of more than one third could be achieved (Amsterdam Climate Office 2008).

The lion's share of electricity consumption in the US occurs in buildings where a very diverse array of cost-effective efficiency investments is possible. For example, replacing appliances such as air conditioners, refrigerators, freezers, and hot water heaters with more efficient models could reduce energy use by 30%. The savings from attaining full deployment of energy-efficient technologies in buildings could eliminate the need for new electricity generation capacity through 2030.

In the US, the Better Buildings Initiative was part of the plan to ensure that America invests in innovative clean energy technologies and doubles the share of electricity from clean energy sources by 2035. In parallel, the Initiative aims at making commercial and industrial buildings 20% more energy efficient by 2020 and accelerating private sector investment in energy efficiency. Through a variety of efficiency improvements, such as improved lighting, greater insulation, more efficient heating and cooling, clear information, and access to financing, buildings can become more energy efficient and better places to live and work while generating employment and fuelling growth.

The White House Better Buildings Challenge is inspiring public and private sector leaders to engage their organisations in saving energy and money and to demonstrate the results of the best strategies. Partners commit to an energy savings pledge, a showcase building, and the presentation of their achievements. They enjoy public recognition, technical assistance, and sharing of best practices through a network of peers. Their leadership can provide real models for emulation and continuous overcoming (White House 2014).

The widespread adoption of energy efficiency technologies has still to overcome many obstacles. The upfront expenditure can be high and can deter investment despite the prospect of long-term cost savings. Volatile energy prices can cause delays in the purchasing of more efficient technology because of a lack of confidence concerning the return on investments. The public authorities that are closest to the citizens must inform on local investments in energy-efficient infrastructure, which can impact patterns of energy use for decades. These authorities are responsible for providing trustworthy information for consumers wishing to learn about the comparative advantages in terms of performance and the costs of efficient technology alternatives. Driving change requires significant public and private support and sustained effort.

Successful voluntary energy efficiency initiatives include the U.S. Department of Energy and Environmental Protection Agency (EPA) Energy Star labelling scheme, which assisted thousands of citizens and industrial, commercial, utility, state, and local organisations to make informed decisions. Energy Star homes are 15–30% more efficient than typical new homes. Most homebuilders go for Energy Star certified homes. Since 1999, more than 30,000 buildings have gained Energy Stars. To raise emulation, EPA lists the US cities with the highest number of Energy Star certified buildings that meet strict energy performance standards and produce limited greenhouse gasses. Fifteen types of commercial buildings can earn the Energy Star, including office buildings, schools, and retail stores. Energy used in commercial buildings accounts for nearly 20% of greenhouse gas emissions. In 2017, Washington, DC topped the list with 790 certified buildings, followed by Los Angeles, New York, San Francisco, and Atlanta (EPA 2016).

Conscious resource management is particularly reflected on eco-buildings for residential, commercial, or industrial use. Eco-habitat, luminous and healthier, with sustainable use of materials and water and low to zero energy requirements, has already endowed many cities with attractive eco-cells. Good integration into the landscape, adaptation to climatic conditions, recourse to ecological materials and renewable energies, and the sustainable management of water and resources, including waste, are critical for the eco-performance of the entire built environment.

Bioclimatic architecture and interconnected smart home design have made great strides during past years. Architects, designers, eco-builders, and eco-citizens are promoting green construction and more ecologically sound techniques in parallel with smart monitoring systems for better resource use. Origin, design, and performance information about the latest low-impact materials and technologies, the best options for the use of renewable energy, water, and resources, and best practices can help make a difference.

Public buildings and privately owned buildings open to the public can act as pioneers and serve as models for intelligent resource-saving and performance-enhancing construction. Landmark buildings have to shine the way. New York's Empire State Building was retrofitted for energy efficiency, reducing its energy consumption by 38%. Public buildings can act as a lighthouse to the energy transition and initiate a widely spread culture of change.

The display of building plans and energy performance certificates and recommended optimal climatic conditions, such as the most favourable indoor temperatures, in all public buildings can promote awareness about the benefits of eco-performance. Public authorities are responsible to implement the recommendations included in the certificate, monitor their performance, and communicate their results to citizens. They can also interact with citizens, involve them in life cycle assessments, and modify conditions and objectives according to public expectations. The "Be Exemplary" Initiative in Brussels calls for all public buildings to use 100% green energy.

Melbourne offers a prime example with the Council House 2 municipal eco-office, which like a sunflower follows the movement of the sun to capture light and collect heat. This multi-award-winning building managed to achieve remarkable reductions in CO_2 emissions and water and energy consumption. This 10-story office building of the Municipality of Melbourne, opened in 2006, also features ground-floor retail spaces and underground parking. The Council House 2 has been designed to enhance the natural 24-h cycle of solar energy, natural light, air and rainwater, to power, heat, cool, and water the building, which purges stale air at night and pulls in pure fresh air during the day. The north façade of the building has 10 dark-coloured air ducts that absorb heat from the sun; the hot air then rises, taking the stale air up and out of the building. The south façade has light-coloured ducts that draw in pure fresh air from the roof and distribute it down through the building. The west façade has louvers made from recycled timber and powered by photovoltaic roof panels that move according to the position of the sun. Council House 2 embraced the best environmental options also in water treatment, with an in situ treatment plant filtering out the water and producing clean water suitable for cooling and irrigation, other municipal buildings, and public fountains.

Public lighting has often been a subject of controversy, as it has been perceived both as a factor of public security and local identity and as a driver of high electricity consumption. Plentiful lighting has a beneficial effect on the atmosphere of a city. It has proved to help reduce both road accidents and insecurity incidents, and the illumination of monuments and public buildings can redesign a city and foster urban identity and belonging. Light festivals in cities gave dramatic results in Lyon. Light designers were frontrunners in a contest to dramatically brighten the city of Montreal. An Iceberg show consisting of illuminated arcs and the day of Eight Suns with projections on the façades of the buildings were some of the innovations, parts of a truly artistic energy adventure.

Lighting typically accounts for about 20% of a city's total electricity consumption. The advent of LED technologies has revolutionised lighting design and longer-term budgets. Intelligent lighting can optimise illumination, increase comfort and urban safety, and reduce accidents and costs. The latest generation street lighting is robust, weatherproof, and economical, often lasting three to five times longer than traditional equipment. In Barcelona, 50% of lighting is remotely controlled. The city has smart LED streetlights equipped with sensors which switch on when they detect motion and dim when the streets are empty; 19,500 smart meters have been installed in targeted areas of the city to monitor and optimise public energy consumption (EPRS 2017).

Responsible public lighting, especially on the waterfront, would benefit from lighting adapted to the degree of darkness. The municipality of Amsterdam replaced older lamps with more efficient ones and entirely switched to electronic components. In 2007, the City Hall of Amsterdam was already fully equipped with energy-efficient lamps, achieving 45% energy savings. Furthermore, new-generation streetlights have been installed and provide huge energy savings throughout the city (Amsterdam Climate Office 2008).The role of cities and local organisations is expected to be much greater in the energy systems of the future. Energy optimisation should be an essential design parameter incorporated into all urban functions performed by municipalities, households, universities, businesses, and civil society organisations.

In the framework of its "Architects Against Climate Change Campaign," the European Council of Architects (ECA) has been seeking to raise the awareness of the public, construction professionals, and decision makers to the role of urban public and private design. Only 4% of all new buildings in the world are expected to be built in Western Europe, against 10% in the Middle East, 15% in the US, and 38% in China. As 75% of the European building stock is deemed energy inefficient, the reduction of GHG emissions in Europe can only come from a radical transformation of existing buildings. The ECA recommended renovating existing buildings, making better use of the buildings that are currently misused, underused, or not used, and regenerating spaces.

Retrofits concentrating solely on technical and fabric improvements have been shown to fall short of the expected reductions in emissions. To realise a step-change in building performance, a genuine renovation must include a functional, spatial, and material reconfiguration of the built environment. Buildings could be fundamentally transformed during the next 50 years in response to developments ranging from maintenance robots to photovoltaic paint, already in progress. Companies are developing promising solutions such as phase-change materials, dynamic glazing, and adaptive facade systems.

The world's first algae-powered building, the BIQ House, opened in Hamburg in 2017. The BIQ House, designed by Arup in cooperation with Germany's Strategic Science Consultants and Austria-based Splitterwerk Architects, has been launched as part of Hamburg's International Building Exhibition, which embraces research and development. Microalgae can thrive in water that is fresh, salty, or dirty. Because of its high growth rate, doubling biomass within a few hours, and higher productivities per unit area, microalgae offer a less resource-consuming production system than plants and therefore important applications in food, feed, industrial, and pharma systems.

The algae-powered building probably initiates a movement towards living buildings that respond and adapt to the surrounding conditions. According to Arup, the facade of the algae-powered building is designed so that algae in the bio-reactor facades grow faster in bright sunlight to provide more internal shading. The bio-reactors not only produce biomass that can subsequently be harvested, but they also capture solar thermal heat, and both energy sources can be used to power the building. Thus, photosynthesis is driving a dynamic response to the required solar shading, while the microalgae growing in the glass louvres provide a clean source of renewable energy.

4.2.3 *The EU Smart Cities and Communities Innovation Partnership*

In the EU, cities can play a crucial role for achieving the Energy Union targets. Cities impact buildings, making them more energy efficient and turning them into micro-power plants. They can make the energy transition more inclusive and address energy poverty by investing in energy-efficient social housing. They can deploy renewable energy technologies and develop smarter grids. They can invest in high-performing municipal transport and low-emission mobility solutions. They can finally serve as living laboratories, where disruptive and innovative technologies can develop through strong engagement in innovation clusters. Taking local ownership is essential.

In 2012, the European Commission launched the Smart Cities and Communities Innovation Partnership to stimulate Europe's capacity for research and innovation, help cities adopt clean energy solutions, integrated with smart mobility systems and sound resource management and business models, while surfing on the waves of digital democratisation. The partnership searched to reduce energy consumption, GHG emissions, air pollution, and road congestion to provide scalable and transferable solutions and contribute to the EU climate action goals. It tried to overcome obstacles, co-fund demonstration projects, and help coordinate existing initiatives. It ultimately established strategic partnerships between industry and European cities to develop the urban systems and infrastructures of the future.

The partnership brought together all actors, cities, academia, industry, civil society, and citizens to improve urban life through more sustainable integrated urban projects. A strategic implementation plan provided a wealth of smart urban initiatives including better sustainable planning, higher energy efficiencies, better transport responses, and a smarter use of digital infrastructure. Selected high-impact lighthouse projects brought industrial consortia together with cities to demonstrate the technologies. A transformation agenda provided the strategic orientation for identifying bottlenecks, to shape incentives and define actions to be undertaken. The replication of successful urban applications and the mainstreaming of innovations are facilitated through the Smart Cities and Communities Stakeholder Platform, bringing, municipal administrations, industry, academia, and civil society organisations together, to exchange and capitalise on the shared experiences (EC 2012b).

Smart applications are expected to play a major role. The digital infrastructure can make a city more intelligent in all its functions including energy consumption for electricity, heat, and transport. Smart buildings and neighbourhood projects, integrating, for example, local and renewable energy sources and expanding the use of highly efficient heating and cooling, are the components of future smart cities. The construction of nearly zero energy buildings and a positive energy-built environment can be scaled up to energy-producing cities. Smart supply and demand service projects can provide data and information to citizens on energy and mobility options and services.

Urban mobility projects could include electric public transport vehicles that are able to exchange surplus energy, braking and accelerating energy, with the rest of the system. Smart technologies would help to manage energy flows or hydrogen as an energy carrier for storing energy and balance the total urban demand. Sustainable digital infrastructures could help reducing the carbon footprint of data centres and equipment in particular, and advance towards intelligent heating and more and better cooling and lighting solutions.

The EU promoted an integrated platform, as a single entry point that can help cities better realise their ambitions concerning energy transition. It hosts all data and information that matter for cities, ensuring that stakeholders and citizens can easily access the best knowledge on city-related European legislative initiatives and strategies, financing instruments such as Horizon 2020, the Structural Funds, or the Juncker Investment Plan, and all EU data and information on exemplary initiatives, such as the Green, Cultural, or Innovation Capitals and Green Leaf Awards.

Since 2014, the EIP Smart Cities Citizen Focus Action Cluster has been involved in intensive mapping, analytical mutual learning, and networking efforts among its 61 Commitments and beyond. In 2015 the Action cluster presented Principles and

Enablers for Citizen Engagement, which was developed with the participation of citizens. The next step focused on the involvement of Regional and Local Authorities in the discussion and implementation of concrete policy measures and tools to achieve responsible, inclusive, and sustainable smart cities through citizen engagement and participation. The process led to the "Inclusive Smart Cities: A European Manifesto on Citizen Engagement," signed in 2016 (European Innovation Partnership on Smart Cities and Communities 2016).

4.2.4 City Networks and the Global Covenant for Climate and Energy

Cooperation among all cities, citizens, and their elected representatives is critical for interconnected global change. The Covenant of Mayors, launched by the European Commission, with the support of the European Parliament, during the 2008 European Sustainable Energy Week, has encouraged local leaders to go beyond the EU objective for 2020 through increased energy efficiency and the rise of renewable energies. The Covenant, open to cities of all size or form, supported the efforts made by local authorities in their sustainable energy journeys and generated thousands of action plans.

The Covenant, a valuable bottom-up movement, has engendered many creative ideas and succeeded in mobilising a great number of local and regional authorities to develop action plans and smart investments towards climate change mitigation measures. Benchmarks of Excellence are examples of initiatives by territorial actors, endorsed as useful and inspiring actions. One of the most significant initiatives is The Mayors Adapt, launched in 2014, that invited cities to make political commitments and take action to anticipate and prepare for the unavoidable impacts of climate change.

The EU Covenant of Mayors inspired other regions across the globe. It was extended towards the east with the Covenant of Mayors East programme as well as towards the southern Mediterranean countries. In 2012, the first EU–China Mayors Forum brought together the EU Covenant of Mayors and the Chinese Mayors' Association for the adoption of the EU–China Mayors' Charter for Sustainable Urbanisation, which engaged the signatories to share experience and knowhow in sustainable energy management.

In 2015, a public consultation process explored the views of stakeholders on the future of the Covenant of Mayors. The response was unanimous: 97% called for a new target beyond 2020 and 80% favoured a longer-term target. The majority of respondents also endorsed the EU 2030 objectives for a minimum 40% emission reduction and supported the integration of mitigation and adaptation to climate change under an integrated framework. This process led to a reinvigorated integrated Covenant of Mayors for Climate & Energy, incorporating also adaptation to climate change and extending it to the 2030 timeframe. The three pillars of the strengthened Covenant, mitigation, adaptation, and sustainable energy, were endorsed and local and regional authorities committed to implement EU climate and energy objectives. The EU Covenant was operated by the Council of European Municipalities and Regions (CEMR), Climate Alliance, Energy Cities, EUROCITIES, and the European Federation of Agencies and Regions for Energy.

In July 2016, the reinvigorated EU Covenant of Mayors created an alliance with the UN Compact of Mayors, launched in September 2014 by the UN Secretary-General and his Special Envoy for Cities and Climate Change, assisting cities to reduce local greenhouse gas emissions, enhance resilience to climate change, and transparently track progress. The Compact was activated under the leadership of the global city networks, Cities Climate Leadership Group (C40), Local Governments for Sustainability (ICLEI), and the United Cities and Local Governments (UCLG), and with support from UN-HABITAT. The alliance of these two major coalitions led to the Global Covenant of Mayors for Climate & Energy, a world movement embracing cities in their transition to a climate-resilient future.

The Global Covenant of Mayors for Climate & Energy is instrumental for scaling up efforts to implement the Paris Agreement and achieve the Sustainable Development Goals, which need a global response from non-state actors, and a credible, reliable non-state interlocutor with the UN system. The more than 7500 world cities, on six continents, engaged with the Global Covenant in 2018 to bring an extraordinary wealth of ideas and actions. They make leading contributions to the climate action agenda, very often setting more ambitious targets to reduce their greenhouse gas emissions than their countries. They also invest in adaptation and wish to make their efforts for resilience visible at the international level. A single global coalition can best promote interactions and attract investments in cities wishing to become smarter and more sustainable inspire each other, and learn from each other's best practices, but also the limitations and failures.

The Global Covenant of Mayors for Climate & Energy is the first global coalition of local leaders pledging to track and reduce GHG emissions. Climate change is a challenge that cannot be resolved by cities or governments alone. The Global Covenant, co-chaired by Former New York City Mayor and UN Secretary-General's Special Envoy for Cities and Climate Change and the Vice President of the European Commission, enjoys the support of global and regional city networks, including C40 Cities Climate Leadership Group, Local Governments for Sustainability, United Cities and Local Governments, Climate Alliance, Energy Cities, and Eurocities. The core principles of the Covenant focus on the key contribution of local governments to global climate solutions, the role of city networks as critical partners, a robust solution agenda with high-impact city actions, and on mitigation and adaptation.

Celebrating the second anniversary of COP21 at the One Planet Summit (Paris, December 2017), the Global Covenant and partners announced the One Planet Charter, a new campaign that will help cities swiftly implement commitments. Through the One Planet Charter, cities invest in action that drives investments, sustainable public procurement, and policy decisions in renewable energy, energy efficiency, electric vehicles, and efforts for zero emissions and zero waste. The Charter underlines commitment to increase demand for sustainable and resilient infrastructure, products, and services, while recognising the importance of working closely with governments and business to mobilise global climate action.

The Global Covenant of Mayors also issued a call for the vertical integration of local authorities in climate investment plans, underlining the need for pathways of active participation and engagement of sub-national governments in the formulation of national climate investment plans, in line with the Paris Agreement. The importance of availing of the necessary funding led to a set of breakthrough partnerships with the EU, the European Investment Bank (EIB), and the European Bank for Reconstruction and Development (EBRD) under the umbrella of "Global Urbis." This global initiative promises to fill the urban financing gap, which is currently preventing many cities from implementing action plans. Each partner has a well-defined role to fulfil. The Global Covenant of Mayors for Climate & Energy is supporting the development and financing of cities' action plans and projects necessary to cut emissions. The EIB, the world's largest climate finance bank with a projected $100 billion in the next 5 years for climate-related lending, aims to develop with the Global Covenant and EBRD an innovative and unprecedented global advisory and financing facility for fostering Climate Action investments in cities.

4.3 The Sustainability Promises of the Maritime Paths of the Planet

4.3.1 Maritime Transport, the Backbone of World Trade, Has to Decarbonise

Twenty-four hours a day and all year round, ships cross seas and connect cities. Global trade and maritime transport are fundamental to connecting people and goods and sustaining economic growth and spreading prosperity and equality throughout the world. Maritime transport is fulfilling a critical social as well as an economic function with key environmental implications.

Seaborne trade accounts for almost 90% of global trade in terms of volume. Maritime connexions are critical for port cities, which increasingly become instrumental for cleaner transport and renewable energy produced offshore. Ocean energy entered the policy debate and promises to contribute to clean and abundant energy and the reduction of greenhouse gas emissions, while respecting the precious ocean ecosystems.

Making shipping greener and preventing marine pollution was one of the original aims of the International Maritime Organisation (IMO), set up in 1948. Environmental rules have intensified in recent years, as pollution became more serious. The IMO reckons that maritime transport emits around 1000 million tonnes of CO_2 annually and is responsible for about 3% of global greenhouse gas emissions. Shipping emissions could increase by 50–250% by 2050, depending on economic choices and energy developments. The organisation is determined to ensure that international shipping remains the most environmentally sound mode of transport. According to IMO, ships' energy consumption and CO_2 emissions could be reduced by as much as 75% by implementing operational measures and existing technologies. Many of these measures are cost-effective and offer net benefits (IMO 2014).

However, a recent report by the International Transport Forum finds that maximum deployment of currently known technologies could make it possible to reach almost complete decarbonisation of maritime shipping by 2035 (ITF 2018). The report identifies four potential decarbonisation pathways resulting in a reduction of CO_2 emissions between 82% and 95% of the currently projected 2035 level. This reduction equals the annual emissions of approximately 185 coal-fired power plants. To advance the decarbonisation of maritime shipping, governments can set a clear and ambitious emissions-reduction target supported by a comprehensive portfolio of policy measures that address technology, operations and alternative energy, and economic incentives.

Regulations governing sulfur oxide emissions from ships are included in the International Convention for the Prevention of Pollution from ships (MARPOL), which sets progressive stricter regulations to control emissions from ships, including sulfur oxides (SO_x) and nitrous oxides (NO_x). In 2016, the IMO set 1 January 2020 as the implementation date for a significant reduction in the sulfur content of the fuel oil used by ships. The decision to implement a global sulfur cap of 0.50% m/m (mass/mass) in 2020 represents a significant cut from the 3.5% m/m global limit currently in place and demonstrates a clear commitment to ensuring that shipping meets its environmental obligations. The reductions in sulfur oxide emissions resulting from the lower global sulfur cap are expected to have a significant beneficial impact on the environment and on human health, particularly that of citizens of port cities and coastal communities, beyond the existing emission control areas.

Exemptions are provided for situations involving the safety of the ship or saving life at sea, or in case of accidents. Ships can meet the requirement by using low sulfur compliant fuel oil. An increasing number of ships are also using gas as a fuel, which has been recognised in the development by IMO of the International Code for Ships using Gases and other Low Flashpoint Fuels, adopted in 2015. Another alternative fuel is methanol, used on some short sea services.

Making shipping greener is subject of technical, organisational, economic, and social innovation. The shipping industry is showing increased interest in the use of liquefied natural gas (LNG) as a cleaner transition fuel towards renewable sources and a decarbonised future. Sulfur and particle emissions would be reduced to almost zero, nitrogen oxide emissions by 85–90%, and net greenhouse gas emissions by 15–20%. Industrial partners and stakeholders, including port authorities, fuel suppliers, and shipping companies, ask port cities to prepare to offer safe storage and bunkering of LNG. Singapore, the world's largest bunkering hub, strives to lead on LNG bunkering by 2020.

Voluntary measures are usually preferred by industry, which can use them in the most flexible way. The Environmental Ship Index (ESI) is a voluntary instrument designed to improve the environmental performance of vessels. It only includes ships that perform better than obliged by the IMO international legislation. It assesses the environmental performance of ships regarding air pollutants and CO_2 and scores NO_X and SO_X emissions directly and proportionally. The Environmental Ship Index is intended to be used by ports to reward ships participating in the scheme and also by business associations wishing to share their environmental performance results and state their engagement with cleaner shipping. The programme is flexible and dynamic and can raise awareness among the global port community and improve the maritime and port environment.

International shipping is addressed by the Paris Agreement's greenhouse gas reduction targets, but because of the eminently international nature of this activity, it is not included in the Nationally Determined Contributions made by countries. In December 2017, at the French One Planet Summit, 34 states pledged for an ambitious contribution of maritime transport in reducing greenhouse gas emissions in the framework of the negotiations at the IMO. The initiative of the Tony de Brum Declaration, named after the former political leader of the Marshall Islands, founder of the high ambition coalition within the UNFCCC at COP21 in Paris, and initiator of the process of negotiation on greenhouse gases within the IMO, exemplifies the mobilisation of states around maritime issues in the global framework of climate action (IMO 2015).

Major port cities cooperate with the maritime industry to reduce air pollution. They have committed themselves to reduce greenhouse gas emissions through the World Port Climate Initiative (WPCI), impacting the sustainability of supply chains, taking into account local circumstances and port management structures. Shipping firms are also under pressure to cut their greenhouse gas emissions.

Marine pollution from shipping is still a serious problem, estimated to about 20% of the total marine pollution. Dumping of garbage from ships is a serious problem. Port waste reception facilities and hosting cities have an important role. The EU Member States have introduced legislation for delivery of waste ashore before a ship leaves port. Sewage pollution from ships is mainly a problem with cruise ships representing the equivalent of a small town and can contribute to local eutrophication problems. The increased requirements under MARPOL are likely to reduce the problems.

4.3.2 *Waterborne Mobility at the Crossroads of the Future*

In 2016, a fleet of trucks from a host of leading manufacturers including Volvo drove their freight across Europe largely without drivers as part of a mission under the Dutch presidency of the EU on the potential of artificial intelligence in the haulage industry. In a report into the driverless car revolution, the World Economic Forum estimated that $2 billion worth of value would shift from traditional cars to those self-driving and that these would make up 5% of vehicle sales by 2025. Urban and waterborne mobility may face great transformations over the common years, as eco-ships and artificial intelligence embrace waterborne traffic.

In early 2018, Rolls-Royce opened a state-of-the-art research facility in Turku to develop the technologies required to shape the future of an increasingly more autonomous global shipping industry. This new R&D Centre for Autonomous Ships includes a Remote and Autonomous Experience Space aimed at showcasing the autonomous ship technologies Rolls-Royce has already introduced as well as those in the development stage, including autonomous navigation, the development of land-based control centres, and the use of artificial intelligence in future remote and autonomous shipping operations.

In 2017, the Norwegian company Yara announced its plans to launch the world's first autonomous zero-emission container vessel *Yara Birkeland*. Initially manned, the vessel will move to remote control in 2019, before becoming totally autonomous in 2020. *Yara Birkeland* will operate on a sea route between Brevik, Herøya, and Larvik ports in southern Norway. The cutting-edge vessel will replace 40,000 road trips from Yara's Porsgrunn fertiliser plant to the ports of Brevik and Larvik, while reducing the level of NO_x and CO_2 gasses emitted by haulage trucks. Norway leads in electric shipping, in part because of cheap electricity and also funds from the NO_x ship pollution tax on shipping to invest in the development of green shipping technology.

The autonomous container ship will demonstrate innovative maritime technologies to enable green shipping operations. It will feature a mooring system that will require no special preparations at docks to enable autonomous berthing and unberthing. The ship's navigation and autonomous operations will be supported by a number of proximity sensors, including radar, a light detection and ranging device, an automatic identification system, an imaging system, and an infrared camera. Cargo loading and offloading will be performed automatically via electric cranes and related equipment. The zero-emission vessel will be equipped with an electric propulsion system powered by a battery pack with a capacity of up to 9 MWh. The vessel's battery pack will act as permanent ballast, eliminating the need for installation of dedicated ballast tanks.

The ship will be monitored and controlled from three operation/control centres, which will handle emergencies and exceptions, observation of conditions, operational monitoring, and decision support. The centres are also expected to manage the surveillance of the autonomous vessel and its surroundings, as well as various safety aspects.

Challenges abound, not least those institutional. The United Nations Convention on the Law of the Sea requires all ships to be "in charge of a master and officers who possess appropriate qualifications," not something an artificial intelligence system on a fully automated ship could fulfil... The International MARPOL Convention and the International Convention on Oil Pollution Preparedness, Response and Co-operation (OPRC) also require that ships have special preparedness plans in place in the event of a pollution emergency. However significant the changes required to allow autonomous ships to operate in the widest possible area, most accept that future regulations must be based on a set of international standards set by the International Maritime Organization. Without a unified code, it is possible that some ports would simply prevent automated ships from doing business.

Cities also can become mobilised. Singapore signed an agreement with the trucking companies Scania and Toyota Tsusho to design, develop, and build self-driving freight lorries to haul containers from one port to another. This idea could lead to a truck with a driver leading a convoy of driverless trucks to reduce congestion and boost productivity in the island's ports.

4.4 Smarter Mobility for Cleaner and More Inclusive Cities

4.4.1 *Competitive, Affordable, Clean, and Safer Mobility and Accessibility*

Transport infrastructures are the arteries for the free movement of socioeconomic vitality in cities. Urban transport is a sociotechnical system with interacting visions, cultural and mobility patterns, technologies and infrastructures, business models, and policies responding to citizen desiderata. As most journeys begin and end in cities, increasing demand for urban mobility has created an unsustainable situation characterised by severe congestion, poor air quality, noise and high levels of CO_2 emissions, and citizen dissatisfaction (UN-HABITAT 2016).

Transport has a central role for societies and economies that strive for sustainable prosperity and benefits more equitably shared among all citizens. The sector has been responsible for the fastest growing source of greenhouse gas emissions and the greatest problems for decoupling economic prosperity from pollutant release. Despite the important technological improvements that transport has witnessed over the last years, the increasing demand outweighed the eco-efficiency gains. Low-carbon transport is an essential element in the transition to sustainable development. The 2016 Summit of the International Transport Forum focused on the changes needed to make transport more inclusive and promote equality of income and opportunity and launched the Decarbonising Transport project together with more than 40 partners. Through a comprehensive computer model of global transport activity, the project aims to assist decision makers in making best choices for decarbonisation (ITF 2016).

In 2013, the European Commission adopted the Urban Mobility Package, reinforcing its measures in the area of urban transport by sharing experiences and best practices, and fostering cooperation, research, and innovation on new and better solutions (EC 2013). 'Europe on the Move,' a wide-ranging set of initiatives to make traffic safer and smarter, aimed at helping reduce CO_2 emissions, air pollution, and congestion, cut red tape for businesses; fight illicit employment, and ensure proper conditions for workers. The long-term benefits of these measures are expected to extend far beyond the transport sector by promoting growth and job creation, strengthening social fairness, and widening citizen choices.

Cleaner mobility is essential for the shift towards post-carbon cities. The EU mobility strategy, launched in 2016, aims at increasing the efficiency of the transport system by making the most of digital technologies, speeding up the deployment of low-emission alternative energy for transport, such as advanced biofuels, electricity, hydrogen, and renewable synthetic fuels and moving towards zero-emission vehicles. Cities and local authorities are already implementing incentives for low-emission alternatives, encouraging modal shift to less unsustainable travel, such as cycling and walking and public transport, and/or shared mobility schemes, such as bikes, car-sharing, and car-pooling, to reduce congestion, pollution, and emissions. Citizens can benefit from more efficient cars that are less energy consuming and from better infrastructure for alternative fuels, improved intermodal links, and reinforced safety. Digital technologies progressively allow citizens making informed choices for daily mobility needs.

Coherent strategies are critical for cultivating the trust needed for the transition to clean mobility. Mobility and accessibility strategies try to reduce the carbon intensity of transport, improve the modal split and give priority to more sustainable modes, that is, walking, cycling, and public transport, while fostering a balanced citizen-friendly multimodal system able to flexibly respond to all needs. Plans usually include integrated sets of technical, political, and social measures to improve the performance and cost-effectiveness of innovative technologies. As all urban sustainability plans, a sustainable mobility plan embraces a transparent and participatory approach and is closely discussed and monitored by all major stakeholders. In particular in coastal cities, the mobility plan has to ensure that all transport hubs linking the maritime, aerial, rail, and road networks are effectively linked and that flows are optimised and seamless.

A comprehensive integrated and coherent approach is a sine qua non condition. Barriers and success factors for sustainable urban transport abound and are often linked to the local context. The main barriers are physical, financial, and institutional hurdles, often related to uncoordinated actions, legal barriers linked for example to the regulations providing the framework conditions, and social and cultural barriers related to public acceptance and side effects. Accomplishments include a well-informed consensual vision and a portfolio of smart, sustainable, and equitable policy measures, public participation, and accountable governance. The International Transport Forum suggested that the lack of cross-industry cooperation, conflicting policies, and missing standards, and the inherent risks of hyperconnectivity, are the main obstacles for seamless transport and travel. It is suggested that these barriers can be overcome through a game-changer combining intermodal travel, a traffic management system for megacities, and the optimisation of the logistic systems (ITF 2012).

Prospective studies are essential to formulate a mobilising vision for the future of a city and its transport models. The mobility projections by the International Transport Forum suggest that global passenger transport volumes in 2050 could be up to 2.5 times as large as in 2010, and that freight volumes could grow by a factor of 4. The strongest growth is expected from emerging and emerging economies. Car use levelled off in OECD countries and passenger transport grows more slowly than the GDP, but this trend could be transitory. In 2000, half of global passenger kilometres were driven in OECD countries, but this share could decline to 20% in 2050 and follow the eastward shift of economic gravity. Emissions of CO_2 grow more slowly because of increasing energy efficiency but could nevertheless more than double (ITF 2012).

The future of mobility and accessibility in cities depends on topography and urban form, culture and lifestyles, local economy, social landscape, and governance. A new culture of sustainable urban mobility/accessibility asks for less car-dependent lifestyles and the promotion of better alternatives. Increasing the part of walking, cycling, and public transport has never been more urgent. Optimal urban sustainability policy portfolios include legislation, the promotion of lower-consumption vehicles and innovative fuels, and propulsion technologies, demand-management schemes, such as parking and access restrictions, fair and efficient pricing, teleworking, and land use and planning. A mixture of compatible land uses is highly important, as this can lead to mutually enriching functions and ensure the accessibility of places (ITF 2012).

The transition to a more sustainable urban mobility system asks for the right policy measures and technologies to influence citizen behaviour. The car fleet may continue in the medium term be dominated by internal combustion engine technology but it is expected to display smaller and lighter vehicles, more hybrid and electric vehicles, and higher levels of autonomy and interconnectedness. Electric city cars could be foldable or adapted to the space requirements and better served by an expanding shared infrastructure. Most fundamental changes are likely to have occurred in the organisation of transport and innovative intermodal mobility services. As the sharing economy expands, car and bike sharing are expected to grow in some urban areas, which could lead to the reduction of urban vehicle stocks and flows. Freight logistics could be optimised thanks to much more nimble ICT applications.

Governance strategies should handle the transport system as an integrated organic whole. Many studies conclude that just using cleaner fuels and propulsion technologies and optimising the efficiency of transport flows is not enough. Successful pathways require not only transformational change in one of the elements of the system, but in several or in all of them, and at the same time (STOA 2012).

The search for a seamless transport system in the urban world helps to identify investment options that provide sustainable value. The upward pressure on mobility demand from larger populations, including a rising middle class with higher income in the emerging world, is crucial. Cities in the emerging world must invest in integrated, less unsustainable transport networks, smarter and greener infrastructures, and the alignment of mobility objectives with socioeconomic and environmental policy objectives. Multi-dividend practices should be highly encouraged, especially in times of crisis and restricted public and private funding.

Access to places, goods, and services is the ultimate objective of transport. Planning and design for sustainable urban mobility should focus on bringing people and places together in healthy environments. The importance of integrated land use and transport planning has never to be underestimated. In 2013, the United Nations' World Habitat Day celebrated urban mobility because access to goods and services is essential to the efficient functioning of expanding cities and the well-being of their citizens. Accessible cities encourage a shift towards more sustainable modes of transport and draw more and more travellers out of cars and onto trains, boats, buses, bike paths, and sidewalks. Mobility is a service depending from far more elements than just the mode of transport. Urban planning and design should focus on creating the conditions for optimal accessibility rather than simply expanding urban transport infrastructure.

Sustainable seamless travel in cities asks for local and regional authorities to reassess the boundaries of geographic and administrative structures and allow for imaginative new mobility services. Infrastructure investment across all modes and innovative funding arrangements are very important. Business processes must be much more open to mode-neutral mobility, and common standards are crucial for interoperability. Sharing risk and liability with public–private partnerships could be crucial. Transparent and clear pricing rules fully aligned with sustainability objectives are fundamental pre-conditions. Public acceptance depends on informed stakeholder involvement and the clear allocation of revenues from pricing schemes to sustainability projects.

Safety is a key question for urban transport. Every minute, a person dies in city traffic. Millions are killed or injured every year, causing great human suffering and costing around 3% of GDP. Crashes also nurture a feeling of insecurity. Among people killed on city streets, eight of ten are pedestrians, cyclists, and other vulnerable road users. Safer streets encourage sustainable forms of transport and help a city reduce pollution, cut emissions, fight congestion, and promote public health. Amsterdam, Auckland, Barcelona, Berlin, Bogotá, Brussels-Capital Region, Buenos Aires, Copenhagen, Dublin, London, Melbourne, New York City, Paris, Rome, São Paulo, Stockholm, and Warsaw created the "Safer City Streets," the global traffic safety network for liveable cities, to improve safety performance by sharing data, experience, and knowledge.

Safer City Streets builds on a 2013 pilot project with nine cities from Europe and North America that agreed to share data on population, mobility, traffic, and accidents. The success of the pilot gave birth to the worldwide network. Safer City Streets is being developed by the ITF and modelled on the global road safety network known as the International Traffic Safety Data and Analysis Group, which has run for more than 25 years. The group has been hailed by the World Health Organization as a model of a multi-country effort. Cities contribute data to the database and enjoy free access to data from their peers, allowing comparisons and benchmarking.

In 2016, La Paz, Bolivia celebrated 15 years of action of the Zebras Urban Educators Project, and won the Guangzhou International Award for Urban Innovation because of the ease of replicability of the project and its dual educational and social function, evidencing the crosswalks or zebra crossings and reducing road chaos. The municipality had wished to work directly with the citizens, given the limited results of impersonal mass campaigns. To reach citizens in a creative and friendly way, a group of young people proposed a character that would emulate a zebra crossing a street at the right place. In a playful way, citizens would notice that there is a space for them to exercise their rights as citizens and to walk safely through the streets. Young street workers, polishing shoes or selling candies, joined to play the animal, with the right training and costumes. The initiative focused on the welfare of citizens and spaces to meet their emotional needs. The zebra has become synonymous to living in harmony with the environment, with gratitude and respect.

Freight transport is pivotal for the socioeconomic functioning of cities, and the efficiency of this sector has a high potential for improvement. Construction works and retail are responsible for much of freight transport. The sector accounts for about 10–12% of vehicle traffic in cities but causes disproportionately higher impacts on congestion and the environment. The importance of urban logistics could become more important in OECD cities, given the rise of e-commerce and the needs of an ageing population. Many cities impose vehicle size or weight restrictions, or limit access in certain areas or during certain times. A balance has to be struck between access requirements for operators and tolerable impact on local populations.

Intelligent traffic management is a growing challenge for world megacities. Seamless transport is being promoted as a smart investment in advanced transport systems that minimise obstacles to interconnection and access to information and services (ITF 2014b). The Condition-based Megacity Traffic Management System integrates and processes information from vehicles, travel infrastructure, citizens, and the social environment. Using real-time data collection and analytics, it helps reduce congestion, greenhouse gas emissions, and fuel consumption and accidents. The total net benefits of the system can range from $2 billion to $10 billion per city (ITF 2014a).

In the European Union, transport infrastructures cover 10–15% of the urban space. Household spending in transport budget amounts to 13% of the total budget. On average, a European citizen makes 1000 trips per year, and half of these are less than 5 km long. According to a 2013 Eurobarometre survey, half of all Europeans use a car every day, which is more than those cycling and using public transport (16%) combined. Across the EU, only a third of respondents use public transport at least once a week. Around four in ten Europeans have problems when travelling within cities, and a substantial majority of Europeans believe that air pollution, road congestion, travelling costs, accidents, and noise pollution are serious problems within cities. The cost of urban trips matters, and more than half of Europeans believe that better public transport and lower prices for public transport would be the best ways to improve urban travel. Less than a quarter of Europeans believe that the urban traffic situation will improve in the future.

The manifesto on sustainable mobility for European regions, towns, and municipalities by Council of European Municipalities and Regions (CEMR), bringing together approximately 100,000 local and regional authorities, suggested some key areas for further reflection and action at all levels. These ideas focus on the optimal use of public transport, the promotion of worthy alternatives to private car use, incentives to manage demand for road space, quality and safety issues, privatisation, public procurement, concessions, and public service obligations, economic instruments, and synergies between urban planning and transport at all levels of governance (CEMR 2005).

Cooperation across networks and governance levels is essential. The European Union has assisted cities to implement and evaluate a broad array of innovations for sustainable urban transport. Since 2002, the CIty-VITAlity-Sustainability (CIVITAS) nurtures political commitment, marketable solutions, and knowledge sharing for more sustainable transport systems. The initiative promoted the sharing of best practice among hundreds of European cities forming a network forum, open to all cities engaged to achieve a sustainable modal shift in their transport systems. By signing the non-binding CIVITAS Declaration, cities exchanged intensively to raise the bar of excellence. The initiative created a network of cities for cities dedicated to better transport in Europe and beyond. Demonstration projects in more than 80 Living Lab cities, Europe-wide, tested and implemented more than 800 measures and possible solutions. The knowledge garnered through these practical experiences is analysed in research and innovation projects towards a more resource efficient, competitive, and inclusive transport system in Europe.

4.4.2 Sustainable Urban Mobility: Walking, Cycling, and Cleaner Public Transport

The concept of sustainable urban mobility plans promoted by the European Commission in 2013 suggests that the ultimate aim is improving accessibility of urban areas and providing sustainable mobility services to, through, and within the city. It considers the 'functioning city' and its hinterland rather than a municipal administrative area. The plan has to be an organic part of a long-term strategy for the future development of the city and its transport and mobility infrastructure and services and contribute to economic, social, and environmental urban sustainability. Typically, a Sustainable Urban Mobility Plan is being developed in close consultation with citizens and stakeholders and features open workshops, surveys, and public awareness campaigns (EC 2013).

Noble pedestrian streets and districts expanded throughout the world and became the hallmarks of many cities. Is the "car-free city" a utopia or a reality? A research by the European Commission at the beginning of the 1990s suggested that the city could be reconceived and redesigned to give priority to pedestrians. A car-free city would be a polycentric one, made up of small islands on a human scale that would be connected by high-speed means of transport. It seems that the car-free city would benefit not only the environment but also to the economy, as it appeared to be two to five times less expensive, dependent on density. Amsterdam, which had gone through a recent referendum on the restriction of the private car, in the early 1990s, organised an emblematic conference "Car-Free Cities?" The question mark is not negligible, but many negative perceptions and reserves have long been overcome (Mega 2005, 2013).

The transformation of European urban hearts into pedestrian areas with cycling paths and human-centred infrastructure celebrated more than five decades of astonishing results. Italian cities were among the first in excluding private cars from their splendid historic centres and introducing "park-and-ride" systems. Some urban mobility plans made an ode to pedestrian areas and streets as shared dignified public space. Despite difficult beginnings, caused by strong opposition from commercial lobbies, car-free schemes gained many cities in the world. Copenhagen was a pioneer city in recognising the sustainable value of pedestrian streets. The creation of central car-free precincts evolved in parallel with the public transport and the bicycle network and with the parking policy and the elimination of 2–3% of the parking spaces per year in the city heart. Gradually and steadily, urban space was freed from traffic and gained by citizens, initiating a tide of return from anonymous peripheries to the historic hub. The pedestrian Stroget area attracted civic architecture, sculptures, fountains, and musical and cultural events, and became the backbone of an archetypal pedestrian precinct (Arcadis 2017, ITF 2011).

Pedestrian vitality of main streets gained also US coastal cities such as New York, Boston, and San Francisco with increasingly safe and inviting streets for pedestrians (NYC 2015a, b). The litmus test of pedestrian-friendly areas include the high attraction of citizens outside office hours and around the clock, a continued expansion of public transit to flexibly respond to the movement of pedestrians in longer trips, restricted parking, and the ending of the mega-garages of the 1950s and 1960s. Widening sidewalks and reorienting and broadening pedestrian walkways across downtown was the first sign of a new era in urban mobility (WRI 2017).

Walking and cycling are the most sustainable transport means. Policy measures and infrastructures to promote cycling expanded in most cities. Since the last decades of the twentieth century, the bicycle has begun to be valued as a component of everyday commuting, urban transport, and recreation patterns. The modal split of urban cycling has grown significantly in all continents. Amsterdam and Copenhagen are endowed with the most elaborate bicycle network, complementing the road and canal routes and integrated into the urban mobility offer. Both cities developed an effective public cycling infrastructure. In Copenhagen, recognised by the International Cycling Union as the first global Bike City, most citizens commute by bicycle.

Is a bicycle the future of urban mobility? In Denmark, the National Bicycle Strategy brought together all stakeholders and created a common language. Investments in bike-friendly infrastructure brought a rise in ridership, with an increase in the feeling of safety. In Copenhagen, bicycles already outnumber cars and are considered by citizens the quickest way to move around the city. The municipality plans to accommodate an additional 60,000 cyclists by 2025 by developing a network of 28 Cycle Superhighways connecting residential areas around Copenhagen with places where people work or study. A Cycle Superhighway is a route where commuters' needs have the highest priority. For instance, cyclists will have a green wave of traffic lights during rush hour. The highways, as direct as possible, cutting down on the number of times cyclists have to stop, offer fast, comfortable, and safe service, with transit areas for public transport. All municipalities were enthusiastic about cooperating in this innovative project and jointly developed a conceptual strategy and a plan for the network of Cycle Superhighways.

In January 2017, a nearly 3-mile bike ride from the World Resources Institute to Washington's National Press Club gave advocates of city cycling the opportunity to give advice on how to make bicycles a healthy, economical, environmentally sustainable mode of urban mobility. Political attractiveness has been suggested as an important step forward, as bike riders can be a mobilising political force making elected representatives sensitive to the cyclists' concerns about safety and infrastructure.

In November 2017, the World Business Council for Sustainable Development (WBCSD) and the European Cyclists' Federation (ECF) announced their partnership within SiMPlify, WBCSD's Sustainable Urban Mobility project, a data-driven, multi-stakeholder approach that helps cities accelerate progress towards multimodal, inclusive, and sustainable mobility. The methodology offers a policy portfolio based on global best practice. ECF, the world's largest cyclist advocacy organisation, and WBCSD engage European cities to develop data-based holistic sustainable urban mobility plans. ECF has helped to optimise the cycling-related indicators and ensure that cycling safety is measured based on exposure data. With 19 indicators identified to describe sustainable mobility in urban areas, the SiMPlify process has been tested in a range of very diverse cities such as Bangkok, Thailand, Campinas, Brazil, Chengdu, China, Hamburg, Lisbon, and Indore, India. In 2018–2019, SiMPlify will be deployed in 50 cities in Europe to support the development of sustainable urban mobility plans. Although the indicators are not focused specifically on cycling, they do look at infrastructure, access to bike sharing, and cycling safety, thus allowing for tailor-made city mobility plans to take into account the needs of cyclists (WBCSD 2016).

Bike-sharing schemes expanded rapidly in hundreds of cities. The "Velib" bike-sharing service has been successful in many European cities and disseminated extensively in the world. Velo-city events and conferences have been held in Copenhagen, Dublin, Barcelona, Perth, and Montreal to promote pedal power in bicycle-friendly coastal cities. Each event offers participants a chance to share best practices for cycling-friendly cities and involve debates on health, economics, and the environment with experts and citizens. More cities associate cycling with sightseeing of historic and cultural places. Bangkok created a cycling path surrounding its historic Rattanakosin Island.

Each city introduces various innovations. Buenos Aires suggested the bicycle's manual, and Hangzhou, one of the most prosperous cities in China, organised a perception survey to understand drivers and barriers to bike sharing. The results suggested that almost one third of the users had incorporated bike sharing into their everyday practices and, most frequently, used a bike-sharing station closest to either home or work. The modal shifts indicated that bike sharing acted both as a competitor and a complement to existing public transport. Recommendations for improving bike sharing included expanding infrastructure and parking services, improving bike maintenance and locking mechanisms, and extending operational hours.

Cycling has critical links with education, health, and safety. In 2012, Boston added 11 miles of bike lanes and more than 100 bike racks, including two new on-street bike parking sites. The New Balance Hubway bike-share system expanded within the city and spread into Cambridge, Brookline, and Somerville. Hub on Wheels, Bike Week, and Bike Fridays were larger than ever, and partnered with Circle the City to offer three Open Streets events. Some of the major objectives concern children and their access to safe pedestrian walkways and bikeways and arts facilities. After extensive collaboration with city departments including the

Boston Police Department and Boston Public Health Commissioner, a substantive safety report has been launched, together with a comprehensive safety and education programme.

The promotion of walking and cycling has often been complemented with car restriction policies, such as road pricing. In Europe, the experience of Oslo, Stockholm, and London, charging motorists entering their central districts, offers many lessons on urban tolls design and execution, public consultation and acceptance, capital and operating costs and pricing, information provision, technological aspects, and environmental impact. In London, the congestion charging scheme introduced in 2003 represented an investment of £100 million in electronic equipment. The annual revenues have been estimated as 20 times higher than the annual cost of the system. Good communication with the citizens and potential users has been crucial.

Park and Ride systems provide a compromise bridging the use of the private car and public transport and have been quite popular in many cities, from Evora to Amsterdam and Warsaw. They offer the advantages of private cars to reach quickly the periphery of cities, where they leave their cars in secure parking places before taking a more competitive transport means to circulate for the inner city. In Warsaw, one of the few cities in which car ownership rates are still growing, the Park and Ride system registered 740,000 cars in 2014. Infrastructure has to be carefully designed to provide cities and citizens the expected shared advantages and reduce traffic pressure on urban ecosystems.

Public transport can only be a worthy alternative to the private car if it is safe, clean, reliable, fast, frequent, noiseless, flexible, easily accessible, well-designed, environment friendly, and affordable. It plays a major role in the larger and denser cities, in which it carries 2.5–3 times as many people as private transport. Public transport is also vital for the important share of households without a car, which reaches 40% in the European Union.

Highly populated world cities such as Tokyo, where only 1% of commuters use their private car, can provide many inspiring lessons. The 2000 urban rail development master plan identified five major policy targets to address the challenges in Tokyo's urban rail market. The first policy target is congestion. The second target relates to the Tokyo Metropolitan Area's decentralised land use policy, which has resulted in satellite business sub-centres. Measures addressed the need for efficient rail connections between sub-centres and reduced commuting time from residential areas to business districts. The third target also relates to the redevelopment of seaside areas near Tokyo Bay and aims to increase rail capacity, particularly in the central business district of Tokyo. Since the 1990s, a number of high-rise buildings were built because of the redevelopment of seaside areas near Tokyo Bay for business and residential use, and the gradual change in preferences of younger generations starting to move from suburban areas to the city centre. The fourth target relates to the globalisation of business and tourism markets. The improvement of rail access both to and from airports, and of high-speed rail, is critical for better business conditions. Finally, the last target reflects Japan's rapidly ageing society and the introduction of new devices and applications for older adults or disabled passengers (ITF 2014b).

All public systems of the world are obliged to become more efficient, smart, green, and inclusive. Many metropolitan underground systems, already more than 100 years old, go through transformational innovations. Driverless operations, state-of-the-art information, ticketing and security devices, mobile communication capabilities, and wireless connections are among the main innovations. Since the 1980s, tramways have returned in many European cities, where they had often been abolished in the 1960s. The tram, abolished in 1960 in Athens, returned to the city in summer 2004, just in time for the Olympic Games. It is a valuable silent transport mode connecting the city centre to the seaside suburbs.

Following the example of Strasbourg, 28 French cities, from Le Havre to Nice, introduced generations of trams. In Marseilles, the original tram system remained operational until the closure of the last line, 68, in 2004. Three years later, a new generation of trams, compared to a boat on rail, started serving the city. More recent developments ensured the links to the rapid train and the old port of Marseilles. Customised Bombardier Flexity Outlook trams, composed of articulated sections, introduced a new element to the urban landscape. The design seems inspired by the hull of a ship, with the driver's cabin resembling the bow. A lighted circle displays the colour of the tram line, and the colour blue dominates the interior design. The City Pass offers access to transport and cultural services, including boats for the Chateau d'If, former royal fortress and notorious prison. The true environmental benefits of tramways have been questioned in many cities, as it seems that it did not impact the modal split, and more cabled and suspended forms (Orsenna and Gilsoul 2018).

Suspended or skytrains have been crowned with success in many developing metropolises. In Bangkok, the skytrain, operating since 1999, offered a great alternative to congestion, linked to the metro and boat connections. In the Ethiopian capital Addis Ababa, the opening of a modern urban light railway or tram service has been met with great public excitement. The Ethiopian Railways Corporation began construction of the double-track electrified light rail transit project in December 2011. A 17-km first line running from the city centre to industrial areas in the south of the city opened in September 2015. It was soon followed by the construction of a second line, and the system became 31.6 km long, with 39 stations and trains expected to be able to reach maximum speeds of 70 km/h.

Buses, including electric and hybrid buses, biofuel and hydrogen buses, and intelligent transport management systems, develop continuously, in parallel with smart transport passes, often linking transport to other public and cultural services. Many local authorities have chosen to contract out public transport services, through tendering processes which incorporate sustainability criteria in the public service obligations. Stockholm managed to achieve a bus fleet that uses 100% renewable fuels. The city promoted the use of clean vehicles running on biogas, ethanol, or hybrids. It also offered incentives to clean vehicles, including free residential parking and cost-free driving in the congestion charging zone (CEMR 2005).

Most urban transport systems are hybrid and need a bold shift from the status quo, one that includes multi-sector collaboration across public and private sectors. Sydney provides an instructive case. Sydney has a fast-growing population and needs to adopt innovative solutions to ensure smart, sustainable, and inclusive mobility for all citizens. Using the World Business Council for Sustainable Development's methodology and globally benchmarked mobility monitoring and evaluation framework, a 2016 report presented a baseline of Sydney's sustainable mobility performance. Sydney's present mobility system performs poorly for GHG emissions, as well as long commuting journeys (an average of 66 min travelling to and from work), exacerbated by congestion that is impacting economic opportunities. Sydney has an efficient mass transit system in places, but also vast urban sprawl, resulting in too heavy reliance on and inefficient use of private vehicles. The lowest levels of community satisfaction (9%) relate to congestion, whereas only 34% of surveyed respondents felt satisfied that transport options help them to reach job opportunities. Advances in mobility technology such as the autonomous vehicle, the shared economy, and innovative transport solutions to congestion would complement the record levels of investment in public transport currently underway in Sydney (Sustainable Business Australia 2017).

In 2013, the City of Oslo decided to make all municipal cars emission free by 2015 and invested in supporting agencies with an interest-free loan when replacing vehicles. More than a quarter of Oslo's municipal fleet is projected to be electric vehicles. Oslo intends to cut emissions from transport by 50% by 2030 compared to 1991, despite the expected population growth, and to achieve emission-free transport by 2050. Paris is seeking to phase out diesel-fuelled buses and move towards a municipal fleet made up of 80% electric buses and 20% biogas buses by 2025. The plan is expected to lead to an 80% reduction of GHG emissions from the bus system and a 50% reduction of the carbon footprint of the whole Paris public transport system (EPRS 2017).

Regulation is an important driver of change. In California, the Zero Emission Bus Rules were instrumental for inventing better buses. The Proterra EcoRide transit bus was the first full-size fuel-cell transit bus to enter service in the US and comply with California's rules. Proterra already introduced a new generation of 100% electric bus in 2014, building on EcoRide's cutting-edge design and engineering. The second-generation bus is built from a lightweight, durable composite that utilises fiberglass and wood. Proterra offers clean, quiet, and citizen- and neighbourhood-friendly vehicles that meet government regulations and local mandates. The company aims at achieving exceptional fuel savings and dramatically reducing maintenance costs.

In 2012, the International Association of Public Transport (UITP) presented the results of a European project on urban buses, aimed to develop a new generation of urban bus systems adapted to the specificities of European cities and achieve key breakthroughs improving the perception of bus transport. The project highlighted the importance of addressing the bus system as a whole, not only vehicles, but also infrastructure and operations. Innovations in digital systems, vehicle design, driver cabins, and bus stations were tested in real city conditions (UITP 2012).

Many cities have adopted fast and clean green lanes exclusively reserved to public transport and, often, priority vehicles or users. Special lanes restricted to all traffic except buses, taxis, emergency vehicles, and cyclists have been adapted in various ways to the local conditions of coastal cities. Priority lanes can also be used by private cars for a price. Dublin introduced high-frequency bus services operated by environment-friendly buses, featuring well-trained staff, high-quality shelters and furniture, kerb alignments, and traffic signals. The system provides a faster, more reliable, safer and cleaner bus service, adequate loading and parking facilities for businesses, and improved safety (CEMR 2005).

Thessaloniki adopted a Sustainable Urban Mobility Plan, in partnership with all key stakeholders, and the mobility priorities for the entire metropolitan area, which consists of 10 municipalities and more than 1 million inhabitants. The Thessaloniki Public Transport Authority benefited from its participation in the EU "Attractive Urban Public Transport for Accessible Cities" project involving urban areas of eight EU countries. The mobilisation and commitment of stakeholders, including central government, regional authorities, municipalities, trade unions, passenger and cycling associations, transport institutes and the university, and an emphasis on the better use of public transport were key elements of the planning process.

A seaborne transport system in the Thermaikos Gulf constituted one of the services, also including an integrated ticketing system, rapid bus transit, bus priority at traffic lights, and awareness campaigns to discourage private cars and promote sustainable mobility modes. The system favours intermodality among the metro, trams, and buses, flexible transit systems including the restructuring of taxi services and integrating parking policy, and a controlled street-parking system, a bike-sharing scheme, and a congestion charge and access to the city centre.

Public waterborne transport, once the primary mode of urban transport in cities, is being rediscovered. Waterways in London, Gothenburg, Oslo, and Hamburg have regained importance. Waterborne transport can become a fully integrated mode of travel that complements other modes in cities. The renewed interest in waterfront development, in parallel with the increased congestion on land-based transport, has provided an opportunity for waterborne transport to play a significant role in integrated urban mobility schemes. Until recently, many waterborne systems operated independently as complementary commuter or leisure service, but started to be designed as organic parts of local urban sustainable planning and mobility schemes (ITF 2009).

The waterborne transport is a unique mode of travel that is often more pleasant and rewarding. Urban cruises are among the most sought after services by sightseers in cities like Venice, Bangkok, Sydney and Amsterdam. Innovative design of boats and stops with dedicated spaces for children, passengers with reduced mobility, and bikes can far increase its attractiveness. In cities where waterborne transport is part of the urban environment, ferry stops are often located in close proximity to land-based transport nodes and city landmarks, encouraging connectivity between the transport modes and inviting tourists to use the public system.

By focussing on service quality and integration with other modes of transport and waterfront development projects, waterborne transport has a bright future in the overall mobility options for coastal cities. Hybrid and solar boats are being more commonly used to reduce emissions, as are fuel cell boats. Nantes and La Rochelle display innovative public transport and promote intermodal mobility. In La Rochelle, the Sea Boat is a regular boat service between the Old Port and Les Minimes, a district built on reclaimed land towards the end of the previous century, with one of the largest marinas in the country. "Yeló" runs an integrated network of public buses, boats, and bikes.

The capacity of urban waterborne transport is not negligible. Venice ranks first with around 110 million passengers per year, the great majority of which are residents. The waterborne system in Lisbon transports around 23 million passengers per year. Waterborne transport is also well established in a number of cities, such as Amsterdam, Istanbul, Stockholm, Gothenburg, Oslo, Hong Kong, and New York. Other cities with much smaller systems also use waterborne transport to promote the attractiveness and accessibility of the urban environment.

Rio de Janeiro enhanced the 2016 Olympic Games as a catalyst to make significant improvements to its transport design. One of Rio's flagship projects is the Morar Carioca initiative, which has the ambition to improve accessibility, health, education, and the environment in the city's informal settlements, the favelas. Favelas, home to the city's poorest residents, are often left outside the city services, often creating a vicious circle of exclusion as the populations that most need social services have no access to the services most needed by them.

Effective links and access to public transport should ensure that residents move and have access to jobs and opportunities within the city. A household travel survey revealed that 90% of residents travel on foot while inside favelas, yet 70% use public transport when outside their settlement. Morar Carioca is the one of the first favela improvement initiatives in Rio which emphasised connecting favela communities to the city via new public lines like the TransCarioca. Effective public transport is expected to provide vital access to jobs and opportunities in the city centre and help ensure that favela residents are connected to the full range of opportunities offered by urban life.

Since 2014, global electric vehicles have more than doubled. More than 2 million electric vehicles are driving on the world's roads. Oslo is often described as the electric vehicle capital of the world. Norway saves approximately 200,000 tonnes of CO_2 emissions annually thanks to its 100,000 electric vehicles, representing just 3% of the total passenger car fleet. The experience of Oslo suggests that to succeed, electric vehicles must bring a clear advantage to the consumer, from buying to using. Therefore, good sale, charging, and parking conditions must be established. The main challenge for the municipality has been to enable citizens wishing to own an electric vehicle and to use it as a daily commuter. During the planning phase, it was crucial to secure a wide political consensus, across all strata of government deciding the long-term investment in charging stations.

Charging stations must be fully integrated within the urban fabric and easily accessed; this requires active interaction and cooperation with the electric vehicle association and private property owners. Oslo established 400 free charging points with reserved parking for electric vehicles in the city over the period 2008–2011. The number of public charging stations soon exceeded 500 and could soon double this number. There are also 758 privately owned charging stations, 323 of which have been implemented with subsidies by the Oslo municipality. Trends in electric car sales illustrate the enthusiasm of the population.

Smart infrastructure for electric cars equipped with a charging battery is gaining ground in cities, as for Zen cars in Brussels. Amsterdam is taking specific measures to make electric driving the norm. The city has the highest density of charging stations in the world, expecting to reach 4000 stations in 2018. Charging points are powered by wind turbines located in the harbour area. Electric car drivers have priority in residential parking permits, businesses willing to switch to electric vehicles are offered purchase subsidies, and electric taxis are granted priority at the Central Station taxi rank (City of Amsterdam 2016).

Automation seems the next frontier, and cities already experiment much. While Dubai is testing many drone taxis, it is important to explore the safety implications. Automated vehicles will not mean zero accidents. But studies have shown that human error accounts for more than 90% of accidents. Google's self-driving cars, for instance, have driven millions of kilometres in the past 10 years. These vehicles have been involved in 17 crashes and found guilty in none of them. As well as saving lives, automated cars can also give back their independence to the elderly and disabled. One of the first test drivers of Google's car was blind. Having the car, and indicating speed and direction with his head, gave him hope of independence. In US, a former racing driver, paralysed from the neck down, was granted the first driving license for a partially autonomous vehicle.

4.4.3 Sustainable Urban Mobility Weeks, Campaigns, and Awards

The European Commission's Sustainable Urban Mobility Campaign sheds light on many innovative alternative mobility initiatives. Since 2002, hundreds of European cities participate, every autumn, in the European Mobility Week, evolving into a global movement inviting citizens to sustainable mobility. The week encourages local authorities and citizens to propose better ways to travel and contribute to making cities more healthy and pleasant places in which to live. The car-free day is the highlight of the week, with the challenge of organising it during a working day! Increasingly popular, mobility weeks have supported cities in creating a more civilised environment with reduced traffic congestion and pollution and more sustainable transport means, including boats. Lasting impact is particularly valued, and participating cities are encouraged to launch at least one permanent practical measure beyond the events of this exceptional week.

The Sustainable Urban Mobility Planning Award aims to reward local authorities that have succeeded in developing a mobility plan addressing the diverse transport needs of citizens and businesses and, at the same time, improved collective quality of life. In 2017, the 5th Award on urban freight focused on integrating the movements of goods and services in sustainable urban mobility planning. The cities that captured the attention of the jury include Brussels for its innovative logistics schemes, Budapest for successfully integrating freight transport within the overall mobility strategy, and Stockholm appreciated for its strategy stimulating the local economy while limiting the adverse impact of urban freight. Brussels, city-region and home to 1.18 million people, received the award for its strategic plan for freight transport that is well integrated within the passenger mobility strategy. Brussels supports private sector initiatives for improved urban logistics, including the use of consolidation centres, to increase efficiency and reduce the number of 200-km trips in the city.

Malmö, the third largest city in Sweden, received the award in 2016 for its efforts in promoting active travel and prioritising pedestrians and cyclists. Malmö has also implemented a series of permanent measures such as turning a major shopping street into a car-free zone, making its city centre safer and more attractive to local residents, and setting up a new bike-sharing system offering 500 bicycles at 50 stations. Malmö is also successful in partnering with local businesses, proving that smart and sustainable mobility can have a positive impact on the local economy.

The Spanish city of Murcia won the European Mobility Week Award 2015 for its endeavours in promoting sustainable urban mobility, and contributing to improved air quality, its proactive approach to enable a safer coexistence between car drivers, pedestrians, and cyclists, and for its intermodal dialogue and solutions, such as promoting bike racks on buses, opening bus lanes for cyclists, and placing a strong focus on electromobility.

Many cities invested in raising public awareness and presenting urban mobility initiatives as part of a plan to improve urban life and not constraining freedom of movement. Many cities such as Porto developed a cutting-edge mobile phone application, 'Move me,' providing travel information to public transport users. The system offers users a recommended travel route based on real-time data on the services by the various public transport providers, local conditions, and intermodal options.

The European Mobility Week 2012 award winner was the Croatian capital Zagreb, praised for its well-organised campaign to promote the week like an urban celebration, a strong link with the 2012 theme (Moving in the right direction) and an array of activities involving more than 150,000 citizens. Citizens took part in urban fitness and educational sessions, rode vintage trams, visited cycle information centres, and helped redesign public parking services for bicycles. The environmental benefits of the week were measured by monitoring air quality and noise intensity and through setting up health check points throughout the city.

Cities in six islands from peripheral European regions are cooperating in a major project that aims to develop innovative initiatives to upgrade local sustainable tourism and eco-responsible transport. DESTINATIONS, a EU-supported sustainable mobility project, involved insular cities in Madeira, Canarias, Malta, Crete, Tuscany (the communities on the small island of Elba), and Cyprus. Participants wished to exchange tools, methods, and smart mobility solutions. The project was invited to join the International Sustainable Tourism Observatory network, and help monitor sustainable tourism mobility.

Network emulation design can reverse a vicious circle in public transport, in which declining demand translates into poor service, into a virtuous circle, in which greater demand leads to higher requirements and more seamless service. Smart ticketing with integrated tariff structures can improve service, reduce fraud, increase capacity, and bring considerable benefits to providers and citizens. Transport for London, the Association of Train Operating Companies and Cubic, took London's Oyster smart ticketing scheme to a new level. Since 2010, customers have enjoyed the benefits of travelling on all major public transport modes with one flexible ticket, covering 250 rail destinations within Greater London. Within 2 years since launch, usage has grown fivefold, from 0.5 million to 2.5 million journeys per week, on surface railways alone. The project demonstrated that well-designed alliances among public and private stakeholders can create a more integrated seamless transport service (ITF 2012).

Public e-ticketing in transport services often integrates entries to other urban and cultural services, such as museums and concerts. A European survey on eco-efficient transport involving consultation of stakeholders and scientists from the transport sector, suggested that respondents see a very high degree of desirability of interoperable electronic ticketing applications for public transport. Still, additional service improvements are necessary to sustain the modal shift. Tourists would particularly profit from integrated ticketing schemes, as they usually value unique experiences that make their visit simpler and memorable (STOA 2014).

The International Association of Public Transport (UITP) advocates for doubling the worldwide market share of public transport by 2025. The initiative "Youth for Public Transport" (Y4PT) cooperates with youths to educate, disseminate, innovate, and improve urban public transport and the cities of the world. The Y4PT began in 2006, when UITP asked young people to report about their experiences and interactions with public transport. Y4PT acts as the voice of young people in urban transport policy making and mentors young citizens wishing to start a career in urban transport services.

Attractive public transport places are important for introducing more citizens, through art, to sustainable noble mobility. Public service companies should not underestimate the importance of beautiful and welcoming urban transport places. Athens and Perugia integrated archaeology galleries in their public transport places. The "Art in Transit" workshops, organised by the UITP, highlight art in public transport and encourage reflection about the impact of quality environments on citizen well-being. Art and architecture can uplift spaces and lives. A most notable example is the Fulton Centre in New York, opened in 2014 as part of the rebuilding of lower Manhattan after 9/11. The creation of the Fulton Centre gave the opportunity to co-create projects completing an extraordinary underground museum of contemporary art that spans the city. The Metropolitan Transportation Authority network suggested that artworks should strengthen an integrated vision and support orientation (UITP 2014).

References

Amsterdam Climate Office (2008) New Amsterdam climate: summary of plans and on-going projects. Amsterdam

Arcadis (2017) Sustainable Cities Mobility Index 2017. Bold Moves. Amsterdam

City of Amsterdam (2016) Plan Amsterdam. The electric city. Amsterdam
City of New York (NYC) (The) (2015a) OneNYC. The plan for a strong and just city. New York
City of New York (NYC) (The) (2015b) Building a smart + equitable city. New York
Council of European Municipalities and Regions (CEMR) (2005) CEMR manifesto achieving sustainable mobility in Europe's towns and municipalities. Paris-Brussels
European Commission (EC) (2012a) Communication on renewable energy: a major player in the European energy market. Brussels
EC (2012b) Smart cities and communities. European innovation partnership. Brussels
EC (2012c) Figures for the future. 20 years of sustainable development in Europe. Brussels
EC (2012d) Blue growth. Brussels
EC (2013) Urban mobility package. Brussels
EC (2014a) Blue energy. Action needed to deliver on the potential of ocean energy in European seas and oceans by 2020 and beyond. Brussels
EC (2014b) Micro-algae-based products for the food and feed sector: an outlook on Europe. JRC Science and policy papers. IPTS, Seville
EC (2015a) A framework strategy for a resilient energy union with a forward-looking climate change policy. Brussels
EC (2015b) Towards an integrated strategic energy technology (SET) plan: accelerating the European energy system transformation. Brussels
EC (2016a) Opportunity now! Europe's mission to innovate. Brussels
EC (2016b) EU Reference Scenario 2016. Brussels
EC (2016c) EU energy in figures. Brussels
Environmental Protection Agency (EPA) (2016) National awareness of energy star for 2016. Analysis of 2016 CEE Household Survey. Office of Air and Radiation. Washington, DC
European Innovation Partnership on Smart Cities and Communities (2016) European manifesto for citizen engagement. Brussels
European Parliament Research Service (EPRS) (2017) Cities: front line of climate action. Briefing PE 608.730. Brussels
European Wind Energy Association (EWEA) (2014) The European offshore wind industry: key trends and statistics 2013. Brussels
Friends of Europe (2017) Cities: the new policy shapers in the energy transition. Brussels
Global Ocean Commission (2016) The future of our ocean. Next steps and priorities. Oxford
Gore A (2013) The future: six drivers of global change. Random House, London
Greenpeace (2005) Decentralising power: an energy revolution for the 21st century. London
International Energy Agency (IEA) (2016a) World energy outlook 2016 special report energy and air pollution. Paris
IEA (2016b) Renewable energy 2016 medium-term market report. Paris
IEA (2017a) Energy technology perspectives. Paris
IEA (2017b) World energy outlook 2017. Paris
International Maritime Organisation (IMO) (2014) Third IMO GHG study 2014. London
IMO (2015) World maritime day. A concept of sustainable maritime transport. London
International Transport Forum (ITF) (2009) Port competition and hinterland connections. Paris
ITF (2011) Pedestrian safety, urban space and health. Paris
ITF (2012) Transport outlook. Seamless transport for greener growth. Paris
ITF (2014a) Connected world: hyperconnected travel and transportation in action. Paris
ITF (2014b) Urban rail development in Tokyo from 2000 to 2010. Paris
ITF (2016) Green and inclusive transport (2016)
ITF (2018) Decarbonising maritime transport: pathways to zero-carbon shipping by 2035. Paris
Mega V (2005) Sustainable development, energy and the city. Springer, New York
Mega V (2013) Quintessential cities, accountable to the future. Innovation, sustainability and leadership. Springer, New York
Mega V (2015) Conscious coastal cities. Sustainability, blue green growth and the politics of imagination. New York: Springer

Orsenna E, Gilsoul N (2018) Désirs de villes. Robert Laffont, Paris
Scientific and Technological Options Assessment (STOA) (2012) Urban transport technology options. Brussels. European Parliament Research
STOA (2014) Integrated urban e-ticketing for public transport and touristic sites. European Parliament Research service, Brussels
Sustainable Business Australia (2017) A business perspective on Sydney's mobility challenge. Melbourne
UITP (2012) European bus system of the future–paving the way for a bus revolution. Brussels
UNEP/ICLEI (2015) District energy in cities: unlocking the potential of energy efficiency and renewable energy. Paris
Union Internationales des Transports Publics (UITP) (2014) Art in transit workshop, London
UN-HABITAT (2016) The New Urban Agenda. Nairobi
UNITED NATIONS (2017) Progress towards the sustainable development goals 2017. Report to the UN Secretary-General. New York
WBCSD (2016) SiMPlify. Geneva
White House (2014) 16 U.S. Communities recognized as climate action champions for leadership on climate change. Washington, DC
World Business Council on Sustainable Development (WBCSD) (2017) Birmingham Smithfield zero emissions city framework, Geneva
World Resources Institute (WRI) (2017) Connected urban growth: public-private collaborations for transforming urban mobility. Washington, DC

Selected Websites and Social Media

www.worldenergy.org
www.citypass.com
www.civitas-initiative.eu
www.compactofmayors.org
www.covenantofmayors.eu
www.denmark.dk/en/green-living/bicycle-culture/cycle-super-highway
www.ec.europa.eu/transport
www.energie-cites.org
www.energystar.gov
www.energystar.gov/topcities
www.erec.org
www.ere-renewables.org
https://eu-smartcities.eu/
ec.europa.eu/energy/index_en.htm
www.ewea.org
www.friendsofeurope.org
www.fubicy.org
www.fuelcell-info.com
www.globalcovenantofmayors.org
www.greenpeace.org
www.iba-hamburg.de/en/iba-hamburg-gmbh.html
www.imo.org
www.ises.org, estif.org
www.itf-oecd.org
www.managenergy.net
www.mobilityweek.eu/emw-award

www.niches-transport.org
www.nyc.gov/html/gbee/html/about/about.shtml
www.nyc.gov/plany
www.savemorethanfuel.eu
www.smartcities
www.smilesprogramme.eu
www.sustainabledevelopment.un.org/sdg7
www.sustenergy.org
www.tfl.gov.uk
www.uitp.com
www.unep.org
www.usmayors.org/cleanenergy
www.usmayors.org/climateprotection/agreement.htm
www.velib.fr
www.velo-city2012.com
www.velocityclub.be
www.wbcsd.org/Projects/Zero-Emissions-Cities/Resources/Birmingham-Smithfield-Zero-Emissions-City-Framework
www.windday.eu
www.worldmayorscouncil.org/the-mexico-city-pact.html
www.wri.org/blog/2017/01/urban-biking-advice-copenhagen-portland-and-beyond
www.zencars.eu

Watercolour 7
Hamburg: with the Philharmonic to the Ocean of the Future

Chapter 5
Towards Purposeful, Circular, and Sustainable Economies

Abstract Cities are places of cross-pollination on a globalising chessboard, the very sites where many diverse economic flows interact and can be converted and intensified to better satisfy the expectations of citizens, businesses, and organisations. Brainports and well-functioning harbours can reinforce both local and global value chains and generate sustainable wealth. Their competitiveness depends on the strength of their functions, their links and networks, and their multiplier effects. Sustainable investments can help reconcile cities and oceans with people and the planet. The evolving ability of cities to introduce new concepts and models and orchestrate the world economy is at the very heart of this chapter, which also sheds light on the burgeoning ocean economy and its prospects for 2030. Generating and capturing synergies and bringing forth sustainability principles are essential for creating better assets, often from liabilities, and services truly fulfilling citizen needs. Eco-responsible blue–green businesses may have a great contribution towards promoting an increasingly zero-waste, post-carbon circular economy and creating sustainable value from ethical citizen and enterprise values.

5.1 Eco-Creative Cities at the Crossroads of Sustainable Shared Values

5.1.1 Cities Positioning to Ride the Next Economy Wave

Cities increasingly gain ground as leaders in the global chessboard. They are not only the stages of socioeconomic operations and environmental impacts, but genuine interconnected players in the theatre of nations and the journey to sustainability. Cities seem on orbit for taking responsibility for the future of territories. Key drivers of change include demographic megatrends, urbanisation, and the emerging middle class in the developing world, the need to tackle inequality and strategic resource scarcity, connectivity, and urban democratic traditions in an era of global democratic governance deficit and declining trust in public authorities (Gore 2013; Mega 2015; OECD 2017; WEF 2014).

V. P. Mega, *Eco-Responsible Cities and the Global Ocean*,
https://doi.org/10.1007/978-3-319-93680-2_5

Cities have long been the world's economic propellers, and, more recently, acted as enablers of the multiple interactions between local and global flows and dynamics, which led to the term 'globalisation.' The collective GDP of world's top ten cities exceeded the total GDP of 162 countries combined, with Tokyo, the world's largest urban economy, generating more wealth than Canada, and followed by New York, producing a larger GDP than India or Mexico. National economies have become networks of metropolitan economies that produce disproportionally more wealth than their population share (EUISS 2012).

Eco-responsible cities can become the brains of the next world economy, an economy with a soul, at the leading edge of the global sustainability agenda. They have to offer citizens and investors convincing security and credibility, infrastructure and services, and quality of work and life. Robust urban infrastructures and leading-edge institutions can greatly improve the position of cities in an equitable globalisation horizon. Conversely, shortcomings in urban hardware and software are important constraints, whose cumulative effects could drain the potential for sustainable development (Mega 2013, 2015; OECD 2017).

The strength of the future world will be as solid as the weakest link. Democratic cities are stepping up for sustainable development and trying to create a recession-proof resilient economy and society in balance with the planet. Networking with their stakeholders and among themselves is of utmost importance. Especially in times of global democratic deficit, visions, ethics, and actions for sustainability offer cities the opportunity to act as enduring democratic spaces between the world macro-regulations and the micro-regulations of the local communities (Florida 2005; Gore 2013; Mega 2013).

Many metropolitan areas are dynamic hubs of production, consumption, and trade. An analysis of per capita income and employment changes during the year 2010–2011 for 200 of the world's largest metropolitan economies, which account for 14% of global population but almost one half of global output, reveals that 90% of the fastest-growing metropolitan economies among the 200 were located outside North America and the European Union. In almost all regions, metropolitan areas generated disproportionately larger shares of wealth and employment. Most of them significantly outperformed the national average on income growth, whereas several others significantly underperformed on employment growth. In the US, the 53 largest metro areas, with populations over 1 million residents, have accounted for 93.3% of the nation's growth since the 2008 crisis and 96.4% since 2014 (Brookings 2016).

During the second decade of the millennium and as more emerging world cities enter the stage of intensive growth, their contribution to the global wealth is impressive. Through a combination of investment in physical and built capital and consumption, rising cities could contribute up to $30 trillion a year to the world economy by 2025. The 600 cities making the largest contribution to a higher global GDP, the "City 600," including more than 440 cities in emerging economies, the "Emerging 440," could generate nearly 65% of world economic growth by 2025. The Emerging 440 could account for close to half of overall growth (MGI 2011).

Leading cities propose powerful models for the future. The concept of a circular city in a globalised world is gaining ground for progressing towards sustainable development, with its economic, social, and environmental dimensions. The World Economic Forum (WEF) proposed a fourfold taxonomy for city competitiveness, including institutions, policies, and regulation of the business environment, hard and soft connectivity. The WEF Future of Urban Development initiative focused on the transformation of cities to enhance their chances arising with advancing globalisation. It worked hand in hand with the Cities of Tianjin, Dalian, and Zhangjiakou to address the most pressing societal challenges and with the Government of India in the development of the planned 100 Smart Cities. The Initiative suggests that cities are the very places to link the production and consumption patterns in a circular way. Cities can orchestrate the transition away from the traditional linear economy model of 'take-use-dispose' and can, as major aggregators of fluxes, ideas, and actors, reuse and recycle resources (WEF 2014, 2018). Pioneer cities have started preparing vision and action agendas in partnerships with all their stakeholders (City of Amsterdam 2013, 2014, 2015; EC 2017b).

Early collaborative engagement with all stakeholders leverages the joint capability to identify innovative and cost-effective solutions to complex, cross-cutting urban sustainability challenges. Major transformations have to be brought to the design, construction, and operation of urban infrastructures, including buildings, energy, mobility, telecommunications, water and resource management services, and the optimisation of the interlinkages between these systems. Businesses that are committed to sustainability and effective solutions can help cities navigate challenges and turn a high-level vision into implementable action plans (WBCSD 2014).

5.1.2 Maritimisation, a Reinvented Source of Wealth

Oceans contribute much to the economic prosperity of cities. Maritimisation has been the first form of globalisation that intensifies exchanges among world citizens and socioeconomic actors. Because 90% of world trade takes place by sea, a great part of the world's prosperity goes through port cities. Many port cities have historically been strongly linked to global value chains. They operate in an international horizon, competing with other world players for investment and opportunities. Their competitiveness depends on a multitude of endogenous and exogenous factors, including education, training, and skills, ability to innovate, flexibility of the labour market, capacity of physical and digital infrastructure, governance and leadership, the macroeconomic horizon, and openness to markets and investment (Port of Rotterdam 2012).

As population, economic activities, and infrastructure concentrate along most coasts, eco-responsible coastal cities have to lead the preservation of the integrity of those natural and marine resources that increase the liveability and value of these places highly desirable to live, work, and visit. From leisure activities to coastal and marine fisheries and nonconventional security issues of food, water, and energy, ports support an array of activities, but are often located in the heart of sensitive coastal and marine ecosystems (NOAA 2013).

The strength of port operations, their maritime links, and hinterland connections can be important strategic assets for a city. The economy of landlocked cities can also benefit much from the ocean industries, as coastal cities also benefit much from efficient links with other sectors operating outside the port region, for example, electronic equipment. Sometimes, only a limited part of the linkages take place in the port region, with a larger share in the main economic centres of the country, especially in countries where the main metropolitan area is relatively distant away from the port, such as Greater Paris from Le Havre.

The economic capital city and the main harbour sometimes form a bipolar conglomeration, as do Athens and Piraeus, with well-defined, often complementary functions and roles. The integration of ports in metropolitan areas and regions is crucial for the mitigation of negative effects, mostly related to the environment, land use, and traffic congestion. These impacts can be very substantial in extended ports such as the port of Antwerp, which occupies more than a third of the surface of the city. In very concentrated dense ports, such as Hong Kong, the majority of the SO_2 emissions are related to shipping. Most of the negative environmental impacts concentrate close to the port area, but air emissions expand on a regional scale. Noise from ships and port operations can also produce a kind of 'acoustic smog' expanding beyond the port, whereas congestion patterns depend much on the local and regional conditions and transport policy (OECD 2013a).

Many port cities went through years of transition, which questioned the future of traditional activities (Economist 2013). Usually, port operations include industry and refineries, shipbuilding and repairing, and passenger and freight transport and services. Port cities increasingly yield more value as environmental policies and incentive schemes have reduced a variety of negative impacts, including air pollution, emissions, noise, and congestion. Port relocations have often freed up central urban space for other public and private, residential and cultural functions. Seafront development has frequently capitalised on maritime heritage and created new opportunities for mixed developments involving housing and noble spaces for urban fulfilment (City of Stockholm 2014).

Traditional ports organised around shipyards went through profound transformations following environmental measures and the digital revolution. At the beginning of the millennium, China has become one of the most influential shipbuilding countries in the world. The restructuring and upgrading of China's shipbuilding industry is expected to focus on accelerating innovation, promoting high-end products and enhancing international market share. Location and specialisation are very important factors for sustainable shipbuilding. An extraordinary site, Gibraltar hosts renowned and extensive shipyard and repairing facilities. Gibraltar claims a broad range of expertise and activities to ensure year-round quality repairs and services.

Cutting-edge shipyards for niche markets have emerged in Europe, and maritime industry continues at the forefront of innovation and excellence. Helsinki became specialised in steamships, car ferries, and icebreakers, Gdansk in ship hulls, and the Kiel region in container vessels, submarines, and luxury yachts. Arctech, Helsinki's shipyard, is focussing on vessels equipped to work in ice conditions. The shipyard has been using Russian facilities to build most of the blocks before the vessels are assembled, painted, and fitted out in Helsinki. Baltika, the icebreaker and oil spill response vessel, is known for its unique asymmetrical aesthetics, linked to its functionality: one side of the vessel is an oblique icebreaker, and the other side has oil recovery equipment for open seas.

Eco-responsible port cities introduce an array of policy instruments such as incentives and free zones, training and education, platform organisations, and knowledge transfer and sharing schemes to attract high added value companies that could make the city an international maritime services centre. Singapore is a good example of integrated socioeconomic, industrial, maritime, and environmental policies attracting investors to a port-city-island. In 1986, it became the first world port, followed in Asia by Yokohama and Busan. In 2017, Shanghai came to occupy the first world place, followed by two ports on the same Asian coast side, Singapore and Shenzhen. In Europe, Rotterdam reinvented itself as an industrial sustainability frontrunner, and Bremerhaven tries to become a global leader in renewable energy. Evolving digital sustainable infrastructure is most necessary to support the sound ecological development of port operations and the prevention and minimisation of waste and pollution.

Metropolitan port cities with high-quality infrastructure, green and public spaces, a sound coastal environment, and healthy residential areas have great potential to attract investors, citizens, and visitors. Effective governance depends on leadership at all levels, from the national government, a government at the metro-regional level, and local authorities and networks that mobilise nongovernmental actors, civil society, and businesses. Many cities promote multi-stakeholder public–private partnerships for public projects that enhance aquatic environmental assets. Spatial planning and financial mechanisms for the sustainable redevelopment of waterfronts can create functional mix, and integrate high-quality residential areas, cultural services, and noble public spaces.

Smart infrastructure is essential for coastal cities to underpin the blue–green economy and attract sustainable investments (WBCSD 2013, 2014). Public policies and public–private partnerships enriched with quintuple helices can be effective in increasing the twofold 'port-city' sustainability performance. The triple helix model is focussed on university-industry-government interactions, whereas a quadruple model also involved civil society, and the quintuple helix models added relationships with the natural environments of society. Several port authorities have also started to monitor environmental trends and impacts and promote balanced environmental and economic development through dedicated sustainability agendas. Transport and land use policies, and multi-stakeholder cooperation, can have great impacts.

Cities must lead in defining long-term policy objectives, businesses have to inject vigour and innovation, and educational institutions must instill a culture of striving for excellence. Maritime clusters have already decades of existence in some maritime countries and many of them can act as innovation ecosystems attracting and integrating new sectors. They most often involve major maritime businesses, including port authorities, dredging companies, research and innovation institutes, and start-ups and supporting services.

The Port of Amsterdam is an example of smart symbioses of different services and utilities, which can reduce energy and costs. Industrial symbiosis is especially capitalising on synergies among material production, infrastructure, transport, and energy. The Port Authority acts as an infrastructure manager for all industries located in the harbour, supported by a specialised infrastructure for the optimisation of incoming and outgoing materials, feedstock, and energy. New forms of cross-sector and cross-actor cooperation and shared infrastructures and resources allowed for new models to emerge (Port of Amsterdam 2009).

Ports often act as magnets for new industries and workers. Across the Mediterranean and Black Sea, some 600–700,000 people are estimated as potentially employed in maritime clusters. Mainstream cluster policy tends to focus on specialisation, whereas diversification appears to be critical for maritime cluster development. Maritime cluster development touches upon an array of sustainable development policies, ranging from transport and physical planning to skills development, education, employment, health, and safety. Proximity underpins multiple helix approaches and business-to-business and research cooperation, and cluster activities can further enhance this process and build new value chains of products and services (Ecorys et al. 2014).

On the European Atlantic front, Brest is at the heart of France's largest maritime cluster, the 'Pôle de competitivité mer,' a network of 300 partners, including large companies, public and private laboratories, and universities located in Brittany. With more than 2700 km of coastline and 95% of the population living less than 60 km from the sea, regional culture, identity, and prospects have a crucial marine dimension. A long-term vision for sustainable development led to regional strength in sea-related activities and created 40,000 jobs. The cluster intends to respond, through innovation, to sustainable development and safety, and make a difference in international markets. A certification scheme allows projects to attract capital for excellence, including maritime safety and security, naval and water activities, fossil and renewable energy resources, marine biology resources, fishing and aquaculture, and port, infrastructure, and maritime transport.

The shipbuilding and repairing sector, the fourth largest industrial sector in Brittany, involves the construction and repair of navy, military, and civilian vessels for fishing, cruising, oceanographym and passenger transport. Leading companies anchor the region with world value chains in naval design and construction, digital equipment, and system integration and associated services. The sector has experienced many waves of world crises and is searching a new future in terms of resources, functions, and markets.

5.1.3 Blue Cities as Frontrunners of the Circular Economy

The linear economy has had its time of glory. It has been powered by the illusion of endless growth. At the beginning of the linear value chain appeared the spectrum of scarcity of resources, and at the end of the chain now there is the piling up of waste, in both cities and the ocean. The circular and shared economy, as sustainable development, asks cities to invest in generating wealth from waste, make assets out of liabilities, and engage in bringing activities and actors together. An EU-supported innovation SME project on "Upscaling the Oceans" intends to produce and sell fabrics and clothes made from marine plastic litter by recycling and converting these plastics into high-quality textiles. Symbolic actions matter much. An island made from plastic bottles was unveiled in the Bay of Cascais, in 2016, with the aim to promote awareness of the huge issue of the Plastic Ocean. This striking work of art, constructed by the artistic group Luzinterruptus, invites social and environmental reflexions. The 10×15 m^2 island, composed of 5000 bottles from a local recycling plant, was placed 100 m off the coast for almost a week and attracted much public attention. At night the island was lit up, making it stand out against the surrounding dark water. In 2014, Luzinterruptus had highlighted the problem of plastic waste and consumerism with a maze made from 6000 discarded bottles in Poland and a Christmas tree composed of 2000 shopping bags in England.

BlueCity, an incubator for circular entrepreneurs in and around Rotterdam, has the aim to help sustainable entrepreneurs to grow from intention to impact. The hub is situated in a deserted former subtropical swimming oasis, Tropicana, an iconic building that has lost its function and magnificence, now open to entrepreneurs, researchers, neighbours, the government, and educators. The purpose is to provide startups and scale-ups with access to circular resources, knowledge, and talent, to turn ideas into action, and ultimately lead to a better society. BlueCity is the breeding ground for 16 innovative, circular companies linking their resource waste streams to create a tangible and inspiring example for the circular economy.

The BlueCity tries to contribute to a resourceful world and create endless circles of value from an interwoven and unbreakable ecosystem. In this playground, sustainable and circular entrepreneurs envision and practice an economy without a particle of waste. They are giving new meaning, function, and value to deserted spaces, transforming the building as circular as possible, reusing old elements and materials both from within the swimming pool and from other abandoned buildings. Inside this ecosystem, waste streams from one activity become material for other actors, and the outputs of some provide the inputs for some others.

In BlueCity, the produced coffee waste serves as nutritious soil for the cultivation of mushrooms, to be tasted in the same local restaurant. The carbon dioxide that is released in the process is used for the creation of spirulina, and in BlueCity Lab, a mycelium is used to develop packaging materials. The beeswax of an urban beekeeper is used by a furniture maker, and plastic neighbourhood waste is recycled into products that the neighbourhood can use. Fruit waste from the market is collected and turned into leather material that can be used for bags and books. A slow fashion label recycles car and bicycle tires into accessories.

Creating new ways to stem the tide of plastic entering the ocean is a noble aim. A circular economy allows plastic, along with many other substances, to be recovered, reused, and recycled, instead of used once and then thrown away. A repurposed new plastics economy is being stimulated through innovation awards, such as those launched in 2017 by The Prince of Wales and Dame Ellen MacArthur.

The leading ideas included approaches which, when combined, could have significant effect in improving resource management and reducing plastic pollution in the ocean.

Rethinking grocery shopping is a prime challenge. Supermarkets are still full of single-use plastic packaging to keep groceries safe and fresh. Yet by rethinking the whole system, innovators can design out waste. An app invited shoppers to order the exact quantities of the groceries needed, which are then delivered in reusable packaging from the producer to their closest store or their home. A social enterprise in Chile offers products in small quantities in reusable containers across a network of 1200 local convenience stores. Redesigning sachets is essential, as hundreds of billions of them are used each year to get small quantities of food and personal care products, mostly in emerging markets. These sachets are not recycled and many end up to the ocean. An Indonesian start-up designed food wrappings and sachets made of a seaweed-based material that can be dissolved and eaten, used for an array of products from instant coffee to flavouring for noodles.

Reinventing coffee-to-go is also crucial, given the fact that, more than 100 billion disposable coffee cups are used globally every year and still almost none of them, nor their lids, are recycled. CupClub, based in the UK, introduced a reusable cup subscription service, in which reusable cups can be dropped off at any participating store. And TrioCup from the United States offered a disposable paper cup made with an origami-like technique that removes the need for a plastic lid. The team has chosen a 100% compostable material and is working on an alternative that is 100% recyclable.

Beyond formal circular projects, cities are the natural incubators for any kind of pioneering startups as, for example, those that emerged from the 2015 Laudato Si' call by the Pope Francis to care for our common home. The creation of Laudato Si' Challenge launching start-ups, which are developing sustainable ways to solve the global climate crisis by 2030, already had some promising results across eight different categories: energy, food, water, human potential, conservation, urban solutions, health, and industry and finance. Of the 300 early-stage companies from all over the world that applied to the Challenge in 2017, 9 were selected to attend the 8-week accelerator programme in Rome. Most of them work in the developing world and 56% of them are created by women (Pope Francis 2015).

The wealth of submitted proposals is extraordinary. AQUS has engineered an affordable, portable water filter that can process 99.9999% of all pathogens, providing a solution to the burden of boiled water and the expense and waste of bottled water. Pāpr creates paperless workflow solutions for enterprises, and Innov8tia's patents microwave pyrolysis transforming toxic sludge into energy and water. Mandulis buys agricultural waste from farmers and transforms it into clean cooking fuel and energy which is deployed to off-grid, rural areas in Uganda. PROTRASH pays community members for recyclables with cash cards that can only be used for food, medicine, or other basic necessities. SmartYields assists farmers to protect their crops and their profits while ensuring food security in their communities. Scooterino is the first ride-sharing platform in Europe using scooters. Nokero, short for 'No Kerosene,' is a durable solar light, built for and affordable to the populations that need it the most. RISE Products up-cycles wasted grain from breweries, turning it into a delicious high-protein, low-carb, and cholesterol-free flour.

The city of Amsterdam has advanced towards becoming one of the world's first circular cities. The municipality organised a 'City Circle Scan' to identify areas in which Amsterdam can make the most significant, tangible, and impactful progress in applying circular business models. The scan concluded that implementation of material re-use actions has the potential to create €85 million of value per year within the construction sector and €150 million of value per year with more efficient organic residual streams. The material savings could bring nearly 900 thousand tonnes per year, against the current annual import of 3.9 million tonnes by the region. Finally, increased productivity levels have the ability to create up to 700 additional jobs in the building sector and 1200 additional jobs in the agriculture and food processing industry. The project provided a more comprehensive overview of flows that move through the city, developing a circular decision framework, and gave a clear direction of priority actions where the next steps can be taken. The strategy promoted a cooperative model that engages cross-sector organisations, companies, city governments, and NGOs to create pilot projects and proven circular concepts in the real world (City of Amsterdam 2014, 2015).

5.2 Established, Reinvented, and Burgeoning Blue Greener Activities

5.2.1 *The Blue–Green Economy Can Only Bloom If Sustainable*

The blue growth concept has attracted much attention in this decade. The European Union, the Organisation for Economic Co-operation and Development (OECD), and important organisations emphasised that, because marine resources and ecosystems are already under much pressure, the blue economy needs to be solidly anchored in sustainable development. In the long term, economic and social development without sustainable management of natural resources, terrestrial and marine, is brittle. Policy makers should set the ground and establish the framework conditions to allow new breakthroughs to emerge and overcome institutional or socioeconomic inertias. Investments in education and innovation are most important, and global governance is necessary to prevent the tragedy of the blue commons (EC 2012, 2017a; OECD 2016).

There has been a rising consensus that blue growth is a marine-based economy that offers social and economic benefits for current and future generations, while preserving the diversity, productivity, resilience, and intrinsic value of marine ecosystems. A sustainable blue economy should be science based, holistic, innovative and proactive, inclusive and participative, transparent and accountable. All stakeholders, including governments, businesses, non-profit organisations, investors, consumers, advocacy groups, and citizens must co-develop or apply gold standards, guidelines, and best practices, to align their activities with sustainable development (WWF 2015a, b, c).

Blue–green growth means fostering economic growth and development while ensuring that natural marine assets are wisely managed to provide the goods and services on which human well-being depends. The European Blue Growth Strategy adopted in 2012 suggested that the seas hold an untapped potential for invaluable new products, drugs, services, and processes. Since then, many actions have been deployed to unlock and realise the potential in a sustainable way through innovation, facilitating the cooperation across frontiers and sectors, and stakeholder involvement to ensure the preservation and sustainable use of the marine environment (EC 2012, 2017a).

Although the term blue economy has been used in different ways, the World Bank suggests that it comprises the array of economic sectors and related policies that together determine whether the use of oceanic resources is sustainable. An important challenge of the blue economy is thus to understand and better manage the many aspects of ocean and coastal sustainability. As an ocean knows no frontiers, a second significant issue is the realisation that the sustainable management of ocean resources requires international collaboration involving all stakeholders (World Bank and UN Department of Economic and Social Affairs 2017).

The acceptable balance among the economy, society, and the environment is a fundamental challenge for the twenty-first century political strategy. 'Business as usual' is unwise and ultimately unsustainable, involving risks that could impose constraints on further growth and development. Cities are ideal hotbeds for blue green–policies, not only through new technologies but also new partnerships and governance models. Risks include increased resource shortages, air, water, and soil pollution, and irreversible climate change and even disasters, accompanied by a myriad of ripple effects. The way forwards requires a balanced policy portfolio to overcome deeply entrenched individual and societal behaviours and to develop the strategy and plans, policy instruments, and indicators to monitor progress (World Bank 2012).

Unsustainable human activities and climate change are at the origin of damaged ocean health and the collapse of critical marine ecosystems. Sea temperatures, ocean acidification, and sea levels are rising and ocean currents are shifting, resulting in biodiversity loss and perturbance of the ecosystems, changes in fish stock composition and migration patterns, and a higher frequency of severe ocean weather events. The prospects for future ocean development are further aggravated by land-based pollution, in particular agricultural runoff, chemicals, and plastic pollutants that feed into the ocean, as well as by overfishing and resulting depleted fisheries stocks in many seas of the world.

Maritime and marine activities are primarily driven by developments in global population, growth, trade, and income levels, climate and environment, science, technology, and innovation. Even if the world enters a lower-growth phase, wealth is expected to rise significantly during the next few decades. Global trade could triple by 2050, with an increasing share of South–South trade, which could bring a new impetus to the shipping and shipyard industry. With the hypothesis that an expanding share of world production will continue to be located in China, India, and Indonesia, reaching almost 40% by 2030 and around 50% by 2050, carrying increases in incomes and wealth especially in burgeoning middle classes, a gradual shift in trade eastwards is expected. Higher incomes and upward consumption trends could bring greater demand for marine tourism and especially for cruise tourism in Asia. Such trends could also generate a need for high-quality protein seafood, which could only be provided through a significant expansion in aquaculture (OECD 2016; OECD-FAO 2017).

Opportunities for blue–green goods and services are fast expanding and can offer the multiple benefits of environmental quality and job creation. Blue growth aims to increase food, energy, and water security, support sustainable management of aquatic resources, and achieve global, regional, and national impact. Most drivers for blue–green growth are linked to the awareness of climate change and the need for preservation of the integrity of ecosystems and habitats. Obstacles mainly involve social and political inertia and a lack of investments. Fiscal instruments, such as environmental taxes, pollution charges, subsidies for renewable energy technologies, coastal and offshore developments, and tax incentives can be crucial in promoting a blue–green economy (World Bank 2012).

A sustainable blue economy has to address specific challenges. The marine environment and ecosystems are more multidimensional and dynamic than land ecosystems. Difficulties include the density of marine water that prevents remote sensing technologies to penetrate deeply, the extent of marine life, from the sea surface down to the deepest ocean trench, the fluidity, mobility, and interconnection of species and habitats, and the lack of ownership and responsibility in the ocean that can have more serious consequences for sustainability than on the land. Marine species can potentially travel much longer distances than terrestrial biota, making the management of related activities very challenging. Animals in the water column can move from one location to another, complicating mapping and protection measures. Nutrients and pollutants can be retained for a long time until returned by ocean circulation, with a longer-term impact (OECD 2016).

To advance blue growth, cities must catalyse investment and innovation which underpin sustainable development and give rise to opportunities from the sea. Many European Green capitals exemplify all the components that make a city a propeller of the blue–green economy. Stockholm, Amsterdam, and Copenhagen provide models at the vanguard of sustainable development that have yielded multiple benefits. Blue–green growth is seen by these capitals as an opportunity to empower and resource cities (OECD 2013b).

The Ocean Innovation Challenge, at the 2017 Economist's World Ocean Summit, called on innovators to share ideas on bridging the gap between sustainability and economic growth. The initiative focused on financing responsible, sustainable investment in the ocean. The award recognised the Seychelles blue bond project by the Government of the Seychelles in one of the world's biodiversity hotspots, an archipelago of 115 granite and coral islands with an exclusive economic zone (EEZ)of approximately 1.4 million km^2, almost 3000 times the size of its land. Fishery is the second most important blue sector after tourism, contributing 20% to the GDP and employing 17% of the population. Seychelles is also a major seafood processing hub. In 2014, the export of consumable fish and fish products constituted 96% of the total value of exports. However, pressures from the fisheries and tourism sectors on coastal and marine natural resources are reaching unsustainable levels.

A sustainable blue economy offers the country the opportunity to preserve and sustainably manage its marine resources, committing to protect 30% of its exclusive economic zone by 2020, ensure food security, diversify the economy, and create high-value jobs. The world's first blue bond is expected to mobilise public and private investments to finance the sustainable fisheries transition and contribute to achieving the country's Blue Economy strategy. The blue bond financing includes €5 million from the International Bank for Reconstruction and Development and a credit of $5 million from the Global Environment Facility (GEF). In September 2017, the World Bank approved more than $20 million to improve fisheries and marine resources management in the Seychelles, which includes support for the blue bond to strengthen value chains and prepare the fisheries sector for investment. The EU flagship Global Climate Change Alliance Plus (GCCA+) Programme, funded under the 10th European Development Fund framework, supports the country's resilience and vital tourism economy. The GCCA+ programme is investing in building sea defences to protect beaches and hotels, as well as helping farmers preserve the scarce rainfall for irrigation.

Ocean and social stewardship should go hand in hand. The 2018 Economist's World Ocean Summit unveiled the Sustainable Blue Economy Finance Principles, developed through a broad consultation led by the European Commission, WWF, the Prince of Wales's International Sustainability Unit, and the European Investment Bank. If widely embraced, these voluntary, long-term-oriented principles provide a framework for transforming the ways in which humanity manages ocean resources.

The "Blue Invest 2018," the first match-making event in Europe between companies and investors in the Blue Economy, is organised by the European Commission, wishing to make technology and creativity meet finance. In #BlueInvest2018, selected entrepreneurs will have the opportunity to present their projects to a panel of investors, to examine promising projects from startups, scale-ups, and small-to-medium enterprise (SMEs), and to build new networks.

5.2.2 Sustainable Ocean-Based Industries and Marine Assets and Ecosystem Services

The ocean economy encompasses ocean-based established and emerging industries and also the marine assets and ecosystem services, offered for free. Current approaches to the ocean economy are underestimating its contribution, particularly the value of non-market goods and services. Blue growth has to be founded on many delicate balances for the sustainable use of seas and the marine ecosystems. The OECD landmark study on the ocean economy for 2030 outlined a range of critical global factors and changes which, individually and in interaction, have a key role in shaping the context within which the balancing act between ocean use and sustainability is likely to evolve during the coming decades. This study distinguished established and emerging ocean-based activities, presented estimates for added value and employment worldwide, and proposed a set of recommendations for governments and stakeholders that could significantly enhance sustainable ocean management.

The ocean economy appears as an increasingly vibrant and vital part of the global economy. Preliminary calculations on the basis of the OECD's Ocean Economy Database valued the ocean economy's output in 2010 at $1.5 trillion, or approximately 2.5% of world gross value added (GVA), the size of the Canadian economy the same year. Offshore oil and gas accounted for one third of the total added value of the ocean-based industries, followed by maritime and coastal tourism, maritime equipment, and port operations. Direct full-time employment in the ocean economy amounted to around 31 million jobs in 2010, roughly the labour force of France that year. Industrial capture fisheries accounted for more than one third of the total, and maritime and coastal tourism for almost one quarter, of global ocean employment (OECD 2016).

The OECD suggests rapid growth of most ocean industries over the next decades. The size of the ocean economy could then be compared to that of Germany, in terms of contribution to both growth and jobs in 2030. Many of the ocean-based industries have the potential to outperform average world economic growth as a whole, in terms of both added value and employment. The projections indicate that between 2010 and 2030 the ocean economy could more than double its size, surpassing $3 trillion. Employment in ocean-based industries is expected to increase by 30% by 2030, outpacing the increase of the global workforce and reaching approximately 40 million full-time equivalent jobs in 2030. But the great economic potential of the ocean economy can only be achieved through responsible and sustainable development. Unsustainable overexploitation of the ocean resources threatens the welfare and prosperity of humankind as well as the future of the oceans. Irresponsible developments place increasing strain on the ocean environment and ecosystems with significant challenges to ocean management (OECD 2016).

The estimates of the ocean industries' contribution to economic output and employment are, in general, very conservative, not least because several important functions of the ocean economy are not included, this in both established and emerging activities. For example, marine business and finance, ocean surveillance, and marine biotechnology are not captured for lack of robust data. All ocean-driven value chains could greatly benefit from advances in satellite and underwater technologies, sensors and imaging, automation, and big data analytics.

Maritime and coastal tourism, including the cruise industry, is expected to be the largest ocean industry (26%), followed by offshore oil and gas exploration and production (21%) and port activity (16%). Among ocean activities, particularly strong growth is expected in marine aquaculture, offshore wind energy, fish processing, and shipbuilding and repair. Less strong growth is expected in fisheries and offshore oil and gas. The fastest growth in jobs is expected to occur in offshore wind energy, marine aquaculture, fish processing, and port activities. Ocean renewable energy and marine biotechnology are considered to hold a high potential; however, the scaling-up of these aspects is unlikely to happen before 2030.

Fisheries constitute the oldest ocean activity. Growth in global production of capture fisheries has more or less stagnated since the mid-1990s, hovering consistently around the 90 million tonnes mark (80 million tonnes for marine fish). The joint OECD-FAO outlook for 2026 suggests that the situation is expected to continue at least for the next 10 years. Since the beginning of the FAO stock assessments, the proportion of assessed marine fish stocks exploited within biologically sustainable levels declined from 90% in 1974 to 71% in 2011, and the proportion of overexploited stocks has risen from around one tenth in 1974 to less than one third in 2011. Fully-fished stocks accounted for 61% and under-fished stocks for slightly less than 10%. In addition, there is the crime of illegal, unreported, and unregulated (IUU) fishing. The OECD suggests that the world would gain 13% more fish in volume and at least $50 billion per year if all depleted stocks were recovered (OECD 2015; OECD-FAO 2017; GOC 2016).

Most dramatic developments in seafood production will be through aquaculture, making this industry an increasingly important component of global food security and a potent driver of change. Looking further ahead, the World Bank's baseline projection also expects aquaculture to continue to grow over the period to 2030, albeit at a decelerating rate, falling to less than 2% per year by 2030. Nonetheless, in terms of food fish production, it could account for 62% of the global supply destined for direct human consumption by 2030. Marine aquaculture is expected to be dominated by Asian countries, in particular China, India, Indonesia, Vietnam, and Thailand. In addition, some other countries, such as Norway, Egypt, and Chile, could significantly expand their national production (World Bank et al. 2014).

World seaborne trade is expected to increase by more than 4% annually from 2016 to 2030. Shipbuilding growth is influenced by a range of factors, such as global trade expansion, existing capacity, energy consumption and prices, vessel profiles, ship retirement/scrapping, and changes in cargo and trade patterns. The world experienced a considerable buildup of overcapacity, as, in the last 10 years, the global shipping fleet grew at an average rate of 7% per year by far outstripping the annual 3.8% rise for world seaborne trade in tonnes. The demand for specialised ships, semi-submersibles, cruise tourism, supply and maintenance vessels for platforms, anchor handling, and offshore wind farms and floating multi-use platforms is expected to grow noticeably through to 2030.

Marine tourism holds booming promises, especially for cruise tourism. International tourist arrivals worldwide are expected to increase by 3.3% per year from 2010 to 2030, to reach 1.4 billion by 2020 and 1.8 billion by 2030; this implies an annual average increase of around 43 million international tourists globally. Up to 2030 at least, international tourist arrivals in the emerging destinations of Asia, Latin America, Central and Eastern Europe, the Eastern Mediterranean, the Middle East, and Africa are expected to grow at double the rate of that in advanced economy destinations (UNWTO 2016).

During the past six decades, tourism has experienced continued expansion and diversification, becoming one of the largest and fastest growing economic sectors in the world. Many new destinations have emerged, challenging old patterns and practices. In spite of occasional shocks, such as those linked to accidents, economic crises, or terrorist events, international tourism has experienced continuous growth, with a steadily rising share of the world's emerging regions (UNWTO 2016).

A healthy land–sea environment is fundamental to blue tourism and favours the growth potential of new forms of sustainable and green tourism in and around coastal cities. High-quality sea waters and coastal and marine landscapes have a high recreation value and contribute to the attractiveness of cities and the potential of activities as varied as nautical tourism and dolphin, whale, or coral reef watching. The sheer variety of blue journeys provides cities an opportunity to enhance their natural and cultural assets and boost their local economy in partnerships with their seas and their surrounding regions.

Sustainable tourism is expanding in all continents. Worldwide, international tourism has rebounded strongly, the increase more than offsetting the decline caused by the economic downturn. Recovery was particularly strong in emerging countries, where arrivals grew faster than in advanced ones. The European Union is the first world destination. Destination management and hospitality are important for all cities wishing to attract high-level visitors and create sustainable heritage tourism. Coastal cities are gateways to the recreational economy and marine tourism, which, if sustainably managed, can provide precious blue–green jobs. Traditionally second to shipping, the maritime and coastal tourism has become the largest single maritime economic activity in the EU. The European Blue Flags for Beaches and Ports label attributed since 1988 denote a high standard of environmental quality, the provision of certain important services, and the promotion of environmental education and information to the public.

The US, France, Spain, Italy, and China, the five first world destinations, have exceptional cities that serve as cultural attractions for sustainable tourism. Sea basins and coastal regions provide a unique resource, especially when associated with heritage treasures. Boston, a destination offering a high-quality heritage and environment, is encouraging the tourism industry to improve services to visitors, including cultural and emerald routes, waterfront activities and visits to the New England Aquarium, and cruises in the Boston harbour. Many Boston Harbor wharves have been converted to residential or cultural tourism services and offer the possibility for visitors to meet the residents. Québec City founds its offer on services provided around the Citadelle, at the epicentre of the historic district of Québec City, which has been designated a World Heritage site.

Cruise tourism is developing in most seas. It is a global business, which depends on geopolitical stability. Each cruise is a multi-destination experience that can impact other blue economies but also threaten the balance of biodiversity hotspots and sensitive sites. Despite misfortunes, the sector weathered crises well and, even in the Mediterranean, 2014 was the only questionable year since 1995. The average growth rate of Caribbean cruising, approaching 3%, can be reached in the Mediterranean. In the future, Asia is set to register quite spectacular cruise tourism growth rates, from 1.3 million in 2013 to 7 million in 2020, and global cruise passenger numbers could almost triple, from 19 million passengers in 2010 to more than 54 million in 2035, signalling annual growth rates of well over 7% (Cruise Industries News 2018).

Yachting tourism develops also for superyachts, of a length greater than 30 m. Yachting is expected to grow by 2–3% per year. EU shipyards and repairing centres have been successful in serving this specialised market, both large cruise ships and small leisure vessels. According to the Cruise Lines International Association (CLIA), the European ocean cruise market has grown to more than 6.9 million passengers in 2017, a 2.5% increase compared with 2016. In 2017, 6,941,000 Europeans took a cruise, for an average duration of 8 nights. Germany leads the source market in Europe with more than 2 million passengers. European cruising has been steadily growing for the past 10 years, led by the German, UK and Ireland markets (CLIA 2016).

The share of returns from cruise and coastal tourism with local communities is a thorny question. Congestion is the main burden, and time management of ports as "chronotopes" is a key issue. Local authorities typically accuse the cruise lines for unfair share of benefits and burdens. The data provided by the cruise lines, suggesting that the average cruise passenger spends €75 per day in each port of call in the Mediterranean in 2015, are considered overestimated for local authorities which do everything to attract cruise ships.

The search of sustainable tourism development models is much advanced in the European Union. Since 2006, the European Destinations of Excellence (EDEN) initiative proposes a selection of a diverse bunch of destinations that highlight the values, diversity, and common features of European tourist destinations. All selected destinations contribute in a balanced way to the economic, social, and environmental dimensions of sustainable tourism. The initiative is developed around an annual theme, and also sheds light on emerging European destinations, creates a platform for sharing good practices, and promotes networking and cross-pollination. Europe's best hidden gems in 2016 include towns with a special relationship with the sea, such as Positano in Italy or Menton in France.

Sustainable tourism destinations and routes offer the possibility to link outstanding marine physical resources and cultural urban assets in quality tourism services for more sustainability-minded tourists. Europe, hosting 37% of the sites of UNESCO's World Heritage list, is already being promoted as a 'unique tourism destination.' It is estimated that four of ten tourists choose their destination on the basis of the cultural offerings. European Cultural Routes further prolong the journey into space and time, as transnational routes help visitors discover the ways that Europeans have lived since ancient times. The concept was launched by the Council of Europe in 1987. In 2016, 32 Cultural Routes illustrated, through a variety of themes, European memory, history, and legacy to the future. The impact of European Cultural Routes on innovation and competitiveness suggested that cultural routes have much potential for small business generation, clustering, and skills, and intercultural dialogue. These Routes can have a major contribution to the creation of entrepreneurial ecosystems, building on local knowledge and talents and often promoting hidden gems destinations.

Sustainable tourism on land and the sea tries to bridge physical and cultural resource conservation and quality of leisure for personal and collective fulfilment. Cultural and marine commons, parks, and itineraries are key components of cultural tourism that involve public heritage spaces and merit particular attention for sustainability. Linked to green ethics, the design of cultural parks has to take into account all natural and cultural assets of an area and support ecosystem-based management. Cultural itineraries can serve as metronomes of journeys. Each city is a world and has its place in the global community as it becomes a cultural reference and a sea of wonder.

Renewable energy from the ocean and marine environments is the third major domain of the blue economy. The ocean energy landscape continues to be dominated by fossil fuels (offshore oil and gas), but the rise of renewables is incontestable. During the past 20 years, the offshore wind sector has progressed from the first small pilots to a dynamic industry with the potential for significant further growth and the most promising employment generator from all ocean industries. The EU is the world leader in offshore wind energy, with almost 90% of the global installed capacity more than 7 gigawatts (GW); projections suggest a global potential for 40–60 GW by 2020 and growth of a further order of magnitude by 2050. Industry efforts to reduce capital and operating costs could make offshore wind highly competitive by 2023. Projections for employment suggest the creation of around 170,000 jobs by 2020 and 300,000 jobs by 2030 (EWEA 2014).

Emerging blue economic sectors with significant potential but not yet at commercial scale by 2030 include ocean renewable energy. The ocean is a massive source of potential energy—tidal, wave, current, osmosis, and ocean thermal energy—to be harnessed for the green power generation transition to a low-carbon future. According to the Ocean Energy Systems Implementing Agreement, functioning within a framework created by the International Energy Agency, there is the potential to develop 337 GW of wave and tidal energy worldwide by 2050.

In Europe, the ocean energy supply chains are organised at a pan-European scale. Large engineering conglomerates, notably from Germany, Austria, and Denmark, are actively involved in many of the large-scale prototype projects and suggest that developments provide significant opportunities to expand markets. However, ocean energy technologies are still in an early demonstration phase, largely involving short-duration testing deployments, with only a few prototypes initiating the first steps towards commercialisation. The challenge is to accelerate the commercialisation of ocean energy through technology development and cost reduction. Geographic and oceanographic conditions and various technologies can offer a more predictable base-load supply of electricity and back up the fluctuating supply from wind. Tidal barrages are used to capture energy from volumes of moving water. Wave power devices are currently being demonstrated, and underwater turbines driven by currents are close to commercialisation. Ocean thermal energy conversion, which enhances the temperature difference between cooler deep and warmer surface waters, could be another option for tropical territories (EC 2012, 2017a).

Algae-based value chains expand and produce food, feed, and fuels. The global market for microalgae-based food and feed supplements is developed and has a great potential for growth. Microalgae are currently used both as dried whole algae and for the extraction of high-value food/feed supplements and colorants. Although the total production volumes and market size of food and feed supplements derived from microalgae are still relatively small, they have increased fivefold since the turn of the century.

Among the emerging blue activities that will have to gain public acceptance, sea mining is boosted by the increases in the price of many non-energy raw materials, mainly in emerging economies and the related scarcity risks. Under the 1982 UN Convention on the Law of the Sea, a country has exclusive rights to explore its continental shelf and exploit the natural resources, including mineral resources. Offshore mining is currently confined to shallow-water coastal regions. About 75% of the world's tin, 11% of gold, and 13% of platinum are extracted from near the surface of the coastal seabed. Aggregates, including sand, coral, gravel, and shell, are also important. Deep-sea mining operations can also be carried out in the international seabed, outside jurisdictional marine areas. The International Seabed Authority (ISA) supervising deep-sea mining in the "Area," beyond national jurisdiction, approved 29 contracts for exploration covering more than 1.3 million km^2 of the seabed. Contractors include public bodies and private corporations sponsored both by developed and developing states, mostly for manganese nodules, and exploration projects in the Clarion Clipperton Zone of the East-Central Pacific.

Seabed mining could have a disproportionate and even irreversible impact on coastal and marine ecosystems and habitats. Commercial-scale seabed mining operations are presently limited to shallow water, mostly for aggregates, sand, and gravel, for the construction industry and the creation of artificial beaches. The technologies mostly used include dredging, vacuum pumps, and remotely operated vehicles. Aggregate extraction has a large environmental footprint and could irreversibly change the seabed in shallow waters and affect other sectors and stakeholders. Social acceptance issues become extremely important. The mineral resource potential of the deep sea is thought to be huge, although extremely difficult to assess with any degree of confidence. Presently, offshore mining only takes place in shallow water on the continental shelf. There is potential for current offshore mining to expand into deeper water, but it is thought unlikely that this type of mining will extend beyond the limits of the continental shelf. Conversely, most targets for deep-sea mining are at far greater water depths.

Major drivers for deep-sea mining include the shortage of metals from land-based mines, but there is still no indication than the latter cannot continue to keep pace with growth. Some small countries with easy access to resources under the seafloor could benefit financially from deep-sea mining but very few have the capacity to properly undertake future mining projects. In all cases, the high degree of automation expected in deep-sea mining is unlikely to result in significant employment. The main constraint for deep-sea mining consists of sustainability concerns as related to the extraction of minerals from the seabed and the potential disturbance and damage that could be inflicted on ocean-floor and deep-water ecosystems. Deep-sea ecosystems are highly vulnerable and interconnected, and environmental assessment and precautionary approaches are highly advocated.

Marine biotechnology is a promising activity with great potential for new products and processes and has the potential to address a range of critical challenges such as human health, food, and energy security and make a significant contribution to green growth. It has generated great hopes for new drugs, cosmetics, chemicals, and pharmaceutical products. The complex marine ecosystem, with its large number of yet undiscovered microbial species and undiscovered properties even of known marine species, presents a rich and largely untapped resource. The capacity of marine organisms to nurture blue economic activities is just beginning to be explored, partly through new gene sequencing technologies for living organisms. Exploration of marine biodiversity could help us understand, for example, organisms that can withstand extremes of light, temperature, and pressure and develop new industrial enzymes or pharmaceuticals.

The global market for marine biotechnology products and processes is a significant and growing opportunity. In the immediate future, blue biotechnology could emerge as a niche market focused on high-value products for the health, cosmetic, and industrial bio-materials sectors. By 2020, it could grow as a medium-sized market, expanding towards the production of compounds as inputs for the food, feed, and chemical industries. In the longer run and subject to technological breakthroughs, the blue biotechnology sector could become a provider of larger markets with a range of high-added-value specialised products (OECD 2016).

5.2.3 From a Fragmented Activity Landscape to Symbiotic Value Chains

Fragmentation and sector-by-sector management of marine activities further increased pressures and compromised the capacity of ecosystems and the performance of marine activities. The integration of emerging activities with traditional framework poses many additional problems. A high degree of potential interactions exists between activities, especially and surprisingly between offshore wind and other ocean activities. Many cross-sectoral synergies can be generated and leveraged, for example, through the multi-use platforms, a form of joint location of offshore activities; the use of these platforms has been multiplied over recent years.

Although regulation of ocean activities is expected to continue to be largely sector driven, at least for the foreseeable future, efforts focusing on the partial integration or anchoring of emerging ocean industries into existing and often fragmented regulatory frameworks are promising (OECD 2016). Crises have the capacity to stimulate and cross-fertilise ideas and markets. They can bring together blue sectors to conceive, create, promote, and diffuse innovative initiatives. There is no single blue–green growth model but there are various possibilities for reconciling the short term with the long term, depending on local contexts and policies for growing cleaner without growing more slowly (World Bank 2012).

A blue–green growth strategy incorporates a longer time horizon and takes into account the full value of natural capital and terrestrial and marine ecosystems.

Many diverse activities could create propitious conditions for the development of lean blue–green innovative SMEs. The essence of coastal urbanisation, the concentration of people and activities close to the sea, offers a range of diverse resources needed by the green maritime industry and sustainability-conscious consumers. The features of a city favourable to green SMEs include access to precious human and natural resources, the availability of other high-quality services, complementary producers, and a market for green maritime services. Port cities can also provide the necessary knowledge basis and advanced services and support the generation or integration of innovative green SMEs in maritime clusters. Many regional economic benefits associated with well-performing ports often spill over the larger area or even other regions. Doubling port efficiency of two countries is found to raise their bilateral trade volume by 32% (OECD 2013a).

Aquaculture provides half the consumed seafood at the global level, and organisations suggest that by 2030 it could exceed two thirds (FAO 2016; World Bank et al. 2014). The main challenges for aquaculture include lack of available space, feed, breeding technology, competition in the global market, and administrative procedures and constraints. Sustainable aquaculture must also consider potential impacts on wild fish stocks and water quality, and also ethical considerations ensuring, for example, gender issues and the creation of decent jobs. More space for marine aquaculture could be freed up by the development of offshore production systems. Progress in feed technology could achieve a higher proportion of feeds of plant and waste origin.

Cities could also have a major role in managing and enhancing the transit of activities taking place offshore. Multi-purpose offshore platforms could act as innovation spearheads for sustainable maritime economy. Some offshore platform consortia are developing innovative design and new business models to become key players of the future cross-sector offshore economy. The governance architecture tries to generate and capture synergies among diverse economic chains, such as aquaculture and offshore wind energy, facilitated by robotic technologies.

The public and private sectors should work in concert for promoting the blue-growth economy, starting with precompetitive associations, in which competitors pool investments in resources, knowledge, and expertise to learn from successes and failures, to invent, deploy, and scale the cutting-edge technologies that may spark new breakthroughs. This approach offers the opportunity for ocean-minded cities to organise the conditions for new actors to enter the innovation chains and invest creativity and vigour in emerging and changing blue–green growth activities.

Maritime helices and clusters can be a powerful resource for policy makers, as unique platforms for business, education, research, and government to connect and engage in substantial dialogue on the future of blue activities; these require a convincing policy framework, a trusted leadership and a large number of both public and private actors willing to cooperate. Policy makers should carefully map and identify relevant local maritime clusters and put in place instruments and tools to support their emerging phase. A coherent longer-term policy framework is vital. It is also crucial that governance levels are aligned and that efforts are made towards well-articulated multilevel governance.

Well-functioning clusters are often engaged in horizon scanning to serve the longer-term needs of their members and partners. The blue–green maritime sectors are interlinked with competing economic activities as well as the marine environment. Benefits from clusters depend strongly on strategies to strike the right balance between local and international activities and care for the coastal marine ecosystems. Competency, skills, and research are prominent areas for international collaboration (Ecorys et al. 2014; Port of Antwerp 2017).

Partnerships can provide a critical contribution to the renewal of cities across the world, especially at times when public resources are under pressure. Many cities are plagued by outdated transport and energy infrastructure and underutilised industrial areas, which need to be dramatically upgraded, often by capitalising on unknown or radically undervalued and underleveraged assets. Most cities could benefit from smarter use of these assets and better performing sustainable infrastructures. Shifting attention on public wealth and resources from short-term spending to longer-term investments can significantly raise the sustainability bar. To revive their city in the late 1980s, a coalition of national and local officials laid the groundwork for the Copenhagen City & Port Development Corporation. Its success provides a model to revitalise cities and finance large-scale infrastructure by increasing the commercial yield of public land and buildings (Detter and Fölster 2017).

5.3 Responsible Corporate Conduct and Co-optimised Investments

5.3.1 Reconciling Quality and Quantity, the Short Term with the Longer Term

All responsible organisations, public or private, seek to promote the pursuit of happiness for all stakeholders. Sustainability and commercial excellence can be mutually reinforced. Reconciling economic objectives with longer-term political goals is at the heart of strong sustainability. Innovation and entrepreneurship can create sustainable value by embedding long-term sustainability considerations into business strategies and by fully disclosing progress to investors, while opening to society. Eco-responsible companies can help strengthen the long-term sustainability of the financial system. Large enterprises are important for pioneering cities, and SMEs have high potential for revitalising urban fabrics. Cities can create high-performing business environments and establish specialised enterprise areas, start-ups, seed and capital risk companies, bio-clusters, and micro-financing mechanisms. They can also engage with cities and citizens from all parts of the world and influence the health and the future of the commons.

Public recognition and bold decision making are essential for the link between sound business ethics and performance aligned with sustainable development. Businesses ask cities for favourable locations and fees, preferential treatment, and social recognition as a significant generator of wealth and jobs. Municipalities seek, from businesses, employment for citizens, tax revenues, and investment in the local economy. At the crossroads of these requirements, cross-fertilised investments can be beneficial for citizens, the city, and the ecosystems which bountifully support them (Mega 2015).

A more resource-conscious economy advocates for new ways of designing business, responding to the fundamental needs of all, and respecting the availability of terrestrial and marine resources. Scarcity can be turned into plenty when the waste of one product or service becomes the resource providing the input to a new cycle in the circular economy. Cascading nutrients and energy can become a valuable process that generates jobs, builds social capital, and increases income for all. Cities in partnership with businesses are best situated to orchestrate such partnerships and processes, including for Corporate Social Responsibility in the Maritime Industry (WMU 2017).

Eco-responsibility asks for eco-efficiency, one of the cornerstones of the "Vision for 2050" conceived by the World Business Council for Sustainable Development (WBCSD), and based on country dialogues with several hundred companies and experts. The vision highlights that humankind can achieve a low-carbon and zero-carbon society, living well and within the limits of the planet, but all stakeholders must radically change their eco-behaviour, their interactions with economic and ecological processes (WBCSD 2010a, b).

A redefinition and renegotiation of notions and values is a precondition for behavioural change. Society has to rethink the notion of prosperity and well-being and economic growth to be decoupled from energy and material use and ecosystem destruction. Markets have to reconsider values, costs, and benefits in a broader horizon. In a complex and interconnected world, many diverse state and non-state actors have to engage in responsible global governance to effectively manage international systems or resources, such as climate, water, or space.

Cities need to provide high-quality space and environment for sound civic and corporate purposes. Interconnected port cities, the capacity of which to handle container traffic may have to rise considerably, have to discuss new models. China may hold a share of nearly 40% of growth in global demand for urban building floor space to 2025. Africa and the Middle East could account for almost 14% of the global rise in municipal water demand in large cities, almost twice their share of urban GDP growth. But cities have also to care about green infrastructures and resource- and energy-efficient developments.

A sustainable city can use resources more efficiently, thrive economically, and create an inclusive community. Businesses are essential for the development and operations of urban infrastructures, which provide the 'hardware,' increasingly incorporating 'software' for a better performance. The main infrastructure, technology, and services to reinforce the sustainable visions and strategies of cities are predominantly developed and implemented by the private sector. The WBCSD's Urban Infrastructure Initiative brought together a diverse group of companies from key sectors, including energy, water, materials, equipment, and support services to assist urban authorities with pragmatic and cost-effective plans for sustainable infrastructures. Urban markets offer companies the opportunity to provide system solutions, products, and services for sustainable buildings, energy, infrastructure, and resource and waste management (WBCSD 2010b, 2014).

These partnerships can only be beneficial. Cities win by getting practical, cost-effective solutions to envision and implement actions towards sustainability. Through collaboration toward common shared goals, business can address some of the critical challenges while strengthening resilience. Business can and should have a key role in addressing marine debris by having more companies involved to scale-up emerging solutions. Based on company insights along the entire plastics value chain, the WBCSD pinpointed four main business drivers for acting on plastic marine debris, including maintaining core business, enhancing brand and reputation, securing competitive advantage, and improving business operations. Solutions to reducing ocean waste can generate additional environmental, social, and economic benefits, creating shared value beyond company walls if seen from a broader perspective. For instance, job creation, inclusion of the informal waste management sector, and better conditions in cities and communities in which companies operate, enhance local stakeholders' trust in business and improve community engagement (WBCSD 2017).

5.3.2 Multi-stakeholder Engagement for Raising the Sustainability Bar

Social responsibility of all public and private actors is essential for the blue–green economy. Cities and enterprises embracing and supporting the sustainability ethos should make optimal use of marine resources and technologies. They have to ensure that the best green products and services come to the market and ultimately improve human lives. Enterprises have to deploy their inventiveness and bring to the market blue technologies, products, and services. Enlightened leaders can only succeed in concert with companies, universities, citizen associations, labour unions, environmental groups, cultural institutions, and philanthropies, often working in overlapping and interlocking networks. As cross-sector multi-stakeholder networks and platforms multiply, it is important for efforts not to be dispersed and synergies to be generated and captured. Codes of Conduct for responsible networks of networks and platforms can be inspired by some businesses.

Improving social and environmental performance of business and disclosure of information on their results is capital. Ethical citizens and vigilant consumers and organisations call for cities and companies to design and prove their contribution to sustainable blue growth. Awareness campaigns try to overcome information deficits and increase consumer knowledge on the potential impacts and the benefits of alternative patterns. Citizen associations and the media tend to create a climate of trust, surrounding sustainable cities and businesses. Companies and cities without a declared commitment to and action for sustainable development could face citizen and consumer boycotts, attacks on fixed assets, failure to attract forward-looking stakeholders, stockholders, and employees, restrictions on port operations, and obstacles in infrastructure development. Pro-action is a must, because damaged reputation, impaired licenses, disillusioned shareholders, and disappointed citizens may strongly impact future investments.

Climate change has added a major and most powerful dimension in the shared social responsibility among cities, business, and citizens. Climate change affects the array of blue economy sectors, from shipping and food production to tourism and service industries. The Carbon Disclosure Project suggests that cities recognise and report climate change threats to business. Businesses and cities should be aligned in addressing climate change risks, and urban adaptation actions should contribute to business resilience. Cities are providing regulations, incentives, and information, as well as investments in infrastructure that support businesses and all stakeholders (CDP 2014).

A growing number of ocean-minded cities and companies make public their intentions, plans, actions, and results in preparing for the post-carbon age. Corporate codes of conduct gained momentum. Businesses adopted a broad range of approaches, from non-reporting to social reporting to increase confidence in their performance. Assessment and public reporting have to be organically integrated components of the business cycles. The voluntary nonbinding nature of most codes is often related to the absence of independent auditing, even if codes spell out the necessity for monitoring, life cycle assessment, and reporting.

Eco-responsible businesses are instrumental in creating value out of values and can be valuable allies of eco-responsible cities. They increasingly integrate social, environmental, ethical, and human rights concerns and principles into their strategies and business operations in close collaboration with their stakeholders. The aim is to achieve results beneficial to both society and the economy, and to minimise and prevent negative impacts on the environment, including marine habitats and ecosystems.

Corporate and multi-stakeholder engagement for sustainable development is highly promoted by many investors, environmental and public interest organisations, and advocacy groups. The FAO's Code of Conduct for Responsible Fisheries suggested that ethical governance is a necessary condition for the sector to fully realise its potential for sustainable fisheries. The voluntary Code of Conduct consists of a set of principles, goals, and actions that took more than 2 years to reach agreement, and it is deemed to represent a global consensus or agreement on a wide range of issues.

The Code of Conduct for Responsible Fisheries advocated that countries should have clear and well-organised fishing policies to manage their fisheries. These policies should be developed with the cooperation of all stakeholders that have an interest in fisheries, including the fishing industry, fish workers, environmental groups, and other interested organisations. Fisheries should be managed to ensure that fishing and fish processing are conducted in ways that minimise negative impacts on the environment, reduce water and energy use and waste, and preserve the quality of caught fish. Governments should have enforceable laws with procedures for determining and punishing violators. In cases that fisheries resources are shared among countries, the Code calls for new regional fisheries organisations to be established or for existing organisations to be strengthened. Cooperation in this way is the only realistic approach to achieving the long-term goals.

The FAO guidelines on aquaculture recommend that, to prevent overfishing, the size of the fishing fleet should not be too large for the natural supply of fish. Fishing methods and gear should be selective, and designed to minimise waste and the catching of fish species that are not wanted or which are endangered. Important fish habitats such as wetlands, mangroves, reefs, and lagoons should be protected from destruction and pollution. When natural disasters harm fisheries resources, countries should be prepared to take emergency conservation and management measures (FAO 2017).

As global sustainability challenges become higher than ever and the urgency for inspired leadership and responses multiply across the globe, the foremost policy makers, investors, and companies connect to accelerate the transition. The Coalition of Environmentally Responsible Economies and Societies (CERES), founded by a small group of investors largely in response to the *Exxon Valdez* oil spill in 1989, wished to bring together environmentalists and capitalists to forge a new sustainable business ethos. The oil spill into Alaska's Prince William Sound devastated one of the world's most pristine habitats. Suddenly and painfully, businesses had to account for the environmental and social impacts of their operations. CERES brought forward a bold vision of a world in which business and the capital markets jointly engage in the well-being of human society and the protection of the precious natural environment. At the heart of the vision, the coalition's principles proposed a code of corporate conduct to be publicly endorsed by companies.

The CERES coalition has been instrumental in weaving environmental and social challenges into company and investor decision making. The Global Reporting Initiative, launched in 1997, proposed a harmonised public disclosure to deliver a structured flow of verifiable and comparable information. The coalition's sustainability reports enabled organisations to 'walk the talk' and serve as a model for the transparent reporting of achievements against commitments. This is more than needed for the accomplishment of the SDGs and the follow-up of commitments surrounding the targets.

The Roadmap for Sustainability, issued in 2010, outlined meaningful steps that companies have to undertake in order to succeed their economic goals, while committed to sustainability principles. The Roadmap intended to bring investors, companies, and other key economic players together to find scalable solutions for integrating sustainability across capital markets and the economy. In 2014, CERES launched the Clean Trillion, highlighting the annual global investments needed to limit global warming to 2 °C and ensure a clean energy future (CERES 2010, 2014).

Public campaigns have been crucial for raising awareness. The CERES 'Connect the Drops' campaign showcases the strong leadership of California companies and policymakers to advance resilient water solutions that ensure sustainable management of the stressed and often unpredictable water supplies. The 'Disclose What Matters' campaign asks investors and companies to disclose material sustainability issues, such as climate change risks, in their financial accounts and to better manage opportunities associated with wide-ranging shorter-term and long-term decisions that impact sustainability performance.

Corporate social responsibility is crucial for mainstreaming ethical principles in business activities and catalysing actions in all sectors. It could link different activities of the blue economy and diffuse to a larger scale of initiatives such as the UN Global Compact, created in 2000, the largest voluntary corporate responsibility initiative in the world. This strategic policy initiative engages businesses that align their operations and strategies with ten universally accepted principles in the areas of human rights, labour, environment, and anticorruption. A commitment to transparency and disclosure is critical.

The Global Compact provided the framework for the development, implementation, and disclosure of cross-sector sustainability policies and practices. It called businesses to support a precautionary approach to environmental challenges, undertake initiatives to promote greater environmental responsibility, and encourage the development and diffusion of environmentally friendly technologies. The Compact bridges the global and local levels, the private and public domains, and activities on land and the sea. Voluntary yet accountable, it searches to combine the UN moral authority and convening power, with the private sector's vigour and the contributions of a range of key stakeholders. Benefits include the sharing of best and emerging practices and world linkages among business units and subsidiaries across the value chain with the Global Compact's Local Networks, including in emerging markets.

The Compact also has an accountability policy component, the Communication on Progress. The annual communication is an important demonstration of engaging with the UN Global Compact and its principles. Failure to communicate may result in the loss of participant status and possible expulsion. In promoting the implementation of sustainable development goals (SDGs), the UN Global Compact tries to identify each year a group of SDG Pioneers to shine a spotlight on individuals championing sustainability and mobilising companies to be a force for good and an example for others to follow (UN 2017).

Guidance and reporting matter and contribute to the circle of learning, thought, active awareness and action. The SDG Compass providing guidance for companies to align their strategies as well as measure and manage their contribution to the SDGs was launched in September 2015 by the WBCSD, in collaboration with the Global Reporting Initiative and the UN Global Compact. WBCSD's Action 2020 platform, fully aligned with the SDG agenda, is driven by a series of business solutions which are well positioned to bring significant contributions across the full spectrum of the Goals.

Corporate social responsibility for blue–green growth and sustainable development is promoted through the Global Partnership for Oceans (GPO), an alliance of governments, international organisations, civil society groups, and private sector interests committed to the health, productivity, and resilience of the ocean. The alliance targeted the priority issues of overfishing, pollution, and habitat loss that are contributing to the depletion of the natural resource offering nutrition, livelihoods and vital ecosystem services, together with the multi-stakeholder leadership, knowledge, and finance (GPO 2013).

The Blue Ribbon panel, convened by the World Bank in 2013 to advise the Global Partnership for Oceans, brought together decision makers, experts, and CEOs of some of the largest seafood companies in the world. It emphasised that without action to restore the declining health of the ocean, the consequences for economies, communities, and ecosystems will be irreversible. Improving ocean health is a complex process that requires leadership and participation across a broad sphere of communities, industries, and governments and the concentrated effort of society, government, and science. Fragmented approaches that fail to consider systemic social, political, economic, and ecological relationships cannot bring viable solutions. The panel calls for an integrated approach to ocean investment and pinpoints the essential role of public–private partnerships that combine the efforts of big and small business, government, and science. Such multi-stakeholder partnerships and panels can bring together cities that care for the ocean and wish to offer citizen support, expertise, and resources.

Eco-responsible investments that advance sustainable livelihoods, social equity, and food security, ocean health, effective governance systems, capacity building, and innovation can only be beneficial in the long term. Policy options must be multidimensional and dynamic and integrate all aspects of the socioecological ocean system. The Blue Ribbon panel's principle-based strategy allowed focusing on highly impactful action. The panel recommends five interlinked principles to be incorporated into all levels of ocean policy reform, from fisheries management to incentives and habitat restoration. These principles include sustainable livelihoods, social equity and food security, a healthy ocean, effective governance systems, and long-term viability, capacity building, and innovation (GPO 2013).

Sustainable fisheries offer ample opportunities for exercising social and environmental responsibility. In the UN Conference "Our Ocean" (New York, June 2017) more than 40 buyers of the tuna sector pledged that, by 2020, all tuna products in their supply chains will be sourced from socially and environmentally responsible producers and will be traceable to the vessel and the date. Ecolabels can amplify progress. The Marine Stewardship Council (MSC), an independent nonprofit organisation, sets a science-based standard for sustainable fishing. Seafood products can display the blue MSC ecolabel only if they can be traced back through the supply chain to fisheries that demonstrate that they are well managed and sustainable compared to the MSC standard. There are more than 20,000 MSC-labelled, certified sustainable seafood products on sale around the world, from fresh fish to prepared products. Back in 2000, Western Australia rock lobster became the first fishery to earn MSC certification. More than one tenth (12%) of global marine wild catch is MSC certified, a figure that has doubled since 2010 (MSC 2017).

5.4 Prosperous and Sustainable City Rankings and Indexes

5.4.1 Place, Location, and Agglomeration Do Matter for Sustainability

Urban liveability is cardinal for all citizens, who may however not share the same desiderata and sustainability values (Arcadis 2016; EC 2016a, b). It is also important for business activity and the attraction of the best world talent and visitors

(Florida 2008, 2011). Composing a universal list of the world's best or most enjoyable cities is a near-impossible task, attempted by some think tanks and organisations. The Economist Intelligence Unit (EIU) endeavours, since 2002, to quantify the liveability of cities, according to 30 criteria related to safety, healthcare, educational resources, infrastructure, and the environment.

The EIU 2017 global liveability report and index, assessing 140 cities, suggested that, after a decade of liveability decline, cities are on the rise again, especially in Canada, the EU, Australia, and New Zealand, which host the best to-live-in cities. They enjoy widespread availability and accessibility of vital goods and essential services. According to the 2017 liveability ranking, Melbourne continues to be the most liveable of cities, followed by Vienna. Hamburg is the only other European city appearing in the list of top ten. Vancouver, which ranked as the most liveable world city until 2011, came in third place in 2017 (EIU 2017).

In 2012, the Economist Intelligence Unit teamed up with the data-sharing company BuzzData to offer users the opportunity to make their own ranking by combining data from the Cost of Living and Liveability surveys with other sources. Opening up the concept of best cities to a diverse set of persons generated an array of innovative approaches. Interactivity allowed respondents to propose additional drivers and suggest priorities towards a collective choice of best city. This approach had the dual benefit of having citizens expressing subjective choices and effectively crowdsourcing responses to inform future methodologies (EIU 2012). It is interesting to note the evolution of the list of criteria, as, for example, with the introduction of a connectivity indicator, which propelled Hong Kong or Dubai to the highest places of the rankings.

At the other end of the spectrum, lists of the worst cities in which to live and invest can be equally impactful as the lists of the best cities, because they indicate the cities that hold the strongest potential for improvement. The Economist Intelligence Unit ranked cities according to the 'tolerability' of living in a given place, relative to crime levels, threat of conflict, quality of medical care, levels of censorship, weather effects, and education and transport services. The ten least liveable cities in 2017 included Kiev, Douala, Harare, Karashi, and Algiers. Conflict seems a key factor in undermining liveability.

Location matters for quality of life and work. The quality and cost of living surveys and the ranking of world cities, conducted by Mercer, the world's largest consultancy in human resources, also offer interesting insights and information. Mercer's surveys help multinational bodies fairly compensate employees on international assignments. Each survey is based on an array of assets and services reflecting the typical expenditure of the expatriated population. For each city, Mercer provides an overview of conditions and hardship premium recommendations. Safety is a key factor considered by multinationals when appointing expatriate workers, both because it raises social apprehensions about personal safety and has a significant impact on the cost of remuneration packages.

Mercer's quality of living list includes 230 cities across five continents, benchmarked against New York, the base city. The 2017 19th annual edition suggested that despite recent security concerns, migration issues, and social unrest, European cities continue to offer some of the world's gold standard quality of living. Auckland

and Vancouver are the only non-European cities in the list of the ten best cities. Vienna continues its reign as the top place for overall quality of living, followed by Zurich, Auckland, and Munich. Vancouver is North America's highest ranking city, and Singapore the highest ranking Asian city, at 25th place. Mercer's survey also identifies the personal safety ranking, based on internal stability, crime incidents, local law enforcement, and openness to the world. Luxembourg tops the personal safety list, followed by Bern, Helsinki, and Zurich. Baghdad and Damascus occupy the lowest positions (Mercer 2017a).

Next to its quality of living survey, Mercer published in 2017 its 23rd annual cost of living survey, focusing on the comparative cost of more than 200 items in each location, including housing, transport, food, clothing, household goods, and entertainment. Exchange rates and Mercer's international basket of goods and services from its quality of living survey have been used as base measurements for more than 375 cities throughout the world. Mercer's cost of living survey suggests that factors including currency fluctuations, cost inflation for goods and services, and instability of housing prices contribute most to the cost of expatriate remuneration packages for employees on international assignments.

According to Mercer's 2017 cost of living survey, Hong Kong tops the list of most expensive cities for expatriates, pushing Luanda to second position. Zurich and Singapore remain in third and fourth positions respectively, whereas Tokyo is in fifth. Kinshasa, ranked sixth, appearing for the first time in the top 10, followed by Shanghai, Geneva, N'Djamena, and Beijing. The survey indicates, as the world's least expensive cities for expatriates, Windhoek, Namibia, Cape Town, South Africa, and Bishkek, Kyrgyzstan. The most expensive European are the Swiss cities, although several cities across Europe remained relatively stable, including Paris, Milan, Vienna, and Rome. London and Birmingham dropped 5 and 16 places, respectively, but the German cities of Munich, Frankfurt, and Dusseldorf climbed in the ranking. In the US, New York is the highest ranked city at the eleventh position, followed by San Francisco, Los Angeles, and Seattle (Mercer 2017b).

Among the rest of international rankings, the Mastercard Global Destination Cities Index is more than just a ranking of the 132 top destination cities; it also provides deeper insights of the reasons that world citizens travel and consume. Since 2009, international travel and spending by overnight visitors to cities have grown faster than the GDP. Asian and European cities, with Bangkok on the top, and including New York, make up the list of the first ten cities in 2017. Tokyo is expecting the largest growth in visitors among the prime destinations (Mastercard 2017).

5.4.2 *Interconnected Global Power and Emerging Cities*

Stellar megacities could emulate on a global scale and bring challenges and opportunities to a new level. Powerful world cities attract promising activities and have a great impact on the world (NYC 2015a, b). The Global Power City Index, first Asian ranking by the Institute for Urban Strategies in Tokyo, which celebrated its tenth anniversary in 2017, evaluates and ranks the major cities of the world according to six major urban functions: economy, research and development, cultural interaction, liveability, environment, and accessibility, and on the basis of 70 indicators. Evaluation is carried out through the lenses of key stakeholders, including managers, researchers, artists, and visitors and local residents. An indicators matrix of city actors and functions helps appraising and ranking cities (Mori Memorial Foundation 2017).

Power also derives through eco-responsibility and connectivity reach. Interconnected eco-responsible cities are best placed to lead action for the global commons. Since the rise of global financial industries in the 1980s, New York, London, and Tokyo form a global triangle of urban power. Despite the ascent of many cities such as Paris and the Asian giants Shanghai, Singapore, Seoul, and Hong Kong in the 1990s, which experienced rapid urban growth led by dramatic economic change, the top triad has changed little. The 2017 ranking of the most powerful cities on the world chessboard reconfirmed the position of London, New York, Tokyo, and Paris as the top four global cities, respectively. The same cities form the league of the World Economic Forum's 'super cities' for real estate investment. Pre-Brexit London is the leading among five European cities in the top ten, also including Paris, Amsterdam, Berlin, and Vienna. Three more Asian cities, Singapore, Seoul, and Hong Kong, complete the top decade (Mori Memorial Foundation 2017).

According to the Global Power City Index, New York City ranked as the first world powerful city until 2012, the year that yielded the top spot to London. Tokyo is always the first in population global megacity and it is expected to remain first in 2030, despite past UN population revisions suggesting that it would have already been dethroned by Mumbai. The Japanese capital is still the top Asian city, but it seems that it represents a mature megacity offering fewer new opportunities to attract people and capital and caring more for its shrinking and ageing population. However, the ranking is dynamic, and several upcoming events, such as the withdrawal of the UK from the EU in 2019, or the 2020 Tokyo Olympics, are expected to heavily influence city power.

The debate on a Perfect Global City is endless. At A.T. Kearney, the Global Cities Index, which examines, since 2008, a comprehensive list of cities on all continents according to their global engagement in business activity, human capital, information exchange, cultural experience, and political engagement, provided in 2014 an interesting direction. The Perfect Global City could be a composite result of 16 cities including New York, Tokyo, Paris, and London (Kearney 2014).

In 2017, the A.T. Kearney's Global Cities (7th edition) testified to the continued strength of many of the world's largest cities with vibrant ocean dimensions. Identifying New York City as this year's leading metropolitan area, this ranking also makes projections about the potential of the frontrunners, suggesting that San Francisco is best positioned to attract and retain global capital, talent, and ideas in the near future (Kearney 2017).

Best to live in and invest or most expensive cities are necessarily not the most powerful global cities, which are necessarily not the most competitive. Blends of more indices that try to capture the economic power of global cities bring New York to the top of the global urban pyramid and its position seems secure, at least for the medium term. It is the world's most open and diverse large city, the only one to come first or second according to all criteria in all major rankings. London seems overall second and Tokyo third, Hong Kong is fifth, and Singapore, Shanghai, and Los Angeles follow in the race to the top.

Another interesting survey, the seventh edition by PricewaterhouseCoopers, is based on the assessment of ten indicators focusing on quality of life, economics, and instruments. London, Singapore, Toronto, and Paris top the 2016 benchmark of the most successful global cities. This latest edition analysed the trajectory of 30 cities, all capitals of finance, commerce, and culture, and identified strengths and weaknesses. The survey suggests that the heart of the matter revolves around balanced social and economic strengths. Quality of life factors seem to be an increasingly important factor of urban success in all cities (PwC 2016).

In 2014, the previous edition brought forward London and New York. The exercises have been instrumental in benchmarking cities from various perspectives including unmet demand and potential for future growth. London had come first with a good margin and marked the highest scores in technology readiness, economic influence, and gateway to the world. New York displayed a balanced performance across domains, while Singapore climbed to the third place as world leader in transport and infrastructure, and the enabling business environment. Australia's largest city scores in the top ten in four indicators and in more than half the variables. The comparative advantages of Sydney shine in demographics and liveability, sustainability and health, safety, and security. Dubai ranked 16th of 30 Cities of Opportunity in 2014, coming in first for airport connections to business districts, fifth most affordable city, and eighth world city as an urban gateway. The city performs remarkably well across several domains, including infrastructure, potential for future investment growth, and safety. It is also very interesting that the survey suggests that the sustainability and intellectual capital of the city hold a high potential for improvement (PwC 2014).

Next to the A.T. Kearney Global Cities Index 2014, an Emerging Cities Outlook shed light on 34 cities located in low- and medium-income countries. The potential of a city depends on its ability to attract, retain, and generate flows of ideas, capital, and people, and the time needed to catch up with the global leaders. The indicators can be grouped into three categories. A first range of indicators includes business activity reflecting the evolution of a city's prosperity, changes in infrastructure, business environment, and perceptions regarding public transparency. A second group focuses on human capital, trends in stability and security, healthcare availability and quality, income equality, and environmental sustainability. Finally, the third category of indicators composing the innovation profile of a city include patent filings per capita, new business generation, volume of venture capital, investment in research and demonstration, and access to finance.

Two vulnerable Southeast Asian mega-capitals, Jakarta and Manila, head the list of the Emerging Cities Outlook, including the cities with the highest potential to progress. Jakarta and its 1000 islands, the beating heart of Indonesia, in the hub of ASEAN, seems to move up quickly in relationship to human capital, but also in addressing income inequality and environmental concerns, as well as across several important innovation indicators. Manila also seems favoured by a relatively sharp increase in human capital, with a particularly notable improvement in healthcare. The Philippine capital, located in the fascinating hotspot of the coral triangle, enjoys an extraordinary biodiversity reflected on its seal, a pearl embedded in a shell. Both cities can only progress in harmony with the planet and their marine environments and if investing in resilience and sustainable development (Kearney 2014; Mega 2013).

References

Arcadis (2016) Sustainable Cities Index 2016. Putting people at the heart of city sustainability. Amsterdam

Brookings (2016) Achieving an advanced economy that works for all: the Brookings Metropolitan Policy Program in 2016 and beyond. Washington, D.C

Carbon Disclosure Project (CDP) (2014) Global corporate use of carbon pricing. Disclosures to investors, London

CERES (2010) The 21st Century Corporation: the CERES roadmap for sustainability. Boston

City of Amsterdam (2013) Towards the Amsterdam Circular Economy. Amsterdam

City of Amsterdam (2014) How Amsterdam goes circular. Amsterdam

City of Amsterdam (2015) Circular Amsterdam. A vision and action agenda for the city and metropolitan area. Amsterdam

City of New York (NYC) (The) (2015a) OneNYC. The plan for a strong and just city. New York

City of New York (NYC) (The) (2015b) Building a smart + equitable city. New York

City of Stockholm (2014) Stockholm Royal Seaport. Stockholm

Coalition of Environmentally Responsible Economies and Societies (CERES) (2014) Investing in the clean trillion: closing the clean energy investment gap. Boston

Cruise Industry News (2018) 2018–19 Annual report. New York

Detter D, Fölster S (2017) The public wealth of cities. How to unlock hidden assets to boost growth and prosperity. Brookings Institution Press, Washington, DC

Economist (The) (2013) The shipping industry sinking under a big green wave. London

Economist Intelligence Unit (The) (2012) Best cities ranking and report. London

Economist Intelligence Unit (The) (2017) A summary of the liveability ranking and overview. London

Ecorys et al (2014) Support activities for the development of maritime clusters in the Mediterranean and Black Sea areas. Brussels

European Commission (EC) (2012) Blue Growth. Brussels

EC (2016a) Opportunity Now! Europe's mission to innovate. Brussels

EC (2016b) Sustainability Now! A European vision for sustainability. Brussels

EC (2017a) Report on the Blue Growth Strategy towards more sustainable growth and jobs in the blue economy. SWD 2017–128. Brussels

EC (2017b) Report from the Commission to the Council on the urban agenda for the European union, COM (2017), 657 final. Brussels

European Union Institute for Security Studies (EUISS) (2012) Global trends 2030 Citizens in an interconnected and polycentric world. Paris

European Wind Energy Association (EWEA) (2014) The European offshore wind industry: key trends and statistics 2013. Brussels

Food and Agriculture Organisation (FAO) (2016) The state of world fisheries and aquaculture. Rome
FAO (2017) Aquaculture development. Aquaculture governance and sector development. Rome
Florida R (2005) Cities and the creative class. Oxford, Routledge
Florida R (2008) Who's your city. Basic Books, New York
Florida R (2011) The great reset: how the post-crash economy will change the way we live and work. Harper Publishers, New York, London, Toronto, Sydney
Global and European Cruise lines (CLIA) (2016) Contribution of cruise tourism to the economies of Europe 2015. London
Global Ocean Commission (GOC) (2016) The future of our ocean. Next steps and priorities. Oxford
Global Partnership for Oceans (GPO) (2013) Indispensable ocean. Blue Ribbon Panel Report. World Bank, Washington, DC
Gore A (2013) The future: six drivers of global change. Random House, London
Kearney AT (2014) Global cities index and emerging cities outlook. Global cities, present and future
Kearney AT (2017) Global cities 2017: leaders in a world of disruptive innovation future
Marine Stewardship Council (MSC) (2017) Global impacts report 2017
Mastercard (2017) Global Destination Cities Index 2017. New York
McKinsey Global Institute (MGI) (2011) Urban world: mapping the economic power of cities
Mega V (2013) Quintessential cities, accountable to the future. Innovation, sustainability and leadership. Springer, New York
Mega V (2015) Conscious coastal cities. Sustainability, blue green growth and the politics of imagination. Springer, New York, Cham, Switzerland
Mercer (2017a) Quality of living survey – worldwide rankings 2016
Mercer (2017b) Cost of living survey – worldwide rankings 2016
Mori Memorial Foundation (The) (2017) Institute for Urban Strategies Global Power City Index Yearbook 2016. Tokyo
National Oceanic and Atmospheric Administration (NOAA) (2013) State of the Coast. National Coastal Population Report . Population trends from 1970 to 2020. Washington, DC
Organisation for Economic Co-operation and Development (OECD) (2013a) The competitiveness of global port cities. Synthesis report. OECD, Paris
OECD (2013b) Green growth in Stockholm, Sweden. OECD, Paris
OECD (2015) Green growth in fisheries and aquaculture. OECD, Paris
OECD (2016) The ocean economy 2030. Paris
OECD (2017) Trust and public policy: how better governance can help rebuild public trust. OECD, Paris
OECD-FAO (2017) Agriculture Outlook 2017–2026. 13th joint edition, Paris
Pope Francis (2015) LAUDATO SI': encyclical letter on care for our common home. Holy See, Vatican
Port of Amsterdam (2009) Port-City. Amsterdam
Port of Antwerp (2017) Sustainability report 2017. Antwerp
Port of Rotterdam (2012) Port Vision 2030. Rotterdam
PricewaterhouseCoopers (PwC) (2014) Cities of opportunity 6: we the urban people. New York
PwC (2016) Cities of opportunity 7: the living city. London
United Nations (UN) (2017) Progress towards the sustainable development goals 2017. Report to the UN Secretary-General New York
United Nations World Tourism Organisations (UNWTO) (2016) Annual report. Madrid
World Bank (2012) Inclusive green growth. The pathway to sustainable development. Washington, DC
World Bank and UN Department of Economic and Social Affairs (2017) The Potential of the blue economy: increasing long-term benefits of the sustainable use of marine resources for small island developing states and coastal least developed countries. Washington, DC

World Bank et al. (2014) Fish to 2030. Prospects for fisheries and aquaculture. Washington, DC
World Business Council on Sustainable Development (WBCSD) (2010a) Vision 2050. Geneva
WBCSD (2010b) A world of sustainable cities. Geneva
WBCSD (2013) A solutions landscape for Kobé. Geneva
WBCSD (2014) The urban infrastructures initiative. Final report. Geneva
WBCSD (2017) The business case for reducing ocean waste. Geneva
World Economic Forum (WEF) (2014) The competitiveness of cities. Geneva
WEF (2018) Circular economy in cities: evolving the model for a sustainable urban future. Geneva
World Maritime University (WMU) (2017) Corporate social responsibility in the maritime industry. Springer, New York
World Wide Fund for Nature (WWF) (2015a) Living Blue Planet Report. Species, habitats and human well-being, 2015
WWF (2015b) Reviving the ocean economy: the case for action. London
WWF (2015c) Principles for a sustainable blue economy. London

Selected Websites and Social Media

#BlueGrowth
#BlueInvest 2018
www.aivp.org
www.allianz.org
www.amsterdamsmartcity.com/knowledgecentre
www.arcadis.com/en/global/our-perspectives/sustainable-cities-index-2016/
www.arup.com/Publications/SlimCity.aspx
www.barcelonaactive.cat
www.bluecity.nl/en/
www.ceres.org
www.circle-economy.com
www.circle-economy.com/how-amsterdam-goes-circular/
www.coastalsocioeconomics.noaa.gov
www.eiab.eib.org
www.isa.org
www.kobenhavn.dk
www.laudatosichallenge.org
www.mckinsey.com
www.mercerhr.com
www.mind-lab.dk/en
www.mori-m-foundation.or.jp/english/research
www.nyc.gov/site/sustainability/onenyc/
www.ocean-energy-systems.org
www.rotterdam.nl
www.sdghub.com/
www.stockholm.se
www.theatlanticcities.com
www.unglobalcompact.org/sdgs/sdgpioneers/2017
www.unwto.org
www.wbcsd.org
www.weforum.org
www.worldbank.org
www.wwf.org

Watercolour 8
Malta, Between the Sea of Hope and the Promised Land

Chapter 6
An Inclusive, Just Society as the Ultimate Frontier on Land and the Sea

Abstract Fairness constitutes the ultimate ethos of the social architecture of sustainable cities that strive to offer citizens a life that is fulfilling and worthy to live. Eco-responsible cities care much about the distribution of sustainable wealth across urban actors, space, and time. Interactions among diverse people and activities are invaluable assets for cities, schools for respecting difference and living harmoniously together. This chapter focusses on the social capital of sustainable cities and their capacity for intergenerational and intercultural equity, public health, safety, and solidarity, including in relationship to surrounding seas. Cities are themselves important direct and indirect sources of investment, education, and employment, the first factor of social integration, followed by housing. They can give the impulse for local to global initiatives on land and in the sea. As migration rises on the global horizon, reinforcing inclusiveness is of increasing importance in cities that feel part of a universal community. Nobody should be left outside or behind; all citizens should have access to the conditions to enjoy a healthy, meaningful, and rewarding life, in balance with the blue planet.

6.1 Cities and Seas Without Borders: Welcoming and Sharing Places

6.1.1 Urbanisation, Migration, and Integration

Migration is a great challenge for urbanisation and inclusiveness in both advanced and developing countries. In ageing societies, as in Japan and Europe, migration is often the only source of net demographic increase, while in developing cities, the migrant population joins the increasing endogenous population in often explosive ways. Migration, especially when chosen and not forced, brings multifold opportunities that can be grasped for the benefit of all. A key feature of the increasingly interdependent world, international migration, although three times less comparing to that internal, attracts the most attention, as it enables the exchange of talents, services, skills, and experiences. If "sustainability is equity extended into the

V. P. Mega, *Eco-Responsible Cities and the Global Ocean*,
https://doi.org/10.1007/978-3-319-93680-2_6

future," a responsible resource-conscious journey to the prosperity and well-being of all, the welcoming and integration of immigrants through activities in line with sustainable development are an important part of the journey. The blue–green economy has a high potential for decent jobs and incomes, food and nutritional security, and public health and well-being (OECD 2016b).

The world population of international migrants reached 258 million in 2017, 3% of the global population. International migration could increase and counterbalance demographic dynamics. Migration offers a means to alleviate pressures on the social security systems of countries with rapidly ageing populations that face workforce shortages. One study suggested, however, that although global migration flows were very likely to rise, they would not suffice to offset the effects of population decline in OECD countries (OECD 2013; EC 2017a; UN 2017).

Most migration journeys choose metropolitan areas, cities and their suburbs, as destination. Nearly one in five of all migrants live in the world's 20 largest cities. In some cities, migrants represent more than a third of the population. Cities are places where inclusive growth policies can be most influential, ensuring the access of migrants to higher incomes and opportunities. In Asia and Africa, rapidly growing cities are expected to absorb almost all the future urban population growth of the world, and this mobility pattern is characterised by the temporality and circularity of the internal migration process. The fast rate of urbanisation, and the rising migration to cities, brings with it both risks and opportunities for migrants, communities, and governments. Migration is a great source of diversity, opening cities to world flows of mobile citizens, ideas, and perspectives. Ethnic, religious, and cultural diversity is extremely important for sustainable cities; it favours interaction, co-creation, and resilience (OECD 2013).

Migration and migrants are shaping cities, being, in turn, shaped by cities and citizens, organisations and institutions, in multiple interacting ways. Much of the discussion about migration trends and policy tends to focus on the national level. Taking the issue to the city level increases understanding of the local political processes of migration and the close connection between migration and urban development. The discussions on migrants and cities, inclined to concentrate primarily on the Global North, are also enriched with the conditions of migrants in the cities of the Global South, as all cities across the development spectrum have to address challenges brought by growing mobile and diverse populations (IOM 2015).

As advancing climate change redesigns the map of the liveable world places, the first waves of climate migrants added around 32 million to displaced world citizens, next to 50 million asylum seekers in 2014. The cost of inaction is likely to be higher than the cost of policy measures, especially if reducing the likelihood of knotty population displacement. Policy attention to migration in the context of environmental change should also be given at the earliest possible stage. New Zealand has already proposed a new humanitarian visa to receive displaced citizens in the Pacific (EC 2017a).

The local environment and experience play a significant role for the integration of migrants. The strength of local networks and their interconnection with world networks can be decisive for cities to welcome migrants and weave ties with their places of origin. Migration towards lower-risk places could pose stimulating questions concerning citizenship and identity and generate societal tensions, which often fuel populist political movements. The conception and deployment of sustainable infrastructure for the meaningful integration of migrants would require partnerships among governments and local authorities, business and civil society organisations, and international institutions. Systemic approaches help to upscale micro-solutions to macro-urban and ultimately world environments (WEF 2016, 2017).

Ports often attract more migrants than do other cities. Access to the sea holds multiple opportunities and signals an urban lifestyle more open to diverse cultures and values. Major coastal cities, such as New York, Boston, and San Francisco, the largest immigrant gateways in the US, and London, Rotterdam, and Amsterdam in Europe are receiving large flows of migrants. Almost half the population of Rotterdam is composed of foreigners of 171 different nationalities. The labour and housing market situations are defining attraction factors, but also local openness and acceptance of new citizens. Long-term unemployment of migrants is, however, a serious challenge in many countries. The average unemployment rate of the foreign residents in OECD countries has increased by 5% between 2008 and 2012, compared with 3% for those native born (OECD 2013).

Many cities owe much of their wealth to foreigners. International statistics suggest that Dubai welcomes the largest migrant population in the world, with more than 70% of its population being foreign nationals. Although this diaspora is predominantly from South Asia, there is also a significant Western population. Vancouver, another notoriously multicultural city, owes much of its success to immigrants, many of whom have claimed Canadian citizenship. The S.U.C.C.E.S.S (Immigrant Settlement & Integration Programme) assists immigrants, refugees, and their families to connect with community resources and government services, develop language and communication skills, find employment, and engage with community development.

As a politically sensitive issue, migration requires public policies to clear up stereotypes and misunderstandings and promote a welcoming culture. From refugees and asylum seekers to expatriates, the spectrum of immigration is increasingly diversified and defies standardised approaches. Cities can have a role in changing the perception of migrants in the societies of both origin and destination, forging solidarity bonds among world communities. Interfaith and intercultural centres and activities, especially in disadvantaged areas, help minorities coexist harmoniously and interact peacefully. The organisation of exceptional events, for example, interreligious open days, can stimulate dialogue and mutual understanding (Cities of Migration 2012).

Many European cities have started seeing immigration as an opportunity to be shaped and grasped. They can learn from Amsterdam, European capital of tolerance, one of the best examples of openness and willingness to accept and integrate newcomers as a source of wealth and progress. Since the city's golden age, political refugees, intellectuals, merchants, and workers of every race and belief have lived side by side with the local population and invested in common activities, not least in port industries. Plurality, a key element in European civilisation, has been important in the prosperity of the city.

6.1.2 Access to Employment and Opportunities for Sustainability

Cities are unique laboratories, where opportunities for prosperity coexist with stark inequalities. Eco-responsible cities must invest in the co-creation and distribution of sustainable wealth. The blue–green economy has an important potential to attract a diverse workforce in cities. The development of skills and activities, integration in the labour and the housing market, and the welcome of mobile offshore workers such as seafarers make the role of cities crucial as central institutions engaging all stakeholders. Europe's maritime sector employs more than 5 million job holders, generating almost €500 billion a year, with a potential for many more jobs. In 2012, 99% of all jobs could be classified under five broad areas: living resources, nonliving resources, transport, shipbuilding, and tourism, but the landscape is changing. Explosive growth in the installation of offshore wind farms brought forward a sixth sector, offshore renewable energy, which accounts for 150,000 jobs (EC 2017b).

Cities are the mirrors of societies, with their strengths and tensions. Training and the development of skills provide a valuable community service and have to be continuously redesigned to support the dynamic path towards sustainable development. The transition from fossil fuels to renewable energies and the bioeconomy is a source of multiple opportunities which should attract professionals of a broad spectrum of qualifications. Unemployment is a major waste of resources and talent that undermines urban futures. The surge of citizen movements is indicative of the political repercussions which find in cities their privileged spaces for protest. The 'Indignados' in Madrid is an outstanding European example that generated new political movements (Harvey 2012).

The youth labour force is extremely important for sustainability and has declined sharply during the past 20 years. Youth unemployment is particularly alarming as young citizens usually are the main victims of crises. Youth are three times as likely as adults to be unemployed (ILO 2017). Structural trends underline the need for investment in human capital. A protracted recovery that does not generate employment has not only important economic and social costs but signals longer-term deficiencies in the labour market which weaken the seeds of growth. The concurrent rise in job vacancies and unemployment indicate labour market mismatches among required skills and education and training. Education and lifelong training policies are essential to equip the unemployed with the skills required for promising sectors that contribute to sustainable development. Training portfolios related to the exploration and the care of the ocean can initiate a new blue–green employment dynamic (EC 2015).

Integration of youth and newcomers through employment for sustainable development is vital and provides many links to the ocean commons. From marine biology and aquaculture to offshore renewable energy and sustainable nautical tourism, many coastal and maritime sectors have much potential for those looking for a job and a meaningful and principled way to serve humanity. Fisheries still employ 4.4% of the estimated total economically active people, of which 84% are in Asia and 10% in Africa. Women are estimated to account for more than 15% of these. Employment in aquaculture seems to have much potential, which has to be sustainably developed. Cities can create a wide array of direct and indirect jobs to stimulate, design, regulate, and support the marine and maritime economy. In the EU, the coastal and maritime tourism sector has been identified as a domain with special potential to create new quality employment in the EU regional sea basins (EC 2017b).

In Liverpool, the School of Engineering, Technology and Maritime Operations at John Moores University produced, in 2008, a 'Skills for growth agreement' that envisaged that the expansion of the port would generate 20,000 new jobs. A course called 'sharp end' tried to reconvert professionals from various sectors into promising maritime activities. At the top end, the University pinpointed hard-to-fill vacancies in fields such as automation engineers. Such top-end jobs bring innovation for new products, processes, and services, which, in turn, create demand for lower-skilled workers. A network organisation for the maritime sector, Mersey Maritime, supporting 1500 knowledge SMEs has proved instrumental for discussing skill agendas.

Shipping and maritime activities that respect the environment can offer rewarding onshore and offshore careers. Maritime jobs can involve working with revolutionary technology, for example, in exploratory ships, open sea laboratories, or platforms, maintaining high quality and safety standards, supporting technical knowledge, and preventing harm to marine ecosystems. The European marine equipment manufacturing and industry, including propulsion, communication, automation, and integrated systems, is composed of about 7500 companies. The European shipbuilding and ship repairing industry is made up of around 300 yards, of which more than 80% can be considered to be small to medium ships of 60–150 m. This sector employs more than 500,000 people.

Employment still remains the best protection against poverty and inclusion. Although employment is considered to be the most impactful factor of social integration, more than 8% of the EU population in employment is estimated to be at risk of poverty, and the trend in income inequalities remains a generally upwards one (EC 2015). This trend exemplifies the need for an integrated approach addressing both social and employment situations. Working on a temporary contract, which is the case for 40% of youths, is another crucial feature that must be seen in the context of rising automation. In times of uncertainty and complexity, social inclusion asks for structural reforms addressing integrated competitiveness, fairness, and sustainability challenges. Cities have to care about the quality of jobs but also the quality of life of those not in employment, education, or training or at risk of poverty. They have to ensure that the risk of poverty is contained and exclusion is prevented, while providing equal opportunities and investing in the interconnected spheres of education and training, labour, and housing markets (Mega 2013, 2015).

In 2016, the year of the new urban agenda at global level, an OECD groundbreaking study provided internationally comparable data on well-being outcomes, socioeconomic segregation, and economic growth and inequality in metropolitan areas of at least 500,000 inhabitants. With the "All on Board for Inclusive Growth" initiative, the OECD has brought inclusive urban growth to the forefront of the global policy agenda. Cities offer targeted services, and form and attract highly educated workers and innovative employers. It is usually easier in cities for individuals to climb up the income, education, or jobs ladder. But cities, especially the largest ones, also concentrate inequalities in both income and opportunities. Urban dynamics have to be taken into account by national and local governments reorienting efforts towards cities that strive for sustainability (OECD 2016a).

Mobile world citizens have good reasons to come to cities. In OECD countries, cities hosting on average 50% of total population contribute to 60% of total employment creation and GDP growth in the past 15 years. On average, household incomes are 18% higher in cities than elsewhere, although higher living costs may partially offset such a premium. Cities offer opportunities to citizens wishing to increase their incomes regardless of their background. In Canada and the US, for example, prospective earnings of urban residents are less correlated to family income than those of nonurban residents. At the same time, cities within countries such as Italy or the US have very diverse profiles.

OECD cities have experienced very different patterns of growth since 2000. There is an array of approaches to inclusive urban growth, such as evaluating the ways in which cities have increased economic prosperity, while engaging more of their residents in generating it. Some cities have seen both an increase in growth and in labour participation (e.g., Tallinn, Santiago, Perth), whereas in others the GDP per capita has increased although labour market participation has declined (e.g., Poznan or Takamatsu), and in others both growth and labour participation have stagnated or declined (e.g., Catania or Las Vegas).

The size of a city and the place in which citizens live within a city may greatly influence their condition and prospects. Larger cities tend to generate greater income inequalities. In Canada, France, the Netherlands, and the US, the shares of the working-age population with a tertiary education vary by more than 15% among cities. Income inequality, which has been rising in the past decades, is higher, on average, in cities than in their respective countries. Life expectancies, for example, differ by a staggering 20 years across neighbourhoods in Baltimore and London. When income, jobs, and health are considered together, differences in overall living standards in the various places within a country are starker than those in terms of income only, showing that different well-being outcomes amplify the concentration of prosperity or exclusion in regions (OECD 2016a).

Often, cities are split across socioeconomic lines, which may reproduce disadvantages across generations. Rich and poor urban residents tend to live in clearly separate neighbourhoods. In Denmark and the Netherlands, spatial segregation in cities affects more the poorest households, whereas in Canada, France, and the US, the richest are more likely to live in separate neighbourhoods. Citizens living in disadvantaged areas often have lower-quality public services, which undermines opportunities. In the Netherlands, a relatively egalitarian country by many standards, children growing up in the poorest neighbourhoods have, on average, adult incomes that are 5% to 6% lower compared to those in the most affluent neighbourhoods. Inclusive growth in cities can become a reality, if national and city governments work together (OECD 2016a).

Opportunities for inclusive growth depend on urban governance, citizen expectations, human capital, and the business environment. The latest evidence suggests that cities characterised by a lower level of administrative fragmentation also display lower spatial segregation. Urban inequality and spatial segregation are, however, not inevitable. National and city governments should work with all stakeholders for inclusive growth. Although there is a broad convergence between the priorities for national urban policy and for cities, policies for transport, education, and jobs are not under the sole responsibility of one level of government. Bridging national and local initiatives at the right scale and moment is essential. Without an integrated approach, policies and regulatory frameworks put in place by different levels of government may shift problems from one area to another.

Spatial segregation can prevent access to opportunities for many low-income urban residents living in distressed neighbourhoods. Averting cities from becoming inequality traps requires investments in adult skills training and entrepreneurship, and creation of activities in locally relevant value chains across the sustainable development spectrum. Improving access to education with a particular focus on disadvantaged groups and increased investment in early childhood education is vital, together with the establishment of vocational training programmes that match local needs. Better targeted housing allowances can make access to housing fairer and promote mixed-income neighbourhoods. Coordination of investments for urban housing and transport, access to education and healthcare and generation and capturing of synergies between national and local policies for urban development can only have beneficial and mutually reinforcing results.

The sustainability transition embraces many initiatives with a high potential for job creation. A prime example is the U.S. Better Buildings Initiative, ensuring that America invests in innovative clean energy technologies and doubles the share of electricity from clean energy sources by 2035. The initiative targets commercial and industrial buildings that must become 20% more energy efficient by 2020 and accelerates private sector investment in energy efficiency. The Better Buildings Initiative brought a great array of green jobs in sectors as diverse as lighting, insulation, heating and cooling, information, and access to green financing.

Decent green jobs are already part of urban strategies towards post-carbon societies. The city of Rotterdam estimated that in 2013, some 3600 jobs were created in its region, in relation to climate change adaptation, infrastructure, and smart monitoring (EPRS 2017). New York's Greener Greater Buildings Plan has been a source of thousands of green construction-related jobs. The Plan embraced four regulatory pieces supported by extensive jobs training. It asks for large buildings to annually benchmark their energy performance, that a local energy code be adopted, that these buildings conduct an energy audit and a retro-commissioning every 10 years, and that, by 2025, the lighting in the nonresidential space be upgraded (NYC 2015a, b, 2016, 2017; UNISDR 2013).

The experience of Boston suggests that climate action can generate many high-quality jobs and bring significant economic benefits. Demand for energy and climate-related services have a potential for thousands of jobs. Energy efficiency and behaviour change are great multi-win action domains with multiplier effects, expected to produce net savings of energy while creating quality employment. From a citizen perspective, choosing such skills and employment pathways is vital for a successful professional and social life. From the perspective of the sustainable urban economy, it is crucial to improve skills prospects, labour market adjustments, adaptability of enterprises and workers to sustainability, and support to aligned value chains. The concept of flexicurity, 'flexibility within security,' can be useful for blue–green cross-sector activities.

The creation of green jobs, central in the transition to a zero/low carbon economy and society, is considered by the World Bank to be the most human face of green solidarity. Blue–green jobs signal the transformation of economies, enterprises, workplaces, and labour markets into a sustainable economy providing decent work in balance with the sea. However, innovative strategies to promote blue–green jobs can only succeed with the full involvement of local authorities and the participation of workers and enterprises (ILO 2012; World Bank 2012).

The emerging blue–green economy may offer multi-sector and multilevel jobs. The ocean economic outlook for 2030 suggests the fastest growth is expected in off-shore wind energy, marine aquaculture, fish processing, and port activities. The OECD modelling exercise suggested that many of the ocean-based industries have the potential to outperform average world economic growth as a whole, in terms of both value added and employment. The projections indicate that employment in ocean-based industries is expected to increase by 30% by 2030, outpacing the increase of the global workforce (OECD 2016b).

6.2 Universal Values, Local Communities, and Intercultural Dialogue

6.2.1 *Equity and the Eradication of Exclusion*

Social cohesion is of utmost value for sustainable cities, places of harmonious economic, gender, social, ethnic, racial, generational, and cultural coexistence and cross-fertilisation. Social justice is a precondition for the creation and sharing of sustainable wealth. Diversity is not an obvious asset, because otherness and difference often raise feelings of suspicion, which may generate incomprehension and engender defiance. Openness, understanding, and mutual recognition and acceptance have been invaluable ingredients for many thriving cities and are considered to be critical factors for prosperity (OECD 2017a, b; Storper 2013).

A strong human and social capital is a high feature of all cities striving for sustainability, and it increases the urban capacity to withstand crises. Given the powerful relationship between urbanisation and capital accumulation, any city can easily become a key arena of confrontation. The role of equity in shaping sustainability processes is unparalleled. Unequal distribution of wealth may result in revolt, unsustainable lifestyles, and obstacles to cultural change. The breath, scale, and impact of these tensions are very diverse. The escalation of conflicts, especially in times of crisis, may transform movements for the right to the city into urban uprisings and revolutionary actions (Harvey 2012).

The main social challenges for cities include the fair distribution of wealth and solidarity, the harmonious coexistence of an increasing number of generations, races, and religions, high-quality employment and fulfilling environments, the creation of a welcoming and friendly environment for newcomers and the most vulnerable and disadvantaged citizens, public well-being and security, and the enhancement of opportunities linked to intercultural diversity. Longer healthy lives can be a great chance for cities wishing to address the expectations of an older population but also benefit from the experience and contribution of all and that wish to bring to the future generations a well-preserved overall natural, human, social, and man-made capital.

Almost 900 million people are at risk to fall into poverty traps. Urban poverty is an utmost problem. According to the United Nations Development Programme Multidimensional Poverty Index, about 1.5 billion people in the 102 developing countries currently covered by the Index, about 29% of their population live in multidimensional poverty, that is, with at least 33% of the indicators reflecting acute deprivation in health, education, and standard of living. Environmental disasters and extreme weather phenomena linked to climate change may further aggravate structural or situational vulnerabilities that reduce human potential for fulfilment.

Universal access to fundamental services, such as health and education, is crucial for strengthening human capital and preventing social exclusion and multiple inequalities. Local services are the best placed to understand the persons behind the statistics. A three-pronged approach has to address the three main dimensions of poverty and exclusion, including income poverty, material deprivation, and exclusion from the labour market. The reduction and eradication of poverty is a strong priority of all sustainable development policies, especially in ageing societies. Poverty-trap conditions of low income and welfare-dependent structures can exacerbate exclusion and seriously compromise chances for upward change. Increased financial pressures, in a complex and fragmented institutional environment, have to be addressed through the horizontal and vertical integration of decision-making systems and the optimisation of the commitment and contribution of the public, private, and social economy sectors.

Sustainable enterprises can have a great role for local development and social cohesion. Cities can develop a favourable environment for enterprises that incorporate sustainability values and offer opportunities for the empowerment of minorities. The social enterprises, rooted in local realities, demonstrated a better resilience to the financial crisis, which is a prime signal of their capacity to sustain principled economic activities.

Social cohesion reinforces the resilience of places and communities. A quartered city mirrors multiple divisions caused by unheeded developments, often resulting in dichotomy between spaces and functions. The decline of many urban port areas has led to distressed areas suffering functional impoverishment, with destitute housing, insufficient equipment and inadequate services, delinquency and crime, high unemployment, low mobility, little access to information, education, and training, and high levels of substance abuse. Very often, transport infrastructures, whether operating or disused, fragment the urban web and further isolate distressed spaces from vibrant city centres. Comprehensive urban policies have to simultaneously address all these aspects and create new momentum for attracting and integrating a diverse population.

Cities have to raise awareness and ensure the regeneration of their spatial and social fabric, and preserve and enhance their social capital, in balance with the economic and environmental capitals. They must offer all citizens access to information, education, and training, adequate housing and noble public, green, and recreational spaces, social services, and the possibility to participate in co-designing the future of the city in harmony with the blue planet. Vulnerable coastal cities must in particular inform citizens about the risks that the proximity of the sea involves and engage them in collaborative prevention design processes. Building social cohesion contributes to the creation of a resilient society, able to endure threats (UNISDR 2013).

The social face of cities is composed of very diverse singular attributes. Its richness springs from the plurality of otherness. It is crucial to promote intercultural dialogue to avoid segregation and "ghettoization" and prevent spirals of identity violence. The blue economy can generate many precious synergies, which can be captured and harvested for the benefit of all citizens. Working together with stakeholders, citizens, and civil society sensitive to public goods and global commons can create a climate that offers everyone the opportunity to share information and knowledge, exchange ideas, and engage in shared action. If cities manage diversity properly, they can benefit hugely from the potential of all minorities for entrepreneurship and innovation. For this, they should review the array of policies, services, and instruments to create the appropriate conditions and governance structures.

Intergenerational solidarity is an essential dimension of sustainable cities. The noble long-term aspiration of sustainable development can offer a shared value to adhere and make a transgenerational contribution. Instead of bringing conflict over scarce resources, ageing could enrich community life and forge more conscious and vibrant relationships across generations. Otherness of cultures could also add various parameters of difference. Older scientists from other locales have a precious capital to transfer to the younger generations, as have older leaders and businessmen. Sea navigators and explorers are particularly appreciated as role models for younger generations.

Intergenerational cities are those in which every generation brings its best to the community. The talents of the elderly are invaluable to the next generation of city leaders, professionals, and citizens. Active ageing policies should include supportive measures such as discouraging early retirement, stimulating learning and training to avoid skills obsolescence, adapting working conditions to the specific characteristics of older adults, valuing volunteering community work, promoting the good health of older workers, and providing care for the elderly, while making them most valuable to the city. Universities for the elderly expand their offer for sustainable development to address a great variety of interests to older adults, and cities can sponsor courses that generate curiosity in the exploration of the oceans and the transmission of healthy seas to future generations.

Nantes boasts the creation of social capital which transcends the generational divide but also many other invalidating notions such as disability and vulnerability. Senior citizens have been introduced to digital creation. Accessibility of all to culture has been essential for revitalising the city, a major Atlantic port for trade with the colonies and the slave trade from the seventeenth to the nineteenth century and a flourishing industrial city in the twentieth century, hit by the crisis of the shipyards and reinvented through art and culture. The Castle of the Dukes of Brittany, the last château on the banks of the Loire before the ocean, has obtained a quality label for disabled visitors in four categories (physical, sight, hearing, and mental). Its wonderful buildings, dating from the fifteenth and seventeenth centuries, promote accessibility on a truly magical voyage through the history of Brittany (Ville de Nantes 2012).

If cities wish to attract talent in a competitive world, they should develop a culture welcoming to newcomers. Boston created a Mayor's Office of New Bostonians to better recognise the resources and skills of immigrants and provide them with the necessary attention and support. The settlement and integration of newcomers is a fundamentally local experience and the quality of the welcome has a huge influence on their future success and ultimately on the prosperity of the cities. Bilbao developed a multicultural resource guide offering information on a wide array of services including healthcare, education, and employment services. With the support of the Maytree Foundation, cities started exchanging their experiences in welcoming migrants and contributed to more than 100 "Good ideas in integration" (Fundación Bertelsmann and Cities of Migration 2012).

Cities are uniquely positioned to learn from each other and to adapt and replicate good practice. In Barcelona, Fundación Bertelsmann, founded in 1995, focused first on developing the reading and media culture in Spain, then one of the top migrant-destination countries. This focus resulted in efforts to benchmark 20 public libraries and promote cooperation between libraries and schools in seven cities. As part of its tenth anniversary, Fundación Bertelsmann shifted its focus to the promotion of civic engagement and committed in intercultural dialogue. In partnership with the international Cities of Migration project, it is identifying good integration and engagement practices in Spain.

All cities can provide a better welcome and create the conditions for newcomers' attachment and belonging, while striving to redefine and enrich community relationships. Involvement can be a powerful generator of integration and development and can be stimulated by cities. Whatever the cultural mindset, legal and jurisdictional framework, or differences in national context and municipal responsibilities, cities around the world are accountable to their constituents and can exchange experiences, successes, and failures. Cities can inspire each other as service providers, as employers and diversity managers, and as the drivers of the local economy, from infrastructure and procurement to support for investment, entrepreneurship, and small business incubation (Cities of Migration 2012).

6.2.2 Universal Languages for Preventing Xenophobia

Openness of cities and their professional actors and bodies can stimulate bottom-up and mutual trust approaches to assist the integration of ethnic minorities. Cultural proximity can create a bond between cities and communities and encourage new talent and creations. The Suikeroom project in Amsterdam is a fund for ethnic start-up companies financed by established companies. The fund, created in 2006, helps ethnic minority entrepreneurs, appraised as potential profit makers, to start a business on solid foundations (Eurocities 2010).

Many cities have adopted active integration strategies and mainstream them in their policies to strengthen their collaborative gene and aim for leadership that reflects the full diversity of the urban society in the public, nonprofit, and private sectors. The number of languages spoken in a given city is becoming an important indicator of cultural diversity. It is estimated that more than 170 languages are spoken in Rotterdam, the first European port city counting 640,000 inhabitants. Schools are the roots of integration, experiencing cultural and religious diversity and the natural open-mindedness of children.

Innovative multi-stakeholder partnerships empowering immigrants and especially women, through micro-enterprise development and initiatives promoting sustainable value chains, are most promising. Antwerp searched to integrate immigrants in its dynamic creativity web. In this innovative fashion city, Betet Skara, a local weaving house, works with Assyrian Christian refugees using traditional weaving techniques. As most newcomers and political refugees, the arriving Assyrian community had to find employment and create wealth. The professional craft of hand weaving, once a declining industry, is being revitalised by local social enterprises, through techniques and patterns passed on from generation to generation. For the design of its port house, the city invited an architect world citizen, Z. Hadid, who beautifully endowed the city with a new diamond.

Education, housing, health, sports, culture and the arts, and local politics are the domains with the highest potential for the integration of migrants in the city (EC 2010a, b). The local environments impact the ways in which immigrants interact with institutions such as schools, cultural associations, and healthcare systems and contribute to the wealth of the city. Cities should invest in preventing tensions between spatial proximity and cultural distance, also linked to the coexistence of migrants with other very diverse established social groups and the prevailing harmonious or conflictual dynamics.

Auckland is a city reputed for its quality of life and the warm welcome to immigrant communities. The city enjoys one of the better-quality environments in the Southern Hemisphere, with well-developed businesses and social infrastructure. Many world citizens are attracted to New Zealand's capital, which honours them through a great number of events to celebrate their origins and cultures. The city enhances the universal languages of arts, culture, and the sports for promoting communication and intercultural dialogue with the newcomers.

Socioeconomic integration can start with music or dance, video or soccer. Arts and sports forge bonds among populations divided by mother tongue and tradition and help the integration of young migrants or refugees while fostering a sense of identity and belonging. In Auckland, soccer has been used as a strategic tool to reduce the social isolation of young refugees and to help them feel connected with each other and become a genuine part of the wider community. The Refugees in Sport Initiative, launched in 2006 by "Refugees as Survivors," a nonprofit refugee mental health agency, enabled young refugees to achieve better access into mainstream sports and have a safe place to meet with others sharing and understanding their experience. The introduction of other sports, such as cricket and martial arts, and the encouragement of girls with refugee backgrounds to participate in both team and individual sports, further expanded opportunities (Fundación Bertelsmann and Cities of Migration 2012).

The Intercultural Cities programme (ICC), introduced by the Council of Europe, supports cities in assessing their visions and actions through an intercultural lens and developing strategies to better identify and realise the potential of the diversity competitive advantage. Compelling research evidence demonstrates the value of an inclusive approach to diversity for human groups and communities. Cities can gain enormously from the entrepreneurship, variety of skills, and creativity associated with diversity, and should adopt policies and practices that facilitate intercultural interaction and inclusion of minorities. More than a hundred cities in Europe and beyond are currently engaged, through the Intercultural Cities network, one of the main tools of the ICC programme, in further developing the intercultural integration knowhow, and exchanging good practice to improve their policies. An impact evaluation of the intercultural integration approach has confirmed its effectiveness for building inclusive, tolerant, and pluralistic communities.

Openness and tolerance, understanding, solidarity, and inclusiveness are key principles for integrating immigrants and enriching cities. The Barcelona Interculturality Plan, published in 2010, after more than a decade of work by the Barcelona City Council, intended to serve as a roadmap for action to address the challenges of "coexistence in diversity." Based on a triangle of principles, equity, recognition of diversity, and interaction, it proposes the establishment of an institutional innovation, Espai Avinyó, to promote interculturalism values, the social use of Catalan, and the historical and cultural understanding of the city. The Interculturality Plan initiated a policy that focuses on the relationships, the dialogue, and the mutual opening of both newcomers and long-term residents. The purpose is to create new foundations for community life which transcends cultural divergences, prevents conflict, and creates the conditions for solidarity.

6.2.3 *Intercultural Dialogue Across Shores, Values, and Vision*

Living on the shores of the same sea can be a great source of intercultural dialogue, underlining the role of seas in uniting rather than dividing. European citizens and those in countries on the Southern and Eastern shores of the Mediterranean (SEM) have a growing desire for mutual knowledge, but face the reality of an alarming rise of xenophobic and sectarian trends impacting social dynamics across the region. This amazing conclusion comes from the Anna Lindh Report 2014, following the results of a regional opinion poll carried out through a representative sample of around 13,000 citizens from 13 different countries of the region, exploring their expectations and wishes for the future (Anna Lindh Foundation 2014).

The Mediterranean Sea is an environmental, climate, economic, social, cultural, and migration hotspot, a space of intense interactions. A first report on intercultural dialogue in 2010, following a first poll conducted in 2009, in several European countries and countries bordering the Southern and Eastern Mediterranean had highlighted the trends. The first survey included eight European countries and five SEM countries. The 2014 survey was specifically designed to identify changes since the Arab Spring. The fieldwork captured the opinions in the autumn of 2012, 3 years after the initial poll. The target population included all individuals older than 15 years, the questionnaire was translated into the major languages of each country, and responses went through quality control procedures. Some of the countries were the same in both studies, although others changed to provide a wide coverage. The results contribute to fill the gap in perceptions and to understand the existing dissonances and divergences between people and communities across the shores of the Mediterranean. Some of the results also help tackle misperceptions and rebuild human and cultural bridges in the region and ultimately facilitate the actions of civil society, decision makers, and opinion leaders. The dialogue has a long-term perspective of shaping the Euro-Mediterranean space as an area of cooperation, exchange, mobility, and peace.

The Mediterranean evokes first and foremost "Mare Nostrum," a sea with a strong common heritage and cultural ramifications that go much deeper and beyond the daily news. Citizens on both sides of the Mediterranean shared an overall positive image of the region, presented with several positive characteristics and associations. About nine in ten respondents in Europe and about eight in ten in SEM countries associated the Mediterranean region with a common cultural heritage and history, an art of life and a culture of hospitality. Both the perceptions and the opinions about the region acknowledge the openness to foreigners, as it appears from the emphasis given to hospitality across the Mediterranean region.

Concerning the cultural perceptions in the two regions, the similarities are more striking than the differences. For Europeans, the Mediterranean region is more often linked to a certain lifestyle and food than for respondents from SEM, and the former are those who are more likely to associate it with turmoil and insecurity. The differences concerning religious beliefs and practices are very instructive as they relate to fundamental values. A good example is the tendency of the majority of Europeans to doubt the existence of absolute moral guidelines. The tendency to believe that what is considered good or bad is not predetermined but can change according to circumstances, a relativist moral standpoint, is probably the most revealing on the ways that different societies observe and judge the world in general. The belief in an absolute truth was by far more widespread in SEM. On the other hand, a majority of respondents in Europe advanced the view that truth was rather relative, and only a quarter of them believed in an absolute truth (Anna Lindh Foundation 2014).

Attitudes and social norms that appeared to be rapidly changing were mainly related to women's role in society. On average, across Europe and on both shores of the Mediterranean, roughly six in ten respondents thought that women's role in society had become more important over the past 5 years, but about three in ten responded that nothing had changed in this regard. Respondents in SEM countries, however, were more likely than their European counterparts to think that the role of women in their society would become even more important in the next 5 years (66% and 56%, respectively). Among respondents in SEM countries, Moroccans were the most likely to believe that women would play a more important role in their society in 5 years' time (89%). The proportion expressing this view was also high in Tunisia (63%), Turkey (75%), Jordan (78%), and Egypt, where 48% of respondents thought that women's role in their society would become more important in the next 5 years. In most European countries, the share of respondents suggesting that women would play a more important role in their society in the years to come remained below 60%, with a most striking exception being Albania (73%).

Expectations for the future are always coloured by the perceptions of the past and the present. Compared to the past, the prospects for change seem much stronger in the South than in the North of the Mediterranean. A slim majority of respondents in SEM countries assessed their life situation as being better than 5 years earlier and only 15% stated that life had become worse. Expectations for further improvement are high. A majority of respondents were optimistic, ranging from between 60% in Egypt and 82% in Morocco, hoping that their life situation would further improve in the coming 5 years. The significant changes that Arab countries were experiencing, and which started with the Arab Spring, were an important part of the general discussions about the future of the region and the North–South relationships. On the European shores, nearly half (45%) of the respondents thought that the ultimate effect of the Arab Spring on Euro-Arab relations would be positive, while almost 30% held an opposite view and 10% did not expect any impact.

A rather similar picture emerged in the SEM countries; although it should be noted that the proportion of 'no effect' responses in this country group was somewhat higher (14% versus 10% in Europe), at the expense of the proportions expressing a positive view (42% versus 45% in Europe). Among respondents in SEM countries, those in Morocco were the most likely to think that the effect of the Arab Spring on Euro-Arab relations would be positive, whereas in Turkey, just 35% of respondents expected a positive outcome. Respondents in Belgium and Poland appeared to be as negative in their evaluation of the Arab Spring as their counterparts in SEM countries, and 39% responded that the effect would be negative. This result was in sharp contrast to the majority of positive responses in Denmark (Anna Lindh Foundation 2014).

Participants in the survey probably ignored that if the Mediterranean Sea was a country, it would be the fifth largest economy in the region with an overall value of natural capital at least $5.6 trillion, but this is at risk of collapse. A semi-closed sea, an ocean in itself, the Mediterranean supports an economy that uses far more resources that the region can sustainable regenerate, as suggested by the WWF in a 2017 study for reviving the sustainable economy of the Mediterranean Sea. Rapidly growing mass tourism and unsustainable fishing are identified as areas of major concern that require critical review. It would be an important step forwards to link the intercultural dialogue questions with a sustainable blue economy approach, to implement ecosystem-based ocean management (WWF 2017).

The Union for the Mediterranean (UfM), the intergovernmental institution bringing together 43 countries, 28 from the EU and 15 SEM, to promote dialogue and cooperation in the Euro-Mediterranean region gave all Mediterraneans a common cause on Earth Day 2018. Joining the movement to #BeatPlasticPollution, UfM's Plastic Busters Project, can bring citizens from all shores together in preventing plastic littering in the Mediterranean.

6.3 Housing and Integration, Diversity, and Solidarity

6.3.1 Healthy Living Cells: Second Factor of Social Integration

Decent and healthy housing is considered to be the second, after employment, most influential factor of social integration. Cities have to provide the conditions for good quality private and social housing allowing personal and collective fulfilment. In the path to sustainable development, housing has to go through radical rethinking to enhance ecological performance and become socially uplifting and architecturally compelling. Innovations include smart housing organised in functional neighbourhoods as competitive organic parts of cities, in harmony with their surrounding areas and the planet.

In the emerging world, global partnerships such as Cities Alliance link housing to the fight against poverty and bring together cities striving for urban poverty reduction and the promotion of sustainable development. The partnership features a broad range of stakeholders, including local authorities, national governments, non-governmental organisations, and multilateral organisations. The Alliance aimed at supporting cities in effective governance, an active citizenship, and an economy characterised by both public and private investment. Its goals are to promote good city governance and management: to analyse the Before the turn of the millennium, the "Cities Without Slums Action Plan" was the first endeavour of the Cities Alliance, which proposed a target of improving the lives of 100 million slum dwellers by the year 2020, introducing for the first time such a measurable target. Former South African President N. Mandela launched the Action Plan, and the target was subsequently incorporated into the UN Millennium Development Goals and the SDGs.

South Africa's cities presented unique challenges when trying to overcome decades of apartheid and again gain access to the seas and the world. Neighbourhoods were racially segregated, with poor non-white areas relegated to the urban fringe. Already severely weakened by the end of the 1980s, segregation ended with the country's first democratic elections in 1994. Hopes ran high for changes in all spheres of society, including urban regeneration. Johannesburg was the first city to start attempting transformational change, while other cities watched and learnt to undertake their reforms. The South African Government chose to create a network that would better enable all of them to flourish, by helping them learn from one another, and building their capabilities for development strategy, planning, and implementation. The South African Cities Network was launched in 2002 as an independent, membership-based organisation comprised of the mayors and city managers of the nine largest South African cities.

Homelessness is a serious challenge even for advanced economies and wealthy cities (Eurocities 2012). In the US, surveys on Hunger and Homelessness by the U.S. Conference of Mayors have illumined the hidden faces of American cities. The 2016 report suggested that 42% of survey cities reported that the number of requests for emergency food assistance had increased over the past year by an average of 2%. Among the citizens requesting emergency food assistance, 63% were persons in families and 8% were homeless. Low income was the main reported cause of hunger, followed by high housing costs and poverty. The cities reported a 5% average increase in the number of pounds of food distributed during the last year and an average 8% increase in the budget for emergency food purchases.

In the first world economy, 544,084 people experienced homelessness on a single night in January 2016, and the phenomenon seems three times more important in cities. There was great variation amongst and between cities, and the reported rate of homelessness for cities was 51 homeless per 10,000 inhabitants, versus 17 people per 10,000 at the national level. And even if homelessness is rather an urban problem, solutions seem also to congregate more in cities. The part of the unsheltered population in the study cities (24.7%) is lower than in the US overall (31.6%). Nationally, homelessness has been declining, 12.9% from 2009 to 2016, including a 2.6% decrease from 2015 to 2016, and this pattern was also seen in two thirds of the study cities. Cities respond to homelessness with a variety of housing and service programmes, including emergency shelters, transitional housing, rapid re-housing, and permanent supportive housing, with a shift since 2013 toward the latter two solutions (U.S. Conference of Mayors 2016).

The two larger US cities, Los Angeles and New York, accounted for almost one in five homeless persons in the US with more than 12% in Los Angeles and more than 6% in New York. In recent years, homelessness in New York City has reached the highest levels since the Great Depression. The Coalition for the Homeless provides up-to-date information on the NYC homeless population and proven solutions to end homelessness. In Europe, some pioneer cities, such as Bologna, have given the homeless the right to use the address of the municipality as their postal details. Other cities allowed solidarity action to extend the possibilities of housing. At the other end of the spectrum, many cities use the criminal justice system to punish those trying to survive on the street. In response to the homelessness crisis, the National Law Centre on Homelessness and Poverty published the report "Homes Not Handcuffs," identifying and pinpointing the criminalisation of homelessness. The list of meanest cities has been proposed to name and shame authorities, which even punish citizens for serving homeless people. Many of these measures appear to have the purpose of moving the homeless out of sight, if not out of the city (National Law Centre on Homelessness and Poverty 2016).

6.3.2 *Housing Sustainable Cities*

Sustainable cities care about eco-responsible housing with high environmental performance, often planned and managed by the inhabitants. Mixed ocean and waterfront residential neighbourhoods expand, often on artificial land. Artificial islands are already common extensions of overcrowded frontline cities, and harbouring airports, but dredging activities can bring much pressure on marine ecosystems. Innovative sustainable housing on artificial islands can be a response to severe housing shortages in coastal cities but the impact on surrounding marine and coastal ecosystems has to be carefully taken into account.

In Amsterdam, Ijburg, now an organic part of the city, has been a unique urban planning experiment that consisted of creating a floating neighbourhood from scratch. Ijburg is composed of an archipelago of seven easily accessible artificial islands responding to the Dutch tradition and culture of living with and on the water. The vision for Ijburg was of a diversified urban neighbourhood in symbiosis with the surrounding natural world and the sea. The initiation of the project faced strong opposition by environmental activists on the grounds that it would be a threat to the ecological balance of the area.

The design adopted an ecologically sensitive approach through the involvement of environmental groups in the conception and planning process and ensured an engaging outcome. Ijburg was granted final approval in 1997. The district was developed using an innovative public–private partnership. The private partners committed to purchase land in advance to guarantee the future success of the neighbourhood, developed with a relatively large number of houses that are suitable for families and children. The diversity of building and architecture styles has been achieved with two islands reserved for plot-based housing development. Parcels were acquired by individual citizens, developing their properties with the architects of their choice.

Public investments were instrumental for the creation of artificial land on the sea and the endowment with infrastructure, new roads, bridges, services, and the tram connection of Ijburg to Amsterdam's city centre. Ijburg residents have access to the fastest public fibreoptic broadband Internet. Amsterdam Smart City is working with citizens to develop new products and services for better education, healthcare, transport, and energy services and to improve the local quality of life.

New residential neighbourhoods, including social housing developments, offer ample opportunities for advanced environmental performance and management. In Athens, the Solar Village for lower-income families has been a first attempt to provide eco-quality housing to 1750 inhabitants, 435 families eligible through the Worker Housing Organization. A model experimental village with bioclimatic architecture has been created on a 9-hectare property under a joint Greek-German initiative and the capital that would have used to construct a conventional settlement. Decades of occupation put forth the importance of good maintenance and active care by the occupants.

In Vienna, 25% of housing stock is social housing and a further 35% is limited-profit housing-association stock in socially mixed communities. Housing is seen as a prime human right, and social housing is often endowed with innovative architecture and good design. The renovation of emblematic housing estates has been done with care and the participation of residents continuing their everyday lives during the work. Subsidised rents are funded by taxes, and tenants' rights are strongly protected.

Stockholm Royal Seaport is the largest urban development area in Sweden with at least 12,000 new housing units and 35,000 workplaces. The overall objective is to meet the city's growing housing needs and to offer a model for sustainable urban development, by converting former industrial land into an attractive and vibrant sustainable district, along the waterline of the Baltic Sea, just 10 min by bicycle from central Stockholm. The industrial site will be fully developed around 2030 and transformed into a resource-efficient area with low environmental impact that interacts with the port activities and the residential areas. The district is promising safeguarding biodiversity and ecological values, is climate friendly, and will be adapted to future climate change. To ensure the optimal construction and achieve the sustainability objectives of the housing units, property developers participate in competence seminars at an early stage and communicate with the civil society (City of Stockholm 2014).

In 2008, Copenhagen launched an open ideas competition for the North Harbour, a new urban development area projected for 40,000 residents and 40,000 workplaces. The vision that emerged for this unique site, which features waterfronts on three sides, is a dynamic, lively, and sustainable urban district with diverse, effervescent life created by a broad mix of residents and enterprises. Concretising the vision has been the task of a broad consortium comprising the City of Copenhagen, the City & Port Development, and many companies and utilities. The project gave rise to an urban laboratory to demonstrate future energy solutions in electricity, heating, and transport, ensuring integration and optimal utilisation of renewable energy. This is the first collaboration of this magnitude bringing together relevant stakeholders from authorities, utility companies, industry, and knowledge institutions with the aim of developing energy solutions. The creation of North Harbour is advancing the integrated multi-faceted model of an eco-friendly city, a vibrant city, a city for all and everyone, a city by the water, a dynamic city, and a city with sustainable energy and mobility.

In the UK, the Beddington Zero Energy Development has been the UK's first designed environmentally friendly housing development in the London Borough of Sutton. A multi-award-winning project, created in the first years of the millennium, owned and managed by one of London's largest housing associations, BedZED was the first large-scale community designed to produce zero carbon emissions. The development encourages energy efficiency and use of renewables, public transport, cycling, and walking, and asks for limits to private cars and parking space.

The residents' car mileage is 65% less than the UK average (Bioregional Solutions for Sustainability 2009).

Bioclimatic architecture was well enhanced in BedZED. All houses face south to take advantage of solar gain, are triple glazed, and have high thermal insulation. Low-impact materials were used for the building, selected from renewable or recycled sources within 35 miles of the site, which uses only decentralised renewable energy. Tree waste fuels the cogeneration plant providing district heating and electricity. Rainwater is collected and reused. Appliances were chosen to be water efficient and to use recycled water when possible. A review of the BedZed development in 2010 drew mainly positive comments from the residents and neighbours. It has also highlighted some negative facts, including the reliability problems of the combined heat and power system as the result of technical failures and the intermittent schedule of operation imposed by the local authority, insufficient cleansing of the water by the 'living machine' water-recycling facility, and the still high ecological footprint of the residents, which is, however, about to become half of the UK average ecological footprint.

Paris is preparing to welcome, in 2021, its first 'nudge' building. Its design, inspired by the work of behavioural economist and Nobel Prize winner R.H. Thaler, aims at fostering human relationships and changing attitudes towards the environment. Nudge is planned as a building of 137 homes, a 'vertical village centred on the commons,' embodied in a canopy of shared spaces, and on the ground floor, an agora promoting the circular, social, and solidarity economy and responsible food. The common spaces will be scattered throughout the residence, including the terrace and vegetable garden on the 11th floor, and will offer a path that will be punctuated by a gym, facilities, and meeting points. Private spaces will offer residents the possibility of organising their apartment according to their preference or the evolution of their families. In addition to these developments, playful measurement systems targeted at the energy expenditure of dwellings will be set up, and it will be easy to compare sustainability performance and improve resource consumption patterns.

A well-designed and accessible urban environment can have a significant impact on citizen's lives. The effects of housing, transport, and environmental investments are inherently interlinked, but difficult to coordinate, particularly at the metropolitan scale. For example, narrowly conceived urban and environmental regeneration initiatives may drive housing prices up and put pressure on the transport network, thereby pushing lower-income households out of regenerated neighbourhoods while attracting wealthier residents and high-end businesses. An integrated public investment strategy can improve access to affordable, equitable, and sustainable infrastructure and services, and expand opportunities for socioeconomic mobility in cities (EC 2016; WRI 2016).

Co-designing sustainable housing in developing cities is a special challenge. Medellin took the initiative of assessing and legalising most of the informal housing units that were found to be structurally sound. The Life Articulated Units programme is another project wherein citizens were invited to be actively involved in the planning of communal recreational and cultural venues on previously inaccessible land. The Circumvent Garden is an intervention along the urban–rural edge of the eastern and western mountain ridges that simultaneously addressed control urban expansion and risks of landslides, conditions of settlements, functional public spaces, and connection of the hillside to the city through multi-modal transport infrastructure, thereby improving accessibility to jobs and education.

The favelas in Brazil are some of the largest slums in the world, housing more than 11 million citizens in less than ideal living conditions. However, one slum, Vidigal in Rio de Janeiro, has been reinvented by its inhabitants and transformed into a biological botanic garden, at the initiative of one of the inhabitants who declared his will to clean up the favela in 2005. Everything is up-scaled, flat tires have become paths, and old bicycles are used as tables. In 2008, the government became involved in the process and introduced new police forces to the area, which helped lower crime. Vidigal became a biodiversity oasis and earned a SEED award, and prestigious urban architecture and design awards, for all efforts involved.

6.4 Healthy and Safe Cities and Seas, Youth, and Citi-Zen

6.4.1 Creating Spaces of Choice and Promise

Healthy and fit cities provide the living cells for a healthy planet. They can co-generate virtuous circles and reverse cycles of environmental degradation, physical isolation, obsolete infrastructures, and neglected public spaces, which go hand in hand with poor education, unemployment, weak health and public participation, crime, and violence. Fulfilling homes and welcoming public spaces engender dynamics for good education and employment patterns, public health, and participation (Hall 2013).

The US place-based programmes of choice and promise neighbourhoods brought a difference to lives in fragile communities stamped by poverty and with a greater risk of exclusion. Policy interest in comprehensive community well-being has surged with two federal programmes, the Department of Education's Promise Neighbourhoods, modelled on the acclaimed Harlem Children's Zone, and the Choice Neighbourhoods by the Department of Housing and Urban Development. The essential aim is to address vital local needs with the active participation of all the generations. Promise neighbourhoods tried to foster a culture that creates a "cradle-through-college-to-career" continuum and promotes multiple positive effects. Neighbourhood students have to be safe, healthy, and improve their life and career prospects through coordinated efforts. Choice neighbourhoods focus on communities dependent on social housing to be transformed into communities of opportunity with affordable good-quality housing and high-performing schools and services. The contextual dynamics are crucial in shaping the transformation effort, creating an ascending spiral and capturing the longer-term benefits (Urban Institute 2011).

Cities can play a major role for the well-being of their residents, through forging healthy lifestyles and preventing and fighting diseases. The contemplation of green and water commons even from a distance is considered a source of mental relaxation, peace, and well-being (Children's Society 2014). Good-quality seafood could help change eating patterns and address, for example, the epidemic of overweight and obesity, a critical public health issue. Obesity can lead to serious and chronic diseases and is a major contributor to increasingly unaffordable public healthcare costs. Studies have concluded that obesity, diabetes, and hypertension seem to increase with standardised urbanisation and indicated some 'obesogenic' predominantly urban environments. Urban models and patterns can promote physical effort instead of convenience and minimal exercise. Dense urban transport networks, well-being services and recreational spaces, parks and athletic facilities should be planned as an invitation for citizens to engage in beneficial exercise (Institute of Medicine 2012).

Coastal cities could become pioneers in promoting nautical sports and seafood nutrition programmes. Public beaches accessible to all could help fit cities making physical activity and swimming an integral part of everyday life. They could also make creative partnerships with the private sector and local businesses to create sound food and nutrition environments that ensure that healthy options are the by default choice in public places. Last but not least, they should partner with schools in which so many health and environmental projects begin but also with local chefs and athletes.

Schools could be perfect focal points for obesity prevention, ensuring that all students have adequate opportunities to engage in physical activity and access to nutrition literacy, and enhancing the access to the sea as a gateway to health and well-being. Marine communities have to be empowered to work for change, and cities can support in many ways their efforts to achieve and maintain a healthy lifestyle. Safe public places to walk or play, outlets offering affordable healthy food, and awareness campaigns for healthy foods and beverages can encourage citizens to engage in behavioural change and achieve and maintain a healthy weight (Institute of Medicine 2012).

Well-being for the ageing populations is very important in cities that prepare for gradually older citizens. Centres of autonomy and longevity are being introduced in many cities. Asian societies particularly attentive to the needs of the elderly ask for a range of spa treatments in urban public health centres. Long-term care programmes for healthy active elderly bring together medical doctors, dentists, scientists, pharmacists, nurses, nutritionists, psychologist, general staff, and elderly care volunteers around the concept of holistic and preventive care. Interested elderly citizens are increasingly involved in the design of the programmes that often have a neighbourhood dimension and foster local communities.

Trust and safety in cities were among the issues given priority focus in the European perception surveys on cities (Eurostat 2016). In about one third of them, less than half of the respondents agreed that most of their fellow citizens were trustworthy. Several Eastern European capitals were at the lower end of the scale. In most Nordic cities, about two thirds of respondents always felt safe in their city. There was a strong correlation between the proportion of respondents who agreed that most of their fellow citizens could be trusted and the proportion who always felt safe in their city. Respondents across all surveyed cities were more likely to say that they always felt safe in their neighbourhood than in their city.

6.4.2 Healthy Cities for Healthier Seas

A healthy city is a city committed in promoting public health for the well-being of its citizens, present and future. Public health is not merely the absence of epidemics or disease; it is a state of complete public physical, mental, and social well-being. The Healthy Cities movement, launched by the World Health Organisation in 1987, brings together cities that have developed and implemented a wide range of policies, including health profiles and strategies through community initiatives that address the needs of citizens, cultural lifestyles, environmental health, and sustainable development.

The Healthy Cities movement aims at improving health trajectories for all and reducing health inequalities, and also improving leadership and participatory governance for health. The Zagreb Declaration expressed a strong commitment of political leaders of cities to bolster and champion action on health, equity, sustainable development, and social justice. It highlighted continuing values and priorities and identified new challenges and approaches for cities to address and adopt as they work to protect and enhance the health and well-being of all their citizens (WHO 2009).

In the framework of the physical activity strategy for the WHO European Region 2016–2025, the WHO advocates urban planning as a means to prevent physical inactivity. Its guiding principles to move towards increased physical activity in urban settings insist on transforming public spaces into invitations for physically active lifestyles. Physical inactivity accounts for an increasing proportion of deaths and disability across the Region. It is also associated with high healthcare costs and lost productivity. Regular physical activity, on the other hand, is a safeguard against a range of chronic diseases, such as type 2 diabetes, cardiovascular diseases, and some cancers. It is also a key determinant of weight control (WHO 2017).

Living by the sea has been associated with health benefits and a higher level of physical and mental well-being (Wheeler et al. 2012). The Blue Gym Initiative, run by a team of researchers working at the European Centre of Environment and Human Health at the University of Exeter, highlights the de-stressing effects of marine environments. Interconnected healthy cities offer a broad and pragmatic framework for concrete cooperation and progress at the local level. Cities promote urban health through projects, training, and seminars, many of which link to the sea. Benchmarking between the network cities is promoted and in relationship to many different public health themes. The "Cities for a Healthier Sea" initiative, introduced in 2010, in close connection with the Baltic Sea Challenge campaign, promoted voluntary water protection measures in Baltic cities.

Well-being centres by the sea often promote thalassotherapy, which involves the use of seawater as a form of therapy. The properties of heated seawater, associated with the reinvigorating shore climate, are believed to have beneficial effects on physical and mental health. The practice of thalassotherapy is old, and its origins can be traced back to antiquity. The therapy has various forms, including application of marine mud or of algae paste, or the inhalation of sea fog. France has a network of sea resorts recognised for high-quality thalassotherapy services.

Urban safety is a critical dimension of well-being in cities. The European Forum of Urban Safety (EFUS) shed light on the interconnected values held by cities, democracy, and safety. The "Cities' Manifesto for Safety and Democracy," adopted in Naples in 2000, expressed the desire for quality cities, defined as safe, vital places of harmonious development and immune to violence and extremism. Exchanges of experience supported cities in overcoming aggression acts against vulnerable social groups and sensitive places, such as schools, deteriorated housing estates and streets, and undertaking innovative prevention measures. Continuous dialogue and cooperation are judged essential for guaranteeing the legitimate right to safety (EFUS 2003, 2006).

The increasing complexity of cities and the deployment of digital technologies have many safety implications. As part of the Internet of Things technologies, sensors collect and wirelessly transmit data from physical environments, offering insights into city operations and allowing remote and efficient management of infrastructure and services. The spread of closed-circuit televisions and webcams, and technologies such as artificial intelligence and data analytics, can greatly enhance the capabilities to combat crime and terrorism. Investments in digital technologies should always be accompanied by commensurate investments in cybersecurity. The consequences of neglecting cybersecurity could be dire. For example, if hackers were to shut down the power or the storm surge emergency system, an entire city would be left in chaos.

Security remains closely linked to wealth, but the rankings of high-income cities are falling, and income is only one of the factors governing city performance on security. The Economist Intelligence Unit (EIU) evaluated the relative safety of 60 cities around the world based on qualitative information and an index composed of 49 indicators, assessing digital security, health security, infrastructure security, and personal security. Tokyo has been named the safest city in the world once again in 2017, a position earned in 2015. Although Tokyo earned the highest score in digital security and made improvements in health security, it did fall out of the top ten in the infrastructure security category. East Asian (Tokyo, Singapore, Osaka, and Hong Kong) and European (Amsterdam, Stockholm, and Zurich) cities dominate the top ten positions in the overall index, and Toronto, Melbourne, and Sydney complete the decade. Safety being a key dimension of liveability in cities, this ranking resonates with the liveability ranking of cities by the EIU (EIU 2017a). Dhaka, Yangon, and Karachi are at the bottom of the list. American cities rank relatively low because of their deteriorating infrastructure (EIU 2017b).

Good practice for urban safety provides powerful cases, especially when action is taken by the most affected who progressively become empowered and become masters of their destiny. In Tanzania, the Dar es Salaam Safer Cities project tried to raise awareness and strengthen the capacity of local authorities to create a culture of crime prevention and safety at the neighbourhood level, initiated by the local communities affected by insecurity. All stakeholders worked together to develop local solutions to prevent crime and build awareness among communities. Youth crime and violence against women were identified as priority concerns. Public surveys provided insights for the conception of an integrated crime prevention strategy.

The strategy encompasses law enforcement, the development of neighbourhood watch groups, and Ward Tribunals for minor offences, job creation and skills training, cultural and recreational activities for youth at risk, and improvements of the local environment (EFUS and ICPC 2014; ICPC 2005).

References

Anna Lindh Foundation (2014) The Anna Lindh report 2014: intercultural trends and social change in the euro-Mediterranean region. Cairo
Bioregional Solutions for Sustainability. 2009. BedZED seven years on. The impact of the UK's best known eco-village and its residents. London
Children's Society (The) (2014) Ways to well-being: exploring the links between children's activities and their subjective well-being. London
Cities of Migration (2012) From practice to policy: lessons from local leadership on immigrant integration. Maytree Foundation, Toronto
City of New York (NYC) (The) (2015a) OneNYC. The plan for a strong and just city. New York
City of New York (NYC) (The) (2015b) Building a smart + equitable city. New York
City of New York (NYC) (The) (2016) OneNYC progress report. New York
City of New York (NYC) (The) (2017) OneNYC progress report. New York
City of Stockholm (2014), Publisher: City of Stockholm. Stockholm
Economist Intelligence Unit (EIU) (The) (2017a) A summary of the Liveability ranking and overview. London
EIU (2017b) Safe Cities Index 2017: Security in a rapidly urbanising world. London
Eurostat (2016) Urban Europe. Statistics on cities, towns and suburbs, Luxembourg
Eurocities (2010) Cities accommodating diversity, Brussels
Eurocities (2012) Cities for active inclusion. Brussels
European Commission (EC) (2010a) Empowering people, driving change: social innovation in the European Union. Brussels
EC (2010b) This is European social innovation. Brussels
EC (2015) Employment and social developments in Europe 2014. Brussels
EC (2016) Sustainability now! A European vision for sustainability. EPSC, Brussels
EC (2017a) 10 trends shaping migration. PSC, Brussels
EC (2017b) Report on the blue growth strategy. Towards more sustainable growth and jobs in the blue economy. Brussels
European Forum for Urban Safety (EFUS) (2003) Secucities, schools and cities. Paris
EFUS (2006) The cities' manifesto for safety and democracy. Paris
EFUS and International Crime Prevention Centre (ICPC) (2014) 100 promising practices on safer cities. Paris
European Parliament Research Service (EPRS) (2017) Cities: front line of climate action. Briefing PE 608.730. Brussels
Fundación Bertelsmann and Cities of Migration (2012) Fifteen good ideas. Diversity and inclusion in cities. Toronto
Hall P (2013) Good cities, better lives: how Europe discovered the lost art of urbanism. Routledge, London
Harvey D (2012) Rebel cities: from the right to the City to the urban revolution. Verso, New York
Institute of Medicine (The) (2012) Accelerating progress in obesity prevention: solving the weight of the nation. Washington, DC
International Centre for the Prevention of Crime (ICPC) (2005) Urban crime prevention and youth at risk. Montréal
International Labour Office (ILO) (2012) Working towards sustainable development. Opportunities for decent work and social inclusion in a green economy. Geneva

ILO (2017) Global employment trends for youth 2017. Geneva
International Organization for Migration (IOM) (2015) World Migration Report: migrants and cities, new partnerships to manage mobility. Geneva
Mega V (2013) Quintessential cities, accountable to the future. Innovation, sustainability and leadership. Springer, New York
Mega V (2015) Conscious coastal cities. Sustainability, blue green growth and the politics of imagination. Springer, New York
National Law Centre on Homelessness and Poverty (2016) Homes not handcuffs. Ending the criminalization of homelessness in U.S. cities. Washington, DC
Organisation for Economic Co-operation and Development (OECD) (2013) International migration outlook. OECD, Paris
OECD (2016a) Making cities work for all. Data and actions for inclusive growth. OECD, Paris
OECD (2016b) The ocean economy 2030. OECD, Paris
OECD (2017a) Beyond shifting wealth: perspectives on development risks and opportunities from the global south. OECD, Paris
OECD (2017b) Trust and public policy: how better governance can help rebuild public trust. OECD, Paris
Storper M (2013) Keys to the City: How Economics, Institutions, Social Interaction, and Politics Shape Development. Princeton University Press, New Jersey
UNITED NATIONS (UN) (2017). International migration report 2017. Highlights. New York
(The) UN Office for Disaster Risk Reduction (UNISDR) (2013) Global assessment report on disaster risk reduction. New York
(The) Urban Institute (2011) How to evaluate choice and promise neighbourhoods. Perspectives brief 19. Washington, DC
(The) U.S. Conference of Mayors' Report on Hunger and Homelessness (2016) A status report on homelessness and hunger in America's cities, 2016. Washington, DC
Ville de Nantes (2012) Le voyage à Nantes. The city turned upside down by art. Nantes
Wheeler BW et al (2012) Does living by the coast improve health and wellbeing? Health and Place 18(5):198–201
World Bank (2012) Inclusive green growth. The pathway to sustainable development. Washington, DC
World Economic Forum (WEF) (2016) The global gender gap report (11th edition). Geneva
WEF (2017) Migration and cities. An insight report. In collaboration with PwC. Geneva
World Health Organisation (WHO) (2009) Zagreb Declaration for Healthy Cities. Copenhagen
WHO (2017) Towards more physical activity in cities. Copenhagen
World Resources Institute (WRI) (2016) Towards a more equal city, framing the challenges and opportunities, Washington D.C.
WWF (2017) Reviving the economy of the Mediterranean Sea. Actions for a sustainable future. Headquarters in Gland, Switzerland

Selected Websites and Social Media

"OurOceanHealth
#OceanOptimism
#EarthDay
www.annalindhfoundation.org
www.c40cities.org
www.citiesalliance.org/
www.citymayors.com
www.coalitionforthehomeless.org
www.csiro.au

www.citiesofmigration.ca
www.coe.org
www.efus.org
www.ec.europa.eu/futurium/en
www.fas.ie
www.globalcompactfoundation.org
www.habitat3.org/the-new-urban-agenda/
www.healthycommunitiesinstitute.com/
www.ilo.org
www.iom.org
www.interculturaldialogue.eu
www.intercultural-europe.org/
www.nycgovparks.org/greening/planyc/schoolyards
www.safecities.economist.com/safe-cities-index-2017
www.socialedge.org
www.successbc.ca/eng/company/locations/vancouver
www.urbansecurity.org
www.undp.org/en/content/multidimensional-poverty-index-mpi
www.union-habitat.org
www.ville.gouv.fr
www.who.dk/healthy-cities
www.youngfoundation.org

Watercolour 9
Toronto, a Welcoming Most-multicultural World Metropolis

Chapter 7
Cities and Oceans at the Crossroads of Knowledge, Ingenuity, and Co-creation

Abstract Cities are extraordinary theatres of humanity at the crossroads of very diverse journeys, individual and collective, entrepreneurial, artistic and cultural, learning and creative, local and global. Port cities are reputed to be more open to innovation and trade, ideas and exchanges. The urban and ocean heritage, including also the underwater cultural heritage, are exceptional forms of human expression. The seas have often inspired scientists and citizens to invest ideas and energy in new scientific discoveries, entrepreneurial action, and cultural expression. Ocean-literate cities have a leading role in reinforcing the bond between cities and the blue planet. A sustainable city has to cultivate the seeds of freedom for science and the arts to bloom and to offer a space to all wishing to communicate and interact. Science and knowledge cities invest in education, innovation, and exploration of the seas, and in international cooperation for excellence. This chapter presents a spectrum of inspiring actions to reinforce urban intellectual and cultural capital and make citizens proud not only of their immediate environment but also of the faraway ocean ecosystems that inspire them and ask them to act as responsible citizens. It also explores the principles of ecosystem-based strategic planning for more forward-looking eco-designed policies in balance with the planet, on land and in the seas.

7.1 Forward Cities: Investing in Science, Technology, and Sustainability

7.1.1 Eco-Cities and Innovation Lighthouses

Eco-responsible cities that are open to the world invest in the power of ideas and the capacity of their citizens to envision a better sustainable future and bring it into being. Magnetic cities attract and develop human talent and offer the conditions for the incubation of innovations and the generation of sustainable wealth. The size and the diversity of a city are very important for the breeding and capturing of synergies, and also for providing culture and governance. It seems that new patents are granted disproportionately to larger urban centres. The most powerful world metropolises,

V. P. Mega, *Eco-Responsible Cities and the Global Ocean*,
https://doi.org/10.1007/978-3-319-93680-2_7

such as New York, London, and Tokyo, are also science strongholds. Future-oriented cities offer very diverse urban mosaics enriched with the presence of enterprises and intense education–research–innovation interactions (Hall 1998).

Vienna, ranked top 2017 Smart City, has set its goal to successfully tackle the challenges of the twenty-first century. In 2013, it engaged with citizens in adopting a long-term and holistic strategy, which has the aim to guarantee the highest quality of life for all Viennese and to save resources through comprehensive innovations. "The Smart City Wien 2050 idem vision introduced the triptych "Innovations–Resources–Quality of Life," which reminds us of the key dimensions of the sustainability trilemma.

Innovating with the ocean in mind has given rise to an interesting array of experiments. In Rotterdam, the Erasmus Smart Port Rotterdam, an interdisciplinary centre of excellence, is active in port-related issues, such as port sustainability, synchromodal transport networks, involving the flexible and integral deployment of various modes of transport on a network to better satisfy demand, sustainable global supply chains, and interorganisational systems in logistics. The centre also acts as a resource for the seaport cluster, including maritime industries, such as shipbuilding, transport, logistics, and offshore industries.

A great city is often praised for its ability to create and attract talent. Successful cities are considered those that provide business and entrepreneurial dynamism, but also educational and cultural opportunity. Barcelona ranks high on these fronts. Through extensive public investment and strong strategic planning, Barcelona has reinvented itself since it reconciled with the sea for the Olympic Games of 1992. The city's Entrepreneurship Centre, created in 2004, serves as a reference for existing and future businesses, as well as a hub to promote entrepreneurship and innovation. The centre provides advice and services, and a pre-incubation space allowing potential entrepreneurs to design robust projects and create innovative companies. Barcelona Activa, the local development agency for employment, business, and the economy, is supporting enterprises in job creation.

In 2014, Barcelona was the first city to be recognised as the EU Capital of Innovation ('iCapital'). The European Commission awarded this prize to Barcelona for the innovative use of new technologies to bring the city closer to its citizens. The "Barcelona as a people city" project, introduced in 2011, promoted the use of new technologies to foster economic growth and citizen welfare through open data initiatives, sustainable energy and mobility, social innovation, alliances between research centres, universities, private and public partners, and better smart services at the frontiers of information and communication. The transformation of physical space was artfully combined with efforts to create a new knowledge and innovation hub and industry clusters, including media, medical technologies, information and communications, energy, and design.

The Barcelona Urban Lab welcomed companies with innovative projects at the pre-commercial stage, wishing to test them in a real city environment. The Urban Lab can help with testing pilot projects promising a potential high impact in public places in Barcelona. The Lab acts as a gateway to the City Council and stimulates emulation and coordination. It allows for the creation of cutting-edge programmes, funded by the companies that conceive them, and demonstrate a need to use public spaces and infrastructures and according to Barcelona City Council's objectives and expectations. The candidate programmes must prove a direct impact on the functioning or on the services provided by the city and ultimately benefit the residents. SynergyS connects micro-, small, and large enterprises with the global knowledge community to provide companies with free access to forefront research and emerging technologies and generate new business opportunities.

Amsterdam was the second European Capital of Innovation in 2016, recognised for its holistic vision of innovation in and for the city. Its "Together we make the city" approach provided a shining example of cities embracing testbeds of innovation to improve the way that people live and businesses work. The one-day Innovation Expo embraced a number of events and also linked to the "Europe by People" project and the search for a new narrative for designing democracy. The objective was to engage interested stakeholders in the open discussion of innovations for living better together in future-proof urban environments. One of the most dreamt-about innovations seemed to be "sustainable floating and hybrid cities" to protect urban life from fierce storms and sea rise and achieve a better reconciliation of cities with the sea, especially in fragile deltaic megacities.

Sharing solutions for a possible better future has been a central element of all experiments, trying to bring forwards the ingenuity of young students and professionals. The pre-fab biobased house, made almost entirely (90%) of biodegradable material, including the home interior, has been promoted as the first nearly fully biobased house bringing together more than 40 professionals. It also featured an algae-based system for water purification. "Fabcity," a coastal urban energy- and waste-neutral neighbourhood in the making, has been a temporary living laboratory at the edge of Java-Eiland, and a departure point for students going on a cycling world tour to help other citizens change their car-dependent lifestyles and move from car-centric cities to cycle-resilient cities. The site showcased biobased houses and houses made from waste, urban vertical farming, bioproduction on demand to prevent food waste, circular water and energy cycles, floating solar panels that are four times more efficient than solar panels on dark roofs, and 3D printing experiments for bioproducts and even steel bridges printed by robots.

FabCity was initiated by the EU2016 Arts & Design programme "Europe by People" and investigated the various future dimensions of everyday living in sustainable cities. As a mini-society inhabited by students, scientists, and social entrepreneurs, it tested innovative ideas for the connected, self-sufficient, and circular city. The campus also functioned as a place to collectively experiment with new ways of creating an improved future urban ecosystem through the commitment of all.

Paris was selected as the third EU innovation capital in 2017. It has been praised for its inclusive innovation strategy and the creation of Station F, the world's largest start-up campus. Opened in Paris 13th arrondissement, in 2017, on a former railway depot, the 34,000 m^2 space is to accommodate more than 1000 startups; the essence of the French startup ecosystem, it counts prestigious companies as founding partners. This amazing workstation kept much of the original architectural elements and atmosphere, including the original concrete and shipping containers transformed into meeting rooms. The French ecosystem has a multitude of great startups, but small and dispersed. The emblematic Station F reminds us of the Silicon Roundabout and Factory in Berlin; it brings the ecosystem together to maximise and optimise its chances, through incubators, spaces, and services.

Tallinn and Tel Aviv were selected as runners-up to the EU iCapital 2017. Tallinn wished to provide a model for an eCity and Tel Aviv showcased a dynamic hi-tech ecosystem and incubators. Tallinn City's Innovation Philosophy 2.0 aims at creating a true eSociety, competitive, attractive, and happy in a healthy and unified city. Paris was also a most popular subject of research. Europeana, Europe's multimedia online library, revealed that its traditional annual ranking of searches for 2017 gave Paris the first place.

Citizens' involvement at the earliest stage is a most common denominator in planning public innovation services. Many cities invest in the mobilisation of citizens' resources when projecting high-quality public services for sustainable development. Copenhagen introduced MindLab, a transformative living lab, acting as a cross-governmental innovation bringing together citizens and businesses to develop public innovations and improve public services. The design of any public service has to place citizens at the centre and anticipate and address social needs.

In Korea, SparkLabs, one of the leading incubators in startup ecosystems, launched its Internet of Things Accelerator in Songdo International Business District, 65 km southwest of Seoul, in early 2015. This ubiquitous city, built from scratch on 600 hectares of reclaimed land along Incheon's waterfront, was developed with a technological infrastructure that supports connected Internet services and products in public spaces and residential environments. A great range of new systems, from security devices for recreational parks to healthcare, are connected through smart applications. The IoT, open to world startups, focuses on enterprise- and consumer-connected devices, sensors, and data analytics. Global leaders are integrating cutting-edge technologies into this smart city built from the ground up and providing a unique opportunity for companies to see how ubiquitous technology can be integrated into everyday lives without creating interference or distractions. Songdo provides IoT participants an unparalleled opportunity to plug them into a community of 35,000 people. For example, the city is filled with sensors that can monitor temperature, energy use, and traffic flow. Working with such a wired city open to innovation allows startups to test out their products at scale.

Science is an endless discovery, continuously nurtured by human and physical links and their interrelated intangible dimensions providing meaningful interconnections to make a difference. The construction of the Øresund Bridge between Copenhagen and Malmö in 2000 brought multiple benefits to the connecting cities and countries, which impacted socioeconomic and cultural cooperation and scientific synergies. The bridge encouraged the establishment of the Øresund region, a network of universities from both shores. Co-authorship between Copenhagen and the southernmost province of Sweden more than doubled in the first 10 years of the millennium.

The rise of new digital instruments, sensors, and connected devices and more empowered citizens initiated a new era of more responsive and efficient government on the basis of real-time data and their intelligent use. The 2016 smart city awards recognised New York City and its efforts to ensure the responsible and equitable deployment of smart technologies. The Smart + Equitable City strategy has four pillars: expanding connectivity for all, piloting and scaling smart technologies, strengthening the innovation economy, and ensuring responsible deployments. The City has committed to bring high-speed internet access to all NYC's residents and businesses (NYC 2016).

Since 2016, LinkNYC, which broke ground as the world's largest and fastest municipal Wi-Fi network, provides access to high-speed Internet and a multitude of services to low-income New Yorkers in public housing. Through a network of neighbourhood innovation labs in public spaces across the five boroughs, New York City is also working with communities to leverage connectivity and the Internet of Things to field test innovative technologies targeting local needs. Urban Tech NYC, an accelerator programme, providing space and equipment, helps entrepreneurs in devising smart and sustainable responses to energy, water, waste, transport, and agriculture requirements.

7.1.2 Oceans of Ideas, Cities of Voices

Innovative cities try to incorporate cutting-edge insights in all their functions and promote public engagement in science and a culture of continuous improvement. Symbolic events, such as Days, Nights, or Weeks devoted to science invite citizens and youths in particular to discover a new world and feel the thrill of the unknown. The Galway Science and Technology Festival, founded in 1998, aimed at bridging the gap between education and industry and encouraging students to study science, technology, engineering, and mathematics (STEM) subjects at third level. The Festival runs annually for 2 weeks and culminates with a demonstration of best practice from primary to post-graduate education, and from local companies to multi-nationals. The Marine Institute participates in the event and holds open days inviting secondary school students to discuss the ocean exhibits on display and the opportunities regarding careers in marine engineering.

Genoa initiated a Science Festival, which offered a stage and a voice to all citizens. Making science accessible is the essence of the festival, inviting all and everyone to a world of excitement and wonder. Finally, the Global Festival of Action for Sustainable Development provides a different perspective for assessing sustainable development goal (SDG) action, with a focus on innovation and disruptive change, citizen engagement, and communicating the SDGs. It can be the starting point for partnerships among cities and seas.

In Boston, the US knowledge capital, scientific distinction is the latest in a series of strengths, from being the largest city in early colonial America, to a port for global shipping and sailing, before attaining its position as a biotech and innovation hub. The Harvard innovation lab tries to foster entrepreneurial activities and deepens interaction among students, faculty, businesses, and professionals of the Greater

Boston community, and the Massachusetts Institute of Technology (MIT) is a leading institution for technology business creation. The democratisation of access to knowledge is considered very important for informed public discourse and the capacity for shared civic projects. Harvard University, the MIT, the Boston Globe, and Massachusetts General Hospital pull together their convening powers to bring thought leaders, innovators, and creative disrupters together in seminars, 'hackathons,' and other thought-provoking events. "Innovate locally, exchange regionally, export globally" became a paradigm for the city, its world-class human capital, and its open and dynamic civic culture (Boston Foundation and the Citistates Group 2004).

The vision for a world-leading Boston has been proposed after consultations with opinion leaders and observers from government, academia, business, and advocacy groups. The vision suggests that it is hard to think of a twenty-first century city-state so ideally positioned as Boston in the century of the intellect. As decisive factors are suggested smart ingenuity and collaborative leadership, necessary for Boston to address the critical challenges of climate change and extreme weather events, public health, energy and water supply, shortfalls of knowledge workers, lack of affordable housing, and weakened civic ties. Recognising the waterfront heritage as an asset to be reinforced is crucial for encouraging selective preservation along with quality urban design.

From science festivals to networking events, cities offer also the settings for marches and help gather world momentum when organised in many diverse cities during the same day. On Earth Day, 22 April 2017, tens of hundreds of thousands of scientists and science enthusiasts participated in a series of rallies and marches held in Washington, DC and more than 600 other cities across the world. The non-partisan march, initiated as a grassroots movement, was endorsed by 240 scientific and academic bodies from the American Association for the Advancement of Science, the largest scientific organisation in the world. Particular issues of science policy raised by the marchers include evidence-based policymaking, government support for scientific research, transparency, and acceptance of the scientific consensus on climate change.

In New York, on 22 April 2017, 40,000 marched for science and the values of objectivity, openness, and international collaboration, calling upon the US President to reconsider proposed government budget cuts that could affect the future of science. In Boston, the rally looked more like a huge outdoor science fair. Several thousand people gathered in the Boston Common to underscore the importance of science. In Los Angeles, 50,000 marchers went from Pershing Square to the City Hall, led by former NASA astronaut G. Reisman in an electric vehicle. The 10,000-strong London march rallied outside the Houses of Parliament and followed a route symbolic of the importance of science to the UK. Marchers were of all ages and many different nationalities and marched outside the Science Museum, near the Natural History and Geology museums, and the Imperial College London. The next Earth Day will focus on mobilising the world to end plastic pollution and support a global effort to eliminate single-use plastics.

AQUAE VENICE 2015 has been described as an experiential, educational, and interactive way to understand and feel how and why water deeply influences lives, science, and culture. Aquae Venice 2015 has been the great universal exhibition dedicated to water, the unique official side event of Milan Universal Exhibition dedicated to food and feeding. For this occasion, the new AQUAE pavilion was designed and built as a celebration of water. The extension of the Expo in the Venetian Lagoon further explored the relationship between humanity and water, through a multimedia path and a contemporary exhibition, where food and experience in food preparation and consumption were used as universal communication codes.

Many awards and alliances celebrate innovation and the role of cities as crucibles of resourcefulness and invention. From structures to the soul of a city, the human and the built environment bear witness of many outstanding efforts for excellence. Bright urban ideas are usually brought forward by individuals trying to extend the breadth of alternative futures. The Guangzhou International Award for Urban Innovation recognised worthy initiatives enhancing sustainable urban development through sharing of insights and rewarded innovations striving to simultaneously improve social, economic, and environmental sustainability. It is interesting to note the power of this and other awards to bring together networks and organisations, such as the United Cities and Local Governments (UCLG), Metropolis, and the Guangzhou Municipal Government, for the recognition of outstanding innovative public projects and practices.

Place-based and human-centred innovations can be extraordinarily powerful and completely transform the status of cities. Efficient but non-creative use of capital or technology can lead to the systematic imitation of innovations that are produced elsewhere, a chronic 'import' of ingenious solutions. Risk taking and trial are limited in the search for the optimal and most efficient conditions for transplanting innovations. Perpetual imitation kills the productive seeds and weakens the creative capacity of cities. A permanent environment for the peaceful incubation of genuinely new ideas and unproven goods and services is the hallmark of many innovative cities (EC 2016).

Innovations may bring impressive leaps from value-creating activities, in a continuous endeavour to uplift the threshold of excellence. They may also lead to a point of no return and affect the cultural equilibrium of a city. Innovations can change a city's sense of what is desirable and possible in the relevant timeframe. Strong leadership is needed to orchestrate cooperation. Citizen participation can communicate vision and purpose, sharing the costs and benefits and public acceptance. Business participation can promote efficiency and dynamic practical solutions. Coalitions with citizens at the centre of a genuine strategy seem to have endless and unparalleled potential. Charismatic leaders, business actors, scientists, or simply local citizens and workers are all potential bearers, initiators or adapters of innovations. A common problem or a shared perspective can often prepare the ground of cooperation. The architecture of coalitions is very diverse and challenges general rules. Flexible but strong alliances are often needed to create the space and the conditions. Alliances based on agreement, mediation, political manoeuvring, and negotiation can best direct the wave towards the desired vision. Consensus and persistent commitment are a necessary front for priceless long-gestation projects.

7.2 Outstanding Urban and Underwater Heritage Commons

7.2.1 Cultural Urban Heritage of the Future

Cultural heritage is a unique asset, invaluable and irreplaceable, and a key element of identity. It is the biography of places written down on the landscape, connecting their physical and spiritual dimension. It can provide a soulful experience for visitors, add character to the quality of life of citizens, and create emotional attachment. Conservation projects have increasingly been geared towards preserving and enhancing cultural landscapes instead of isolated sites and favoured citizen participation. The digital era provided further opportunities also linked to the evolution in museums and digitalisation of collections, open networks and platforms, and the interests of virtual or online communities. Citizen- and user-generated information and knowledge on monuments and sites can be collected via crowdsourcing, which can be extremely helpful in cases of conflict or illegal trafficking.

Cultural heritage has high existence and bequest values, as inheritance of the past and legacy to the future. As common goods, heritage assets need an evolving framework of collective, multi-stakeholder, and multi-level governance. The benefits of cultural heritage, most commonly seen in terms of attraction of visitors and businesses, are increasingly recognised as contributors to innovation, social cohesion, and sustainable development. Fashion tycoons and philanthropists often choose to invest in the safeguarding of monuments and sites. Governments and citizens are increasingly considering heritage as a means of improving citizen well-being, urban wealth and quality of life, and community engagement (EC 2015a, b).

The UN 2030 Sustainable Development Agenda acknowledges, for the first time, the key role of culture, creativity and cultural diversity to address sustainable development, advance economic growth, and foster social inclusion. Many experts had already suggested that culture is the glue that keeps everything together (Landry and Bianchini 1995). This idea resonates with the UNESCO Convention on the Protection and Promotion of the Diversity of Cultural Expressions (2005), which underlined that investment in culture and creativity is an investment in sustainable development and can forge new bonds with the citizens and pathways for development. Cities are places where the cultural diversity is the highest, and this is often reflected in the historic heritage.

In 1972, UNESCO's World Heritage Convention had already integrated the concepts of nature conservation and the preservation of cultural heritage. The World Heritage emblem has been designed to reflect the interdependence of the world's natural and cultural diversity. Europe is the continent with the highest density of heritage sites, and Italy and France concentrate more than 10% of the 1058 outstanding listed world sites in 2017. The World Heritage Convention is a very powerful tool to mobilise partners for the preservation of the integrity and authenticity of the sites and to rally international attention and actions through safeguarding campaigns.

The World Heritage list encapsulates many fascinating historic urban cores in close bond with the sea. The Old City of Dubrovnik has endured the passing of centuries and survived many natural disasters. In 1991, seriously damaged by artillery fire, the city was immediately included in the list of World Heritage in Danger, which included 55 threatened sites in 2017. With UNESCO's technical advice and financial assistance, the Croatian Government restored the facades of cloisters, rehabilitated roofs, and rebuilt palaces. As a result, in December 1998, it became possible to remove the city from the list of World Heritage in Danger. The millennium started with new promises for this exceptional city in a unique symbiosis with the Adriatic Sea.

Le Havre, major port city at the estuary of the Seine River, has been shaped by and marked maritime history. Following WWII, the city's devastated heart was boldly redesigned by A. Perret as a model of concrete functionality. A genuine port city, it had developed much ingenuity for managing the crisis of the 1970s and the end of the golden age of ocean liners. Some exceptional modern interventions, buildings, and cultural centres, including the MuMa André Malraux Fine Art Museum, and the Volcano arts centre by leading architect O. Niemeyer, are key signs of the renaissance of the city. The Port 2000 project increased the container capacity to compete with modern ports of Northern Europe, transformed the southern districts of the city, and again welcomed ocean liners. In 2005, the central core of Le Havre has been the first French modern city to be a World Heritage Site. The city can boast a morphologically shaped nineteenth-century fort turned into public gardens with a view.

Cordoba, on the Andalusian route leading to the Atlantic Ocean, is a perfect example of an historic city with a diverse heritage that confers local pride and landmarks. Its historic centre has conserved its medieval plan and the irregular layout of its narrow streets. The squares, promenades, and gardens are integrated into the urban landscape, on the bank of a meandering part of the river. The urban fabric includes houses, built around patios enclosed by grillwork, and public monuments that testify to the historical phases. The mosque–cathedral and the Roman bridge remain the principal anchor points in the cross-epochs horizon that comprises Roman vestiges, Moorish minarets, the Moorish Almodovar Gate, the Jewish synagogue, Christian monuments, the reconstructed Alcázar, the Calahorra Tower, and numerous churches of Mudéjar and Gothic inspiration.

The city received a 2014 EU Prize for Cultural Heritage/Europa Nostra Award, one of the three major EU projects on cultural heritage, together the European Heritage Label given since 2011 to symbolic sites of European significance and integration, and the Open Cultural Heritage Days, that each September offer 20 million Europeans the possibility to access cultural gems, monuments, and sites. In Europe, four of ten tourists chose their destination based on cultural offer. UNESCO and the European Commission cooperated in 2016 in proposing heritage routes linking in innovative ways the star power of heritage sites.

The Council of Europe Convention on the Value of Cultural Heritage for Society (Faro Convention 2005) emphasised the dimensions of heritage relating to human rights and democracy, communities and society, meanings and values. Urban cultural heritage bears witness of the ways that citizens and communities interact and transform their surrounding environment and continuously etch this into the landscape of the future. Volunteering is probably the most widespread example of public engagement with cultural heritage. Organisations such as the National Trust in the UK train and motivate volunteers. Their contribution has in many cases to be more visible and valued to create an example.

In 2011, UNESCO's Recommendation on the Historic Urban Landscape has been the culmination of many standard-setting efforts in monumental and cultural cities. It introduced the Historic Urban Landscape approach, and recommended that the Member States may wish to undertake comprehensive surveys and mapping of urban natural, cultural and human resources, participatory planning and stakeholder consultations on the bonds to the values to protect and transmit to future generations, to assess vulnerability to socioeconomic stresses and the impacts of climate change, and to integrate heritage values into a wider framework of sustainable urban development, taking heritage risks into account.

The UNESCO Hangzhou Declaration on "Placing Culture at the Heart of Sustainable Development Policies, which resulted from a 2013 International Congress, suggests that culture should be considered as a fundamental enabler of sustainability, being a source of meaning and wonder, a wellspring of creativity and innovation, and a resource to address challenges and find adequate solutions. The outstanding force of culture to foster and enable truly sustainable development has been highlighted as more impactful in people-centred and place-based approaches integrated into development programmes and peace-building initiatives.

Integrated urban landscape management, public–private partnerships, crowd-sourced funding, philanthropy, and new forms of governance and participation can help realise the potential of cultural heritage. A study by an advisory group to the European Commission highlighted lessons from places where cultural heritage has been an economic, social, and environmental driver. Cross-pollination with many arts and contribution to the sustainable development agenda can further expand the limits of the possible. Heritage-led urban regeneration seems to bring great returns from investment. The iconic Ancient Roman city of Pompeii, jewel of European heritage and World Heritage Site, benefited from a contribution of the European Regional Development Fund for the continuation of the renovation and preservation works, consolidation of the structures and ancient buildings to restore its urban character, the construction of a water canalisation and drainage system, training, and skills. The Management Plan of the Historic Centre of Porto World Heritage initiated actions either through public-driven interventions, or by private-driven interventions that mirror the recovery and rehabilitation of the built fabric and the integration of commercial activities and tourist infrastructure and services. Actions focused on the physical regeneration of private and public buildings, as well as the public space, and also intangible actions.

In Newcastle-Upon-Tyne, Grainger Town had already become an award-winning example of heritage-led regeneration at the turn of the century. The Grainger Town, a Georgian development built in the mediaeval city in the 1830s, became the subject of a holistic regeneration programme with a strategic approach driven by a 'Civic Vision.' It was led by the Grainger Town Partnership, a public–private partnership established by the city council, and supported by several bodies, regeneration and conservation organisations as well as businesses and citizens. The project addressed the underlying causes of deterioration and promoted change, generating positive dynamics. It also demonstrated that investment in cultural heritage can bring significant social and economic benefits, including employment and skills, business development as well as housing and civic places. The regeneration exceeded almost all targets creating 2300 jobs in 330 new businesses. The €67 m public investment brought a high leverage, equal to an additional €288 m of private investment (EC 2015a).

At the intersections of culture and the environment, cultural landscapes are often multiply endowed with a rich biodiversity as well as extensive archaeological and historical values. Many World Heritage Cities are among the most vulnerable places experiencing rapid changes, not least caused by advancing climate change. They can serve as laboratories for monitoring impacts and testing mitigation and adaptation measures to enhance their physical and socioeconomic resilience, while ensuring the conservation of precious heritage for an engaging humanity. Exchanges among cities can be even more far reaching when youth is involved. The Youth on the Trail of World Heritage programme consists in pairing schools located in world heritage cities. Exchanges allow students to discover another world heritage city and to interact with fellow students to share ideas, knowledge, and work for heritage. They thus become more aware of the universal value of their heritage and the common challenges involved in the management of world sites.

Sustainable cities have to transmit their cultural capital enhanced, or at least intact, to future generations. The Organisation of the World Heritage Cities (OWHC) involving cities with monuments and sites listed as World Heritage by UNESCO supports shared commitments in actively preserving unique and universal heritage. The OWHC was created on 8 September 1993, in Fez. In commemoration of this event, September 8 was declared "Solidarity Day of the World Heritage Cities," and all cities are invited to celebrate it through diverse activities highlighting their uniqueness but also their links. More than half of these are located in Europe, and the listed monuments and sites span a long period from prehistoric to modern times. Governments must preserve the characteristics, values, and qualities that justified the inscription of the cultural sites on the World Heritage list. The OWHC "City2City" programme aims at strengthening exchanges and dialogues on preservation among its World Heritage Cities.

Hamburg's long-time aspiration for the "Speicherstadt and Kontorhaus District with Chilehaus" to be included in UNESCO's World Heritage list finally has had a happy outcome. The emblematic value of the two outstanding districts with a connecting network of roads, canals, and bridges dating from 1885 to 1927 and modern red-brick building ensembles for port-related use, as warehouses and offices, from the 1920s to 1940 has been recognised as of universal value for maritime architecture of the early twentieth century. The Speicherstadt was highlighted as the largest and most advanced logistics centre of its time and is still the world's largest coherent uniform set of warehouses, whereas the Kontorhaus District was acknowledged as an impressive modern office architecture paradigm. This extraordinary ensemble serves as a lively link between the historic old town and the new HafenCity district, creating an attractive city centre with a living historical identity and diverse cultural options, including the Elbe Philharmonic.

7.2.2 *The Seafloor Is a Cultural Heritage Treasure Trove*

Underwater cultural landscapes include sunken cities and submerged sites that have been subsequently covered by water from storm surge and rising sea levels, but also wrecks from ships and aircraft, port-related structures, and even weapons. UNESCO estimates that there are more than 3,000,000 shipwrecks on the floors of the world's oceans to be investigated. Underwater cultural heritage is endangered and still very insufficiently known, researched, protected, or enhanced. Vestiges of human existence underwater 100 or more years old are protected by the UNESCO Convention on the Protection of the Underwater Cultural Heritage, signed in 2001 and entered into force in 2009, which provides the international legal framework. Human interference with the underwater cultural heritage raises the question of sovereign rights, ethics of commercial exploitation, and conflicts and synergies with offshore and underwater economic activities (UNESCO World Heritage Centre 2013).

The UNESCO Convention on the Protection of the Underwater Cultural Heritage is the foremost international treaty concerning submerged heritage and the international community's response to the destruction of archaeological sites under the water by treasure hunters and industrial activities. It reflects the growing recognition of the need to ensure the same protection to underwater heritage as to land-based heritage, and provides legal protection, enables common approaches to preservation, and offers effective professional guidelines for research and action on underwater heritage. The Convention is strongly supported by underwater archaeologists, actively applying and enforcing its regulations to indeed put a definitive end to the brutal treasure hunting and industrial impact currently menacing major submerged archaeological sites. Underwater cultural heritage holds not only information about the history of humankind and the social importance of the oceans but also about the history of climate change and its impact on humanity.

Underwater archaeology can make a significant contribution to the cultural understanding of places and address threats to destruction or loss of historic information and looting. It involves much sophisticated equipment and techniques to support diagnosis, analysis, and evidence making. Latest advances in the protection and enhancement of submerged archaeological sites range from 3D modelling of sites, better understanding of climate change impacts and submerged prehistory, to the development of robot archaeologists. Underwater marine sites are often subject to unpredictable tidal flows or unexpected storm damage. The dynamic nature of the environment may make in situ conservation infeasible. In addition, underwater sites can remain chemically active, and iron can be leached from metal structures. Artefacts recovered from underwater sites need special care.

Most sites need a topographical survey and a plan indicating the locations of artefacts and other archaeological assets. Environmental assessment of archaeological sites requires the recording of site conditions as well as marine organisms. Access to underwater sites is often restricted because of inadequate safety conditions. A variety of techniques are now available to divers to record findings underwater. Photography is the mainstay of recording, and the advent of digital photography multiplied opportunities. Underwater photography and photomosaic images can also greatly assist the recording process.

The European Union Prize for Cultural Heritage/Europa Nostra Awards rewards, each year, the best of achievements linked to cultural heritage. Since 2002, the Awards have raised awareness through the power of the example and celebrate excellence and creativity in conservation, research, dedication, and education. In 2016, the EU/Europa Nostra award rewarded a research consortium from Denmark, Germany, Greece, Italy, the Netherlands, Sweden, and the UK for their ambitious endeavour to promote the state of marine archaeology and conservation in Europe. The project SASMAP (Development of Tools and Techniques to Survey, Assess, Stabilise, Monitor and Preserve Underwater Archaeological Sites) advanced the development of new methods with the objective of locating, assessing, and safeguarding Europe's underwater cultural heritage: included are the ancient harbour of Sounion in Greece, the submerged prehistoric settlement site of Tudse Hage in Denmark, the Roman town of Baia in Italy, and a seventeenth-century shipwreck in the Netherlands. The protection of the sites is one of the main concerns when developing new tools and methods. The research spanned various disciplines including marine archaeology and conservation, in situ preservation, wood degradation, marine geophysics, and geochemistry.

The Greek seas proved to be an inexhaustible source of hidden treasures and the Antikythera mechanism has probably been one of the most enigmatic objects, the first mechanical computer known to humanity. In 2012, the national archaeological museum of Athens presented the mechanism alongside the remains of the original vessel and treasure, which have been reunited for the first time since their discovery in 1901. The device, more than 1000 years ahead of the first astronomical clocks, could have been designed in Rhodes, home to a community of astronomers and mechanical engineers. The Antikythera mechanism seems a genuine cosmograph, describing the cosmos, and also a selenograph, describing the movements of the moon. The astonishing volume of astronomical data needed for such a mechanism challenges current assumptions about the conceptual abilities of the ancient scholars and engineers.

7.3 Tangible and Intangible Co-created Assets and Innovations

7.3.1 Public Goods and Living Heritage

The urban and the world archipelagos contain only prototypes. Every city constitutes a unique microcosm, involving places, visions, facts, notions, concepts, perceptions, smells, colours, sounds, expressions, signs, and symbols. It also has its subconscious, its interwoven bonds and conflicts, convergences and divergences, and myths and legends. Urban culture is composed of invaluable tangible and intangible assets, continually co-reinvented and co-enriched by all. Monuments and sites of local, national, and international importance, unique and universal, are invaluable resources, and it is the responsibility of world citizens to ensure their preservation. But culture is also created through the intangible traditions, the gastronomic experiences and social events that are the source of joy and wonder. Culture constitutes a precious capital reinvigorated daily by citizens' projections of hopes and desires into the urban reality and marvel (Mega 2013).

The cultural understanding and enjoyment of public goods is critical for personal and collective well-being. Symbolic disruptions of everyday routines help discover the ignored assets of neighbourhoods and bring local imagination and life to an apogee. Exceptional events such as heritage nights or light festivals reveal unexpected jewels and invite citizens and children to rediscover urban environments with new eyes and endless wonder. Culture makes the city a space of identity, attachment, and belonging. The city can serve as the mirror in which diverse cultures observe each other, discover their convergences and divergences, interpenetrate and fecundate. As strongholds of civilisation, solidly anchored in local traditions but open to the world, cities can promote intercultural understanding and multicultural cooperation.

Public heritage sites often are exceptional cultural spaces, closely linked to the identity of a city. They have the potential to stimulate the collective memory. They belong, by definition, equally to all and must be accessible to all. R. Koolhaas describes them as fortresses of freedom. They offer great opportunities as islands of civilisation in the archipelago of the city. Streets, parks, and public spaces should be the object of special attention as the nerve centres that ensure the fluidity and vibrancy of the city.

Next to the World Heritage Convention, the Convention for the Safeguarding of Intangible Cultural Heritage was proposed by the UNESCO General Conference in 2003. The international community recognised the need to raise awareness about cultural expressions and traditions as an integral part of cultural heritage. Oral traditions, performing arts, social practices, rituals, festive events, and traditional techniques form an equally important heritage. Exchanges based on tangible or intangible cultural heritage are important for mutual recognition and the shared journey to sustainability.

At the beginning of the European Year of Cultural Heritage 2018, the European Ministers of Culture and stakeholders from the heritage and cultural sectors adopted a Declaration, which highlights the benefits of a high-quality Baukultur to society and pathways for promoting and achieving it in Europe. Considering a trend towards a loss of quality in the built environment all over Europe, the Declaration underlines the urgent need to develop new approaches to protecting and advancing the cultural values of the European built environment. It also underlines the need for a holistic, culture-centred approach to the built environment and for a humanistic view of the ways that collectively shape the living places. An ambitious vision for a high-quality Baukultur embraces a conscious approach which actively builds social cohesion, ensures environmental sustainability, and contributes to the health and well-being of all citizens (Conference of Ministers of Culture 2018).

Baukultur reached amazing heights in Venice, an eternal city of cultural intensity, an open-air theatre in symbiosis with its lagoon and the seas. The Grand Canal, considered to be the most attractive urban avenue in the world, winds across the city, which also takes care of its green spaces, not least the Green theatre. Crumbling waterside palaces or exquisite churches in perilous state are being renovated to combat physical decay. Venice reminds us that the built environment is a testament to the permanent evolution of human ingenuity and appreciation of the world. Architecture involves diverse and often contradictory concepts, but also common values and selective affinities that create a vibrant and fertile dialogue amongst architects, city planners, and citizens, but also epochs and places, history and geography, and individual and civic life. New high-quality exhibition venues, such as the monumental triangle Punta della Dogana, compete in aesthetics with majestic palaces and enchanting open-air public spaces.

Many exceptional coastal heritage cities enjoy a strategic location by the sea and have inherited a majestic protective infrastructure. In Dubrovnik, the city walls are considered to be one of the most magnificent fortification monuments in Europe and a fine example of fort architecture. The Old Town of Corfu, on a strategic position at the entrance of the Adriatic Sea, is emblematic of fortified Mediterranean ports, demonstrating a high level of integrity and authenticity. The three forts of the town, designed by renowned Venetian engineers, were used for four centuries to defend the maritime trading interests of the Republic of Venice against the Ottoman Empire. In the course of time, the forts were repaired and partly rebuilt, more recently under British rule in the nineteenth century.

The physical setting and remarkable relationship with the surrounding environment determined much of Genoa's development. Its historic centre dates from the late sixteenth and early seventeenth centuries, when the Republic of Genoa was a major financial and seafaring power. The city represents the first example in Europe of urban development parcelled out by a public authority. An original ensemble of Renaissance and Baroque palaces along the so-called new streets is composing a public networked commons. Genoa's harbour, the most important in Italy and one of the most eminent in the Mediterranean, has become the centre of cultural, political, tourist, and commercial activities. In 1992, on the occasion of the Columbian Celebrations, the waterfront was redeveloped and the ancient port rehabilitated and opened to the sea. Native architect R. Piano restored the buildings, including the cotton warehouses, and created new landmarks such as the Aquarium, the Bigo trademark, and a Sphere made of metal and glass, installed in the water, and unveiled in 2001 at the occasion of the decisive G8 Summit.

The science and art of city planning reached a new height in Tel Aviv, literally "Hill of the Spring," built as an urban ecosystem with physical, economic, social, and human dimensions. Founded in 1909 on the Eastern coast of the Mediterranean Sea, it is an emblematic city, conceived according to the European model of a garden city. During the era of British rule in Palestine, Tel Aviv developed into a thriving urban centre, becoming Israel's foremost economic and metropolitan nucleus. It can be seen as an outstanding example of the innovative approaches to town planning during the first part of the twentieth century. The architecture and planning reflect some of the most significant trends, taking into account local traditions and climatic conditions. The white city forms the heart of Tel Aviv, and is based on the urban master plan by Sir P. Geddes, a pioneer in considering the urban landscape as an organism in constant evolution.

Among the most spectacular examples of the ways that cities and citizens imprinted the landscape, Rio de Janeiro benefits from an extraordinary setting, a theatre surrounding the sea. From the highest mountains down to the waterfronts, rich natural ecosystems have contributed to making a fascinating city. Rio de Janeiro is also recognised for the inspiration that it has provided to artists, architects, and city planners. O. Neimeyer's masterpieces offer the opportunity for an uplifting walk along the coast. The curvaceous landscapes inspired the pure curves of Brazilia, the first planned capital city in the world.

Urban cultural symbiosis with the sea is a strong characteristic of Valparaíso. The city was the first and most important port on the sea routes of the Pacific coast of South America that linked the Atlantic and Pacific oceans. It had a major commercial influence on the region from the 1880s until the opening of the Panama Canal in 1914. Since then its development has slowed, and the city attracts more attention for its unique physiognomy in a natural amphitheatre-like setting embracing the sea, as a vast orchestra for the dramas of life. The historic Quarter of Valparaíso is an excellent example of late nineteenth-century urban development in Latin America with well-preserved early industrial infrastructures. It is composed of five interlaced neighbourhoods, public spaces, squares, promenades, stairways, and the top stations of some of the city's distinctive funicular elevators. Adaptation of the built environment to the geographic conditions produced an innovative urban composition which attracts many world citizens.

Each major port is a city of confluences, especially at the geographic situation of Panama. Panama City, the oldest continuously occupied European settlement in the Pacific coast of the Americas, was founded in 1519, 6 years after the discovery by the Spanish of the South Sea. The archaeological remains of the original place include the Pre-Columbian vestiges of the aboriginal occupation and a protected heritage site. The settlement was a major colonial site and seat of a Royal Court of Justice during the sixteenth and seventeenth centuries, until Panama consolidated its position as an intercontinental hub. The city was laid out on a rectilinear grid according to the European ideals of planned towns and was later abandoned and replaced by a new town, known as a historic district, which has preserved its original street plan and architecture, and an unusual mix of Spanish, French, and early American styles. The imposing stone architecture of its public and religious buildings reflects the past prestige of the city at the crossroads of the oceans.

Colombia's port of Cartagena is a unique urban site of the coastal World Heritage and an eminent example of the military architecture of the sixteenth, seventeenth, and eighteenth centuries. Situated on a protected bay facing the Caribbean Sea, the city boasts the most extensive and one of the most complete systems of military fortifications in South America. Because of its strategic location, the city was also one of the most important ports of the Caribbean, an essential link in the route of the West Indies and in the history of world exploration and the great maritime routes.

Cartagena was for several centuries the scene of confrontation among the principal European powers fighting for control of the New World. Defensive fortifications were built by the Spanish in 1586 and strengthened and extended to their current dimensions in the eighteenth century. All the harbour's natural ways were protected by fortresses. Within the security of the walls, the narrow streets host civil, religious, and residential monuments which form the historic heart of the city and seal its identity (UNESCO World Heritage Centre 2014).

From architecture to philosophy and science, urban dynamics have contributed to firing collective imagination and nurturing human civilisation. Cities impacted the genesis of many arts, but have also been impacted by the arts. Art is the aesthetic expression of individual, collective, and universal ethics, the ultimate expression of the creative intelligence of a city, a most dynamic component of its cultural capital (Jimenez 2002). It enriches and heightens the human and social capital but also influences the man-made environment of cities. Each city should be enjoyed as a collective masterpiece of art (Olsen 1987).

Cities are theatres of all expressions of human creativity, and dramatic art is a powerful expression. The dawn of the art of theatre in Greece has been a prominent urban and cultural innovation of all times. Initially consisting of a choral group, part of a rural festival in the honour of the god Dionysus, drama gained importance when brought to the city. The spark of the innovation was the injection of the critical word in the lyrical songs. The chorus became the voice of the citizens, a strong expression of public opinion. Tragedy, comedy, and the satyr play were the three dramatic genres to emerge, and most plays received inspiration from Homer's' Iliad and Odyssey. Athens exported theatre to its numerous colonies and allies to promote a common cultural identity. This development had a significant and sustained impact on the world culture as a majestic anatomy of all human passions and the role of citizens in the city. The theatre of Epidaurus, the best preserved ancient theatre of the world, still hosts during the Greek summers, after more than two millennia, the same unparalleled art form that was born there.

Arts and culture can be notable catalysts of urban regeneration and waterfront developments. Efforts to revitalise urban areas often focus on cultural hubs or arts districts, which contribute to making the urban core friendly to the creative class (Florida 2008). UNESCO's design cities provide many examples of future-oriented interventions, making cities more attractive to citizens and visitors. Labelled as a city of design in 2014, Bilbao made a wise investment with the creation of the Guggenheim Museum, which acted as a catalyst for urban transformation. The city had to overcome a past of declining industry and polluting wastelands and needed to be projected into a dynamic trajectory on the world stage. The museum lifted Bilbao into the twenty-first century. It is a temple dedicated to modern art, which gave to the city a sense of permanent wonder. The titanium vessel designed by F. Gehry was instantly hailed as a most significant accomplishment, a radically new design that challenged museographic approaches and changed assumptions about art, architecture, and collections. The museum revolves around a broad atrium bathed in light. Its structure undulates smoothly and captures every ray of sunshine. F. Gerhy stated that light is free of charge and cities should use and abuse it to the maximum.

Exceptional design has been used as the business card of cities. The Sydney Opera House is a great masterpiece and an icon in an extraordinary waterscape. The unparalleled design bears witness of creativity and innovation and seals a visionary conversation among the city, the environment, and the future. A great urban sculpture set at the tip of a peninsula projecting into Sydney Harbour, the building has had an enduring influence on architecture. It encapsulates three groups of interlocking vaulted "shells or sails" that roof two main performance halls. These structures are set upon a vast platform and are surrounded by terraces that function as pedestrian concourses. The original design concept and unique approach to building gave impetus to a collective creativity of architects, engineers, and supporters and influenced its inclusion in the UNESCO World Heritage Sites list (UNESCO 2014; UNESCO World Heritage Centre 2013).

Landmarks and cultural public spaces, often in unexpected surroundings, are invitation to a journey of art and creativity. The development of High Line in New York is a remarkable example, built on an old railway raised to 7.5 m above the ground, which was used to transport goods from 1934 to 1980. At the heart of the West Side of Manhattan, the cultural passage of High Line alludes to the industrial past of a city in eternal transformation. The walk is designed as an urban choreography with performers, musicians, and walkers, suspended in space and time and with access to experimental drawing rooms in industrial buildings and the Whitney Museum. The High Line is an oasis hosting 220 species of plants. It quickly became a reference inviting New Yorkers and visitors to a more open social life.

Citizen imagination could transform waterfronts into places of civic engagement. Attractive public spaces can build the frame of social exchange and promote democracy. Instead, deteriorated public places, victims of neglect or standardisation, easily become places of confrontation and exclusion that generate violence. Many cities invest in the creation of public spaces that combine aesthetics and functionality, such as the development of quality urban beaches during the summer and fountains at bus stops. The development of public spaces has no other limits apart from the collective imagination.

A demonstration of unlimited innovation and art can be seen in Singapore's Gardens by the Bay, a highly digitalised park spanning 101 hectares of reclaimed land adjacent to the Marina Reservoir. The park is becoming an outstanding urban recreation space and a national icon. It consists of three waterfront gardens, conceived as an integral part of a strategy to transform Singapore from a Garden City to a City in a Garden and raise the quality of life by enhancing local ecosystems. The Supertrees Grove, with tree-like structures, dominates the landscape with heights that range from 25 to 50 m. Supertrees emulate vertical gardens that perform a multitude of functions, which include planting, shading, and environmental protection.

The universal language of creativity endowed Supertrees with enclaves of unique and exotic ferns, orchids, and a vast collection of green plants. They are fitted with environmental technologies that mimic the ecological function of trees, especially photovoltaic cells that harness solar energy which can be used for some of the functions of the Supertrees, such as lighting, and absorbs ions of rainwater for use in irrigation and fountain displays. A 3-km waterfront elevated promenade, between two of the larger Supertrees, allows for breathtaking views by the Bay. At night, the Supertrees further expand wonder as they come alive with light and music.

7.3.2 European Capitals of Culture and Reinvented Worlds

The Capitals of Culture had a great contribution in raising awareness about the diversity and richness of the European Union, shared values and traditions. Each European Capital of Culture intended to meaningfully engage all those living in the city and welcome visitors in uplifting celebrations. Athens has been the first engaged title-holder in 1985 paving the way to a fascinating array of Capitals of Culture. Since then, each year has seen one or two cities put in the spotlight for 12 cultural months, which help to raise the visibility, pride, and prestige of places, people, and events, foster citizen engagement, and bring about urban renaissance. Liverpool 2008 had around 10,000 registered volunteers, while all schoolchildren participated in at least one activity during the year. Bridging the ephemeral celebration with the longer-term regeneration has become one of the most valued aspects of the capitals that wish a transformative legacy. Many annual titles acted as catalysts of profound urban transformation. Lille and Glasgow are often cited as the most outstanding examples of cities that enhanced the opportunities offered by their year as European capital of culture as a long-term investment in urban regeneration.

For the millennium year, as many as seven European Union cities were chosen, together with two cities of the countries that were to join the EU in 2004. Brussels introduced the Zinneke Parade, which evolved into a biennial event, with a different theme chosen for each parade. 'Zinneke,' a word originally referring to the city stray dogs which hung around the streets by the Little Zenne, a tangent canal of the River Zenne which ran round Brussels along the city walls until the end of the nineteenth century, became a nickname representing Brussels, the subject of the parade, connecting people beyond cultures, communities, and districts within Brussels. The organisers worked with a large variety of institutions, schools, cultural centres, organisations, and societies to create the ideas and prepare the projects that eventually make up the parade. Other cities searched new ways to bring forth the culture of their diverse mosaics. Linz shared the title with 12 neighbourhoods as cultural capitals of the 12 months of the year, promoting intra-city cultural dialogue.

The competition among cities wishing to become capitals of culture starts at least 6 years before the treasured year, so that bidding cities can avail of time to engage with citizens, envision, plan, and co-create. As shared title among EU cities and with cities from outside the European Union, the European Capital of Culture has already led to prestigious exchanges. Istanbul was the largest ever non-EU city to share the title of the European City of Culture in 2010 with Pécs in Hungary and Essen for the Ruhr, which focused on mythology, metropolis, and Europe. Istanbul, former capital of four empires, featuring architecture from the Roman, Byzantine, Latin, and Ottoman times, centred on modern and contemporary culture reflecting the diversity of cultures and ethnic groups as well as its young, dynamic population. The event generated transformative energy, built capacity to address the urban and cultural dimensions of citizenship, and created a bridge connecting Europe to the East (EP 2013).

The Estonian capital of Tallinn, Cultural Capital for 2011, honoured culture in its wealth and diversity, ranging from traditional to contemporary art. The city hosted many large-scale events such as the Tallinn Maritime Days, a song and dance festival, the concert "Song of Freedom," the "Happy End" during the last day of the event, and the Tallinn marathon. The greatest ambition of the Foundation Tallinn 2011 was to give life to smart ideas and fascinating, unique events that place culture at the heart of the city. Cultural startups, especially by local designers, were encouraged and embraced unexpected ideas. Tallinn 2011 has also initiated a sustained movement of volunteers to become a more bike-friendly city, presented a programme of hospitality and launched a project to attract young people and children from outside Tallinn to the cultural events.

Guimarães, Portugal and Maribor, Slovenia, were designated as the European Capital of Culture in 2012. Guimarães, often referred to as the "birthplace of the Portuguese nationality" or "the cradle city," has a significant historical importance because of its role in the foundation of Portugal. Maribor, the largest city in traditional Lower Styria, is known for culture, including wine. Every June, the two-week Lent Festival brings together hundreds of musical, theatrical ballet performers, classical, modern, and jazz musicians, and dancers from all over the world (Ecorys et al. 2013).

Marseilles with the Mediterranean marked the 2013 European Capital of Culture and attracted 11 million visitors. The oldest city in France, founded 2600 years ago on Mare Nostrum, offers the richness of a unique heritage and an intense sociocultural life. The car-free redesigned old harbour has been at the crossroads of all events and the beating heart of the capital of culture. The Silo, a concert arena in a converted silo on the docks, and a huge hangar at the head of the old port were among the most fascinating venues. Exceptional events included an exhibition bridging science, mathematics, and experimental art and a fictional journey around the Mediterranean Sea, tracing its true and mythological history through the eyes of residents and artists. The La Marseillaise newspaper presented a serialised novel through the weekly installments by 12 authors in residence in symbolic locations. The city celebrated also the sea through the inauguration of the Museum of Civilisations of Europe and the Mediterranean, and the Villa, an international centre for cultural and artistic exchange largely built underwater.

In 2014, Umeå and Riga shared the title of European Capital of Culture and promoted cultural richness and diversity. In Riga, the festivities started with a tribute to the 500th anniversary of the first printed book. The move of thousands of books by a chain of volunteers, hand to hand, from the old National Library to the new site has been an extraordinary event, suggesting that a public library is not just a place to borrow a book but a repository of true masterpieces and a communication node for the local community. Riga also organised the world select games, a kind of World Championship for choirs and 20,000 singers. The Festival of Light featured a large number of international multimedia art projects and encouraged public participation and co-action. As part of the festival, the largest graffiti in the Baltic countries "The Sun, The Thunder, The Daugava," created for the Capital of Culture, was also shown in a new light. The public participation campaign "Light Up Your City!" created colourful light installations on the facades of buildings.

Co-creation was the key idea in Umeå 2014's Capital of Culture. The city wished to be a window to Europe and show that it is one of the foremost culture-driven cities in the continent. Before, during, and after 2014, cultural exchanges between Umeå and the rest of Europe intensified and new cultural bonds were formed. Raising Sweden's European profile was another cultural ambition of the capital of culture that insisted on the role of cultural exchanges for regional development and identity. To promote a long-term perspective and encourage the continuation of capital of culture activities beyond 2014, projects were run by the cultural operators themselves.

2015 marked the 30th anniversary of the European Capitals of Culture initiative results, and offered the occasion for celebration of its significant cultural, social, and economic benefits, in terms of vibrancy, attractiveness, and regeneration (EP 2013). Mons, one of the two European Capitals of Culture of 2015, proposed the theme of illumination with more than 300 events and a thousand cultural and artistic activities, including unusual public spaces, five new museums, two new concert halls, and new types of civil participation and digital opportunities. The returns to the local economy were great, ranging from 5.5 to 6 € for each € of public money invested in the year. Co-capital Pilsen 2015, under the motto "Open up," staged more than 600 events, including plays, concerts, exhibitions, public space happenings, and artistic interventions. The Hidden City project allowed both residents and visitors to explore the history of Pilsen with the help of innovative tools that combine the personal memories of locals with the latest digital mobile applications. Another flagship project is the DEPO2015 Centre for creative enterprise, opened as the first incubator for future cultural and creative industries in a former tram and bus depot. This Pilsen 2015+ volunteering project created a new space where start-ups, established businesses, and cultural actors can come together, in a hub of collaboration, innovation, and creativity that has great potential to bring long-term benefits to citizens. Many hundreds of visitors took part in neighbourhood and architecture walks.

From Spain to Poland and beyond, San Sebastián and Wroclaw were the European Capital of Culture in 2016. "Waves of energy" captured the soul of San Sebastián 2016 with citizen movements the genuine driving force behind transformative action. The Donostia-San Sebastian 2016 programme, "Culture for co-existence," promoted better ways of living together through art and culture. Three lighthouses reflected the values underpinning the programme, including the Lighthouse of Life (the individual and society), the Lighthouse of Peace (respect and co-existence), and the Lighthouse of Voices (diversity and mutual understanding). Wroclaw 2016 introduced "Spaces of Beauty" to celebrate diversity and metamorphosis and to bring forth the "forgotten city" with unconventional artistic events and concerts taking place in marginalised and overlooked spaces, such as townhouse stairwells, historic courtyards, and poor underpasses. Citizens were invited to run part of the programme through a micro-grant scheme. During the year, the city also became the UNESCO World Book Capital, with a special edition of the European Literature Night.

Some of the European Capitals of Culture later became Green Capitals of Europe, similar to Stockholm and Copenhagen. Amsterdam has been a Capital of Culture long before becoming the Innovation Capital in 2016; this may highlight the interdependencies among natural, cultural, and innovation capital and also the dynamics of local governance and openness of cities and willingness to engage with a higher level. In 2017, the environmentally minded Aarhus and the historic city of Pafos were the European Capitals of Culture. Aarhus proposed "Rethink" as central theme and a Creative World Forum that invited hundreds of children to imagine the future in a series of events entitled "Land of wishes." Pafos committed to "Linking Continents, Bridging Cultures" as the common thread. The Open Air Factory invited artists, citizens, and visitors to come together and to co-create in different projects and networks exploring new paths, mindsets and experiences.

Art and sea in cities can have a transformational impact, as culture frees the community and liberates individuals, and the sea inspires artistic expressions. This was the case in Nantes, a formerly thriving shipbuilding port, which was hit hard by the recession of the 1980s and re-emerged to become the 2013 Green Capital of Europe. When the dominating naval industry sank, the city went down with it. The affluent industrial port with its proud shipbuilding heritage, employing 60,000 people at its mid-1970 peak, closed down in the late 1980s, hosting barely 2,000 workers, with abandoned warehouses and empty wasteland. This crisis not only affected the economic web of the city but also shook its very identity and image. Civic pride was to be regained thanks to culture and the orchestration of art projects that captured the imagination and thrilled citizens. Incorporating new art forms has been vital for Nantes, which made available walls, pillars, and fences as open air canvases for the art form of graffiti and transformed the long estuary of the La Loire River into a contemporary art track (City of Nantes 2010).

Thanks to and through art, Nantes turned upside down. Each summer, the "journey to Nantes" promised new adventures. A crane artwork projected into the skyline became a symbol of the city's history as an important shipyard centre. A puppet theatre bringing to life extraordinary puppets up to 12 m tall was a groundbreaking project kick-starting urban regeneration. A 'crazy arts' festival' intended to showcase music, dance, drama, theatre, and art of a chosen city. Another project transformed the city's former biscuit factory into the city's performing arts 'unique place.' Seas were omnipresent in the art epidemic sweeping Nantes, which hosted a 130-m sea-serpent skeleton and unprecedented projects inspired by the work of Jules Verne (Orsenna and Gilsoul 2018).

Many initiatives try to strengthen the bond between artistic creativity and citizenship. In Brussels, the festival "Moussem Cities" is dedicated every year to a different Arab city. The project "Cities on Stage," initiated by the National Theatre in Brussels, invited six European theatres to exchange and interact on the challenge of living together in Brussels, Gothenburg, Paris, Sibiu, Naples, and Madrid. During the years 2011–2016, theatre artists posed original, critical, and poetic eyes on this urban world in transition. "Moving cities" invited groups of citizens, young and old, to question the challenges of contemporary cities in the present and future world. Citizen workshops, accompanied by artists of multiple disciplines, explore through theatre themes such as migration, diversity, multiculturalism, and cross-fertilised urban identity. In parallel, each theatre invites a group of young actors of partner countries to work for a month with a director and associate citizens and support young artists in their urban discovery.

7.3.3 The Ocean as a Prime Source of Inspiration

As cultural initiatives try to reinforce cities on the world scene, underwater cultural heritage and creation can strengthen sites on the ocean atlas. In 2016, Europe's first underwater museum, Museo Atlántico, in Lanzarote, hosted an amazing exhibition, the "Raft of Lampedusa", by artist J. deCaires Taylor, proposed as a modern version of Géricault's 1818 painting "The Raft of the Medusa." Fifteen meters under the surface of the water, a sculpture park represents a human botanical garden with people and plants drawn from the flora and fauna of Lanzarote: this is the latest in a succession of Taylor's underwater installations. The first was in Grenada in 2006, involving "Vicissitudes," a ring of sculpted schoolchildren, placed on the edge of an ocean shelf. That work was instrumental in creating a protected marine park presently listed as one of National Geographic's 25 wonders of the world.

Questioning the relationships between humankind and the sea is at the heart of the underwater exhibitions. The same artist had already sunk, in the Bahamas, the world's largest underwater sculpture, the 40-tonne Ocean Atlas, and exhibited his largest scale work in the only underwater museum in the world, Museo Subacuático de Arte in Cancun, Mexico. His work "The Silent Evolution," comprising nearly 500 statues, was cast by locals in the nearby fishing village of Puerto Morelos. It is significant that he has named his underwater sculpture parks as museums, emphasising the sacred nature of the oceans. The cluster of vibrant structures was smothered in marine life. Many species including corals have been spotted in and around the sculptures, creating a happy symbiosis between nature and art.

The oceans have been a powerful source of inspiration for many scientists and artists, from Homer's *Odyssey* to *The Sea* by Debussy, *The Blue Agreement* by Y. Klein, and the contemporary video art by Bill Viola. Foundations such as Parley for the Oceans use art and creativity to raise awareness about the beauty and fragility of the ocean. Artists, musicians, actors, filmmakers, fashion designers, journalists, architects, scientists, and innovators try to invent ecologically sensible ways to enjoy and protect marine ecosystems. They invite us to turn the ship around before we lose a treasure that we have only just started to explore: we know very little about the extraordinary blue universe of the oceans.

The Good Planet Foundation, directed by Y. Arthus-Bertrand and M. Pitiot, was intended to be a plea for humanity to respect the blue world supporting life. It tried to explain in simple words the greatest natural mystery of the planet, also after the scientific missions of *Tara*, a schooner acting as a platform for state-of-the-art research on marine ecosystems and biodiversity. From 2009 to 2012, the schooner *Tara*, supported by the UN and many institutions and foundations, sailed around the globe, sampling plankton communities from viruses to fish larvae. During the expedition, high-quality and standardised genetic, morphological, and physicochemical samples from 210 stations across the world oceans were collected. Meta-genomes and meta-barcodes from stations are being built as well as quantitative and high-resolution image databases. After results published in a special number of the journal *Science*, the TARA OCEANS database could be considered the largest open access database, clarifying the role of plankton in the earth's life support system, and the effects of climate change on this critical element for the marine food chain (Tara Oceans Consortium Coordinators et al. 2015).

Very diverse artistic productions bear witness of a vulnerable ocean and world. Well before the modern tsunamis devastated cities, the awesome waves pictured by the Japanese artist Hokusai in his "Thirty-six Views of Mount Fuji" had shaken humanity. Mount Fuji was traditionally seen as the source of the secret of immortality, intrinsically linked to the future of humanity. "Wrapped shores" and "wrapped bridges" by Christo may signal the need to protect cityscapes and seascapes. Geopolitics on the exploitation of the Arctic already inspired theatrical creation and the play "Arctic" over the *Arctic Serenity* cruise ship. The "Red Turtle," a 2016 animated film directed by M. Dudok de Wit, can be seen as a pure fable of survival, "a brilliant shard of sea glass rendered smooth and elegant through generations of retelling." And "Blue," celebrating in 2018 the 10 years of Disney Natura, invited people to visit the deep ocean and to be mobilised about the safeguarding of its wonders.

7.4 The Science and Art of Ecosystem-Based Strategic Planning

7.4.1 *Urban Strategic Vision, Objectives, and Actions*

Strategic planning and budgeting are important tools towards the preferred sustainable future of cities and impact the flows of economic, social, and cultural energy. Increasingly embedded in the political culture of cities, planning demands creative rigour to co-define objectives for the longer term, identify the optimal location of activities and infrastructures, and prioritise actions towards the execution of political strategies. Cities striving to reach their intended destination need to be well aware of their starting position, their strengths and weaknesses. They have to decide together with their citizens where they want to be in the future and understand the significant trends that may influence the direction in which the future unfolds.

Increased macro-complexity and uncertainty create a space for participatory innovation and cooperative processes. Paradigm shifts and quantum leaps are possible if integrated in a coherent vision for the desired future. Strategic foresight, which has always to be accompanied by hindsight, deliberately cuts across the traditional boundaries of sustainability science disciplines and policy areas. It can act as a driver of social interactions that stimulate the generation of common public visions (Hall 1998, 2013).

Strategic anticipation includes the exploration of future possible prospects through the early identification of emerging challenges, opportunities and threats, and horizon scanning for unforeseen turbulences and weak signals and "black swans," low-probability/high-impact events. Setting the strategic questions, identifying driving forces of change, determining main issues and trends, clarifying levels of impact and risks and degree of uncertainty, creating probable scenario narratives, assessing policy options, and identifying inflection points are important elements of the process. Ex ante impact assessment of possible, probable, and preferred futures, exploration of alternative scenarios, and consensus building can lead to an ambitious, collective, and engaging vision (City of Amsterdam 2014, 2015).

Governments and cities integrated futures thinking and strategic foresight in their decision-making process and developed policy portfolios to achieve the desired visions according to sustainability criteria and financial constraints and perceived risks. Scenario planning is often complemented by risk assessment approaches, especially for cities facing important storm surge and sea level rising. The process could help strike a balance between bureaucratic effectiveness and creative thinking and lead to forceful goals backed by all stakeholders. Cities together with their citizens should evaluate their resources, respect the geophysical and cultural local limits, seek a symbiosis with the bioregion, and mobilise invisible hands and appraise investment and opportunities costs. Urban planning for sustainability requires a comprehensive interdisciplinary assessment of urban assets, a natural resource information system, and an identification and analysis of the policy distortions and bottlenecks. Tactical decisions have to be inscribed in the overall policy framework, and prevention should be considered an investment.

Vision building is the first mobilising element for creating the sustainable cities of the future. It has to be compelling, and it is increasingly subject to consensus and participative governance. The discussion of many possible alternative futures, envisioned by all stakeholders, is a key element in this process for the future not to result as an unwished or linear continuation of the past. Multi- and interdisciplinary, scientific, technical, and social approaches are crucial, as many challenges are global, multi faceted, and interdependent. For a vision to come true, discussion and mediation of multi-stakeholder approaches at the earliest possible stage are most important as the range of stakeholders progressively opens to include the whole society.

The blue, and inevitably green, city is one of the latest concepts in a chain of concepts for the possible and desirable sustainable city of the future. Visions had already embraced many interrelated concepts such as the green city, the liveable city, the viable city, the affordable city, and the ecological city. Blue cities join the eco-cities and eco-societies league of cities that are simultaneously ecologically, socially and economically conscious. Integrated urban policy and planning are instrumental for reducing the ecological footprint and improving urban performance towards a blue–green city that can offer good living conditions and sound ecosystems (Beatley 2011).

New York developed a fourfold vision for 2030 and the OneNYC, City's blueprint for the future. OneNYC continues to focus on the core themes of past plans, encapsulating growth, sustainability, and resilience, and additional attention on equity and a regional perspective. The first part of the vision sees NYC as a growing, thriving City, the world's most dynamic urban economy. The second part suggests a Just and Equitable City, with an inclusive, equitable economy that offers well-paying jobs and opportunity for all to live with dignity and security. The third sees

NYC through sustainability lenses, as the most sustainable big city in the world and a global leader against climate change. Last but not least, resiliency is the last major criterion, with neighbourhoods, economy, and public services ready to withstand and emerge stronger from twenty-first-century threats (NYC 2017).

Hippodamus and the first grid plans, the collective search of the optimal urban forms for the desired future, generated interesting concepts. Strategic integrated planning is an important instrument for the future of sustainable cities and impacts the flows of energy towards programming the various actions. It defines the location and distribution of various diverse activities and infrastructures. In the era of sustainability, it has to be an open, transparent, and inclusive process and involve all stakeholders. Strategic plans for sustainability focus upon bringing abandoned urban land into mixed-use development and seek to restrain peripheral growth to key locations served by public transport.

Urban strategic plans have to reconcile thematic and territorial policy objectives and be analysed and discussed, co-decided, co-implemented, and co-evaluated. New principles can be injected into old plans and ensure change within continuity. The Tokyo government proposed the concept of an eco-society towards a clean, sound, and citizen-friendly metropolis. Comprehensive actions focused on resource and water management, transport, consumption patterns, and promotion of environmental education and awareness. Rising sea levels and seismic and increased typhoon risk have raised consensus on the need for a strategy that offers protection to the low-elevation coastal zones surrounding Tokyo Bay. In 2016, 'Next Tokyo,' an exercise in sustainable big-thinking for the future, allowed imagining a resurgent mega-city adapted to climate change with the realisation of a high-density eco-district built on a resilient infrastructure. The archipelago of reclaimed land supports transit-oriented development for a half-million occupants while improving the resilience of Tokyo Bay against water-born risks. Next Tokyo addresses the city-wide vulnerability by providing coastal defence infrastructure and protection to the shoreline of Upper Tokyo Bay. These resilient infrastructural elements have a dual function as the foundations for clusters of recreational open spaces and high-density development across the bay, including the 'Sky Mile Tower,' reaching a height above 1600 m.

The Kobé earthquake in 1995 provoked a whole range of innovative managerial responses across the region. The WBCSD Urban Infrastructure Initiative proposed several solutions to address economic, environmental and social challenges, including an ageing population. The 5-year master plan emphasised creativity and innovation, local economic and sustainable community development. Six major corporations and the WBCSD worked with officials from several city departments to identify key sustainability issues and propose policy options and priorities for action, including energy efficiency and renewable energy production, sustainable mobility, knowledge networks, and leadership in disaster resilience. A transformation study led to the formulation of 14 practical solutions including energy efficiency improvements, sustainable mobility measures, and knowledge-based economic development. The multi-sector contributions early in the city's strategic development process were invaluable and enabled city officials to consider a variety of ideas and engage with the private sector collectively. Such holistic approaches are increasingly important for inclusive and sustainable urban development (WBCSD 2013, 2014).

In Singapore, the Concept Plan, introduced as early as 1971, laid the foundation for Singapore's better quality of life with new towns, transport infrastructure, and access to recreation. The Concept Plan was subsequently reviewed in 1991, and in 2001 and 2011 taking into account strategic insights after evolving local and global trends, in an effort to anticipate and address future needs. The plan is vital in helping balance the diverse urban functions. Singapore's Centre for Liveable Cities and the Urban Redevelopment Authority organises the biennial World Cities Summit (WCS), as an integrated global platform since 2012 with the Water Week and CleanEnviro Summit Singapore. The WCS hosts the Mayors Forum, inviting mayors from cities that are highly regarded for their achievements in the fields of civic engagement, urban solutions, international development, and nonprofit work, as well as the dynamic World Cities Summit Young Leaders.

Singapore could have a population between 6.5 and 6.9 million by 2030. The plan outlined the strategies to support population and economic growth for all. The reviews benefited much from extensive public consultations carried out through various channels, such as surveys, focus group discussions, and public forums. Through engaging all stakeholders and the public at large, the city can better understand citizen aspirations and respond to expectations (Urban Redevelopment Authority of Singapore 2014).

Transport and land use policy and planning constitute two interconnected instruments for sustainability. Transportation plans have to be integrated organic parts of the visions and the strategies to implement them. Suburban garden cities or satellite new towns, which have developed around subway stations, have often been transformed into cloned suburbs without any independent identity. Sustainability imperatives ask for these suburbs to become consolidated vibrant neighbourhoods with new offices and jobs created near public transport stations, cultural facilities, and noble public spaces.

Urban planning has evolved much over the past century and gradually embraced sustainability ethics and principles. The New Charter of Athens, issued by the European Council of Spatial Planners (ECSP), signalled a clear shift in planning values since the post-war functionalistic principles and the segregation of spaces for work, living, leisure, and communication. The charter for the third millennium advocated for sustainable human settlements for all, based on true involvement, and responsible planning, which promotes socioeconomic and environmental enhancement and safeguards cultural identity. The 2013 "Charter of European Planning" promoted a more pro-active role for planners in shaping informed public debate and co-creating the future (ECSP 2013).

Strategic urban coastal planning has to be comprehensive and integrated and involve the full array of urban and maritime actors and activities. To ensure the environmental health of all functions, it has to embrace an ecosystem-based approach for the protection of the natural resources and habitats that provide the basis for urban activities. A public and transparent process for analysing, planning, monitoring, and reviewing the spatial and temporal distribution of human activities is necessary to achieve the agreed economic, social, and environmental objectives. The ultimate aim is the optimal use of terrestrial and maritime space for different public, residential, industrial, economic, and cultural activities. Transport infrastructure is important and should be addressed in an integrated way for all modes, including maritime transport.

Sustainable development can only be founded on healthy ecosystems, terrestrial and marine. On land, sustainable cities opt for consolidation and renewal rather than expansion and urbanisation of green fields. Concentrated and intensified use of space in a well-defined urban territory provides multiple advantages for the integration of structures that minimise flows of strategic resources and transport, local pollution, and greenhouse emissions. New visions for synergies stemming from the mix of compatible urban uses affected the evolution of many ports and their articulation to the city and the eco-region.

Until recently, cities have rarely accepted responsibility for marine resources, often because of the apparent immensity of the seas and the difficulty to visualise or quantify the offshore effects of urban life. Cities and citizens started to link the plastics in the sea with unbridled urban consumption. Urban coastal maps usually stop at the water's edge, even though the activities that support urban systems extend further into the sea. But many cities embraced innovative approaches. In Boston, the New England Aquarium produced a fascinating map of terrestrial watersheds and offshore waters on the Atlantic Coast, presenting areas of a wide spectrum of urban activities, including high boat traffic, shipping, fishing, and dredging (Beatley 2011).

A modern port is a very complex environment involving many risks, and safety is a priority for the port community. Much attention has to be given to local nuisances affecting either citizens or the environment or both, and to communication with society. The port of Antwerp has provided a good example since the publication of its first biannual Sustainability Report in 2012. With this benchmarking exercise, the Antwerp port community gave form to its ambition of being a sustainability leader. It was the first time that the many interlinked sustainability aspects of the port and of the hinterland and foreland were reviewed by the entire array of the port community. Transparent, open, and systematic consultation is seen by the various stakeholders as particularly useful. Scrupulous attention is being paid to safety, with the emphasis on accident prevention. A comprehensive survey of the perceptions of nuisances in and around the Antwerp port offers useful complementary information about the negative effects of port activities and the ways that are experienced by the local population (Port of Antwerp 2017).

In an industrial environment such as the port of Antwerp with one of the world's largest chemical clusters, there is a huge demand for energy, and the port community strives to maximise the use of green energy. The Port Authority tries to lead by example through purchasing 100% green electricity. Significant expansion of the wind power capacity within the port is also planned, and the possibilities of bio-based energy sources are being examined. Further research in energy efficiency is continuing, including a heat distribution network. The study of the feasibility and desirability of urban–maritime combinations along the river paid particular attention to multiple factors such as maritime value, housing demands, metropolitan accessibility, landscaping, investment costs, and overall sustainability prospects. From the point of view of sustainability, the growth potential and land demand of both city and port are in balance, and impulse is given to better connectivity through urban public transport.

Dense, compact coastal cities manage resources at remarkably lower levels of material and energy consumption compared to diffuse settlements and dispersed populations and tentacle-shaped, low-rise, car-dependent suburbs. Compact settlements encourage functional diversification and integration of land uses at the neighbourhood level and prevent urban sprawl and resource overconsumption. The Dutch compact city policy is based on the principle of spatial functional mix, in balance with nature. Creating multifunctional urban cells is a guiding principle for many city plans (OECD 2012).

Amsterdam considers functional mix as a valuable attribute of inner-city heritage and tries to strike a balance among spaces for housing, offices, commerce, services, tourism, and leisure. The compact city policy, already introduced in 1985, aimed at enhancing scarce space as efficiently as possible, introducing more scrupulous resource management by creating multiple residential environments and curbing the overall ecological burden. Diversity and mixed land uses are linked to enhance the city's unrivalled character as a cultural melting pot. The Eastern Dock Island at the east of Amsterdam's Central Station is a component of the large-scale project to develop the south bank of the bay. The move of harbour functions has presented new opportunities to transform the area into an intensive mixed urban neighbourhood with residential, commercial, recreational, and cultural functions, including the new Public Library and the Conservatorium. The Library, designed by architect J. Coenen, includes a central underground system for long-term energy storage for air conditioning and heat.

The "Port-City" of Amsterdam and the municipal Spatial Planning Department explored the possibilities for the future of Western Dockland, a busy port area within the city centre of Amsterdam. Both the city and the port had expressed in the past conflicting needs for more space and, after a period of divergence, they tried to work on synergies. They called for reorientation of the relationship between the city and the port towards a better urban–maritime symbiosis along the river. A key question was whether the existing port activities can coexist with housing and civic life (Port of Amsterdam 2009).

Sustainable coastal cities can highly benefit from a genuine urban, marine, and social intermixture. Symbols help reinforce identity but also improve communication and openness to citizens and the world. Many cities experience a desire for "a real urban gene, an organic part of the city, everywhere in the city." The "open block" proposed by the architect Ch. de Portzamparc advocates for urban blocks that open to the world. The concept could favour the interactions and social life at the scale of each neighbourhood in the ocean of each city and beyond.

Beacon sustainability projects can offer significant landmarks to a city, structure urban and ocean debates, and enhance magnificent skylines and shorelines.

They can act as strong catalysts for the future of the cities and the regions. These projects range from ambitious government plans or unique international events, such as Olympic Games, Universal Exhibitions, and high-level fairs, and European cultural, green, or innovation capitals to local maritime and vessel museums. Their planning and construction with respect to sustainability principles are fundamental given their emblematic role and manifold effects on land and the sea.

The 1992 Olympics was a key catalyst for the renaissance of Barcelona. The city, which had long lived with its back to the Mediterranean, invented a new waterfront and was definitely reconciled with the sea. Enriched with new and better places and functions, including a public beach, and integrating the Olympic village, the city was enhanced with thorough restoration projects and noble public places. The rehabilitation of the Ciutat Vella has been an unparalleled event, in terms of investment, effort, and civic spirit. It reflects the ideals of an orderly yet vibrant city, nestled by the sea. The Eixample, designed by the visionary urban planner I. Cerdà Las Ramblas, offers the delights of animated public artery and seals the link to the sea. The four historic quarters of Ciutat Vella, including the gothic quarter, have been thoroughly transformed through selective renovation, rehabilitation, and eco-constructions, civic centres, pedestrian precincts, and green public spaces. Thoroughly designed small interventions acted as stem cells that when injected in the body of the city led to a positive metastasis and overall renewal. The civic groups cooperated with the authorities, having a pioneering role in the allocation of housing and the vibrancy of public life.

Lisbon seized the opportunity offered by two major end-of-the-millennium events, the Cultural Capital of Europe in 1994 and the Universal Exhibition in 1998 on "The Oceans," to reinvent and transform itself and redesign its map enriched with a new quarter to the East, the Oriente. Oceanic vocation, one of the city's essential attributes, has offered a vital thread. Expo 1998 transformed an abandoned former industrial area into a vibrant innovation site and created a whole new resourceful city of 330 hectares. Investments in bioclimatic architecture, quality design, and advanced energy concepts have led to the aesthetic and functional metamorphosis of the area. An eco-efficient distribution network for thermal energy, heat and cold air, was set up, together with an observation and monitoring system. The standards were higher than those required by the national regulations, and the performance of the area has been exemplary. The Vasco de Gama Tower is among the emblematic legacy buildings.

Characteristic of a new millennium and its waves of change, the 2000 first Green Olympic in Sydney, endowed the city with a sustainable new organic park by the sea. A 640-hectare industrial wasteland in Homebush Bay, previously intended for an urban renewal project after 100 years of industrial and military uses, was fully reinvented and transformed into the Sydney Olympic Park. After the end of the 2000 Olympics, the Park was converted to a multipurpose resource continuing to host sporting events. The Sydney Olympic Park Master Plan attracted a broad range of residential, commercial, recreational, and cultural activities. Urban design and landscaping emphasised excellence and efficiency (Mega 2012).

The Blue Olympics organised in Athens in 2004 bestowed the Hellenic capital with an array of major infrastructure projects. The Olympic works endowed the metropolis with a state-of-the-art multiple mobility network, highly appreciated by commuters, citizens, and visitors. The most central underground crossing hosts archaeological galleries with objects discovered during the excavation works that accompanied the creation of the new metro. The Olympic village became a high-quality residential area for low-income families. However, the post-Olympics use of the athletic installations has been largely problematic.

The Olympic dream has sometimes morphed into a nightmare overnight just after the world turned off the spotlights. Rio di Janeiro and Athens have been pinpointed as having promoted Olympics that they could not sustainably afford. In Athens, the Helliniko site went unused until 2015 when it served for accommodating migrants, while other sites have been abandoned. This scenario indicates the importance for all infrastructures to be thought and planned with a long-term vision and benefit for all.

The legacy of the 2010 World Cup has been very beneficial for Cape Town. Improvements to the city centre, the downtown area known as the city bowl, and the public transport system attracted citizens and visitors. Concerns about safety and the haunting legacy of apartheid have been successfully addressed, and sports were used as a means to overcome racial fears and tensions. The short distance to the Cape of Good Hope was truly filled with lofty hope.

A triple Olympic city, London has long been a multi-faceted city and a great intercultural and financial centre. Sustainability was a prominent criterion in the organisation of the city's third Olympics. The Olympic stadium featured flooring made from recycled tennis balls, while the wave-shaped Aquatics centre designed by Z. Hadid demonstrated sea-inspired design. From Munich to Beijing, bio-mimetics had already given amazing architectural venues for previous Olympic venues. The 2012 Olympics brought a revival to London's East End and enriched the city with world-class infrastructures, business hubs, and new landmarks. The new iconic tallest building in Europe, the Shard, has been both praised for its creative and bold design and criticised for disrupting the historic views of the London skyline (Mega 2015).

Eco-responsible cities are conscious chronotopes, with interconnected spatial and temporal dimensions and interrelated historical and geographic aspects. Like space and water, time is a scarce resource for cities. The time dimension does greatly matter in advancing towards sustainable development. It introduces concerns about inter-generation distribution of assets and serves as a litmus test for the well-being of individuals and societies. Time management has a potential for extending the limits of spatial planning. Some governments and cities have been pioneers in promoting plans that strengthen places through the harmonisation of time budgets (INU-Politecnico di Milano 1997).

The concept of a "24-hour city" gained attraction, even if it most often refers to the always active commercial districts of some cities 'that never sleep.' Time plans in Italian cities, such as Milan, Florence, and Bolzano, deployed efforts to optimise public services, opening them at hours most demanded by citizens. They have been linked to mobility plans and led to the modification of timetables of municipal services. Classical urban time planning examples first related to the restriction of the major cities to cars during certain days and hours per week. In Rome, the municipality, the trade unions, the city time office, and the office for citizen rights had then signed an agreement for the reform of the timetables of municipal services. Women's associations have been very active in extending the opening hours of the social services and bringing them in harmony with the times available to citizens and the pace of everyday life.

The reconfiguration of urban spaces and times can have a substantial resounding impact on citizens' quality of life. Strong sustainability asks for actions maximising and optimising the possibilities of time and space. Crowdsourcing urban design can help create new typologies for urban commons to be collectively enjoyed. In Copenhagen, the creation of the extraordinary Superkilen, an 1-mile-long urban park co-designed by architect B. Ingels running through the multicultural Norrebro quarter, has been a laboratory for public participation. Citizens were asked to identify essential elements of their culture that they wished to find there, and they proposed Chinese palm trees, a Moroccan fountain, and benches.

From floating spirals to deep-sea novel concepts, the future of blue urbanism may include floating cities or other forms of permanent or semipermanent habitation of the ocean environments, either on the surface or underwater. The American designer J. Fresco, even before the Second World War, had proposed model ocean cities, maximising efficiency, independently and in constellations. French visionary architect J. Rougerie proposed many concepts, such as SeaOrbiter, to realise the tremendous potential of inhabiting the oceans, a subaquatic archaeological museum for a journey to sunken cities, a floating scientific city, and an underwater habitat-observatory. For 5 years, the Seasteading Institute has been conducting research into the potential for innovative communities, floating at sea. It has proceeded towards the design and the selection of the world's first city at sea.

In 2017, the Institute signed a Memorandum of Understanding with French Polynesia to cooperate on the creation of a sea zone with a unique governing framework: it could be the 254th state. An independent company, Blue Frontiers, was created to develop and construct floating islands and to operate the sea zone. The icompany Shimizu proposed the ocean spiral, a deep-sea future city concept. Many more urban utopias have been proposed such as self-contained cities in the shape of a lily pad or a lotus flower and fertile cities such as the "Recycled City," of half a million residents, that would be created from recycled plastic from the North Pacific Garbage Patch. Other concepts include amphibious tourist resorts and the extension and expansion of cities in vertical and submarine spaces. Human populations already living on ships and multi-purpose marine platforms, in a close connection to oceans, could be the precursors of the citizens of the floating cities of the future (Beatley 2011; Orsenna and Gilsoul 2018).

7.4.2 *Ecosystem-Based Coastal and Marine Spatial Planning*

Ensuring the optimal distribution of space among relevant functions and stakeholders is essential to enable concurrent activities not simply to coexist but to achieve their full potential and generate synergies that can be captured for the benefit of all.

The concentration of diverse activities can lead to competition for space and attention among sectoral interests, such as port development, shipping lines and pipelines, wind energy operators, tidal and ocean energy infrastructures, fisheries and aquaculture sites, emerging marine biotechnology, and seabed mining activities. It can also raise multiple environmental concerns at the crossroads of pollutants and emissions and the requirements for healthy ecosystems. Conversely, healthy coastal and marine ecosystems are essential to continue deliver substantial services, in terms of food security, recreation, well-being and tourism, climate change mitigation and adaptation, shoreline dynamics control, and disaster prevention (Elmqvist et al. 2013).

Competition for maritime space is intense among various coastal and offshore uses, such as renewable energy equipment, aquaculture, and tourism, and this intensifies the need to manage marine waters more coherently. Fragmentation and sector-by-sector management of marine activities further increases pressures and compromises the capacity of ecosystems and the performance of marine activities. A high degree of potential interactions exists between activities, especially and surprisingly between offshore wind and other ocean activities. Multi-use platforms represent a more complex form of joint location of offshore activities, and their use has multiplied over the past years (OECD 2016).

Collective eco-responsibility is important to reduce conflicts among sectors and create synergies among activities, to encourage investment, through clarity, predictability, and transparency, and to increase cross-border cooperation to develop energy grids, shipping lanes, submarine cables, and other activities, but also coherent networks of protected areas, through early identification of impact and exploration of opportunities for multiple use of space. Ecosystem-based management and marine spatial planning (MSP) respond to the special conditions of complex and connected, changing and mobile marine and coastal ecosystems and the challenges to address for all value chains and sectors to reach their sustainability potential while coexisting harmoniously. This plan poses additional demands for data collection, knowledge integration, and process management, not the least collaboration across levels and scales, borders, and sectors to ensure that human activities at sea take place in an efficient, safe, and sustainable way.

Community and ecosystem-based management approaches have proven particularly important. Integrated approaches to the management of marine and coastal resources increasingly incorporate systems-oriented approaches based on precautionary and ecosystem management principles. Implementation of Integrated Marine and Coastal Area Management (IMCAM) gathers pace for the sustainable use of marine and coastal ecosystems and habitats. IMCAM is a participatory transparent process for decision making to prevent, control, or mitigate adverse impacts from human activities in the marine and coastal environment. Strategic plans can provide legitimacy for activities to support the achievement of the SDGs, such as the development of renewable energy sources and grids, and the establishment of Marine Protected Areas. The early identification of impacts and opportunities for multiple use of space and the coordination of land–sea interactions of coastal activities is a major advantage. A holistic approach and multi-stakeholder debates are always the most critical parts of the process.

Much as a response to growing pressures, recent years have seen a significant increase in the number of countries and regions putting in place ecosystem-based strategic frameworks for better ocean management. Presently, more than 50 countries have some form of government spatial ocean management and planning, covering 8% of the exclusive economic zones (EEZs), a percentage expected to be multiplied threefold by 2025. However, many obstacles stand in the way of more effective integrated ocean management for the right balances to be achieved (OECD 2016).

Climate change, including the rise in sea level, acidification and increasing water temperatures, and the frequency of extreme weather events, is likely to affect maritime economic activities and alter marine ecosystems. Strategic planning can be important in prevention and mitigation by promoting the efficient use of maritime space and renewable energy and cost-efficient adaptation to the impact of climate change. Cooperation beyond administrative and sectoral borders and administrations is a major issue and must often overcome local divisions and invest dissent into an integrated and inclusive approach. Integrated risk management has to take into account the perception of risks and the tolerable levels by all stakeholders.

Marine spatial planning and integrated coastal management create a most suitable framework for addressing human impacts on marine and coastal biological diversity and for promoting sustainable use. Crucial components include industrial activities such as construction and mining in coastal areas, mariculture, mangrove management, tourism, recreation, fishing practices, and land-based activities, including watershed protection. Destruction of vital habitats should be prevented, and degraded habitats, including spawning areas and nurseries of stocks of living marine resources, should be restored.

Guidance provided by UNESCO for marine spatial planning highlights a systematic process, beginning with preparatory steps, such as defining the objectives of a marine plan, analysing existing conditions, including the mapping of maritime activities, deciding on a preferred spatial scenario, and approving and implementing the plan. Stakeholder engagement is integrated into the overall process. Sea-based models include the integrated coastal zone management (ICZM) protocol for the Mediterranean, which is being adapted and applied to other sea basins (OECD 2016).

Integrated urban coastal management should empower, enable, and invite all stakeholders to have more control over the natural resources upon which their future depends. Improved understanding of the ocean commons and the invaluable marine ecosystems, their functions, and the ways that human activities impact them is a first critical step. Coastal cities should undertake life cycle resource assessments, appraise investments and opportunities, risks, and threats, respect the geophysical and cultural local limits, seek a symbiosis with the bioregion, and mobilise all visible and invisible societal hands. Strategic urban planning for sustainability requires a comprehensive interdisciplinary assessment of urban assets, a natural resource information system, and identification and analysis of the policy distortions and bottlenecks across the sustainable development agenda.

An increasing number of countries and regions have introduced ecosystem-based strategic frameworks for better ocean management. Presently, more than 50 countries have some form of government spatial ocean management and planning covering 8% of the EEZs, a percentage expected to be multiplied threefold by 2025 (OECD 2016). The EU blue growth strategy underlined the importance of coordinated efforts from EU, national and regional authorities, business and civil society on issues such as research and development, access to financing, and promotion of education and innovation. In 2014, the EU adopted the Marine Spatial Planning Directive aimed at introducing Maritime Spatial Planning and Integrated Coastal Zone Management, methods to promote the optimal location and distribution of marine activities throughout the blue economy in a coordinated and coherent way. Spatial planning provides investors and operators more predictability about future directions and the impact of maritime activities (EP and Council 2014).

References

Beatley T (2011) Blue urbanism: the city and the ocean. Island Press, Washington, DC

Boston Foundation & The Citistates Group (2004) Boston unbound. Tapping greater Boston's assets and talents to create a world leading citistate. Boston

City of Amsterdam (2014) Amsterdam definitely sustainable, 2011–2014. Amsterdam

City of Amsterdam (2015) Circular Amsterdam. A vision and action agenda for the city and metropolitan area. Amsterdam

City of Nantes (2010) Le voyage à Nantes. The city turned upside down by art. Nantes

(The) City of New York (NYC) (2016) OneNYC 2016 Progress Report. New York

City of New York (NYC) (2017) Progress report: OneNYC, New York

Conference of Ministers of Culture (2018) 2018 Declaration. Towards a European vision of high-quality baukultur. Davos

Ecorys et al. (2013) Ex-post evaluation of 2012 European capitals of culture. Brussels

Elmqvist et al. (2013) Urbanization, biodiversity and ecosystem services: challenges and opportunities. A global assessment. Springer, New York

European Commission (EC) (2015a) Getting cultural heritage to work for Europe. Report of the Horizon 2020 Expert Group on Cultural Heritage, Brussels

EC (2015b) Cultural heritage counts for Europe. Brussels

EC (2016) Opportunity now. Europe's mission to innovate. Brussels

European Council of Spatial Planners (ECSP) (2013) The charter of European planning, Brussels

European Parliament (EP) (2013) European capitals of culture. Successful strategies and long-term effects. Brussels

EP & Council (2014) Directive 2014/89/EU of the European Parliament and of the Council of 23 July 2014 establishing a framework for Maritime Spatial Planning

Florida R (2008) Who's your city. Basic Books, New York

Hall P (1998) Cities in civilisation. Weidenfeld and Nicholson, London

Hall P (2013) Good cities, better lives: how Europe discovered the lost art of urbanism. Routledge, London

INU – Politecnico di Milano (1997) Il Tempo e la città tra natura e storia. Atlante di progetti sui tempi della città. Milan

Jimenez J (2002) Teoria de arte. Tecnos, Madrid

Landry C, Bianchini F (1995) The creative city. Demos/Comedia, London
Mega V (2012) Bio-diver-cités. Editions Universitaires Européennes, Saarbrucken
Mega V (2013) Quintessential cities, accountable to the future. Innovation, sustainability and leadership. Springer, New York
Mega V (2015) Conscious coastal cities. Sustainability, blue green growth and the politics of imagination. Springer, Cham
Organisation for Economic Co-operation and Development (OECD) (2012) Compact city policies. A comparative assessment. OECD publishing, Paris
OECD (2016) The ocean economy 2030. Paris
Olsen D (1987) La città come opera d'arte. Sena e Riva, Milano
Orsenna E, Gilsoul N (2018) Désir de villes. Robert Laffont, Paris
Port of Antwerp (2017) Sustainability report 2017. Antwerp
Port of Amsterdam (2009) Port-City. 2009
Tara Oceans Consortium coordinators et al (2015) Open science resources for the discovery and analysis of Tara Oceans data. In: Nature Scientific Data, vol. 2 Article number: 150023
UNESCO (2014) World Heritage and our protected planet. Paris
UNESCO World Heritage Centre (2013) Safeguarding the crown jewels of the ocean. Paris
UNESCO World Heritage Centre (2014) The list. Global strategy. Paris
Urban Redevelopment Authority (2014) Master plan. Singapore
World Business Council on Sustainable Development (WBCSD) (2013) A solutions landscape for Kobé. Geneva
WBCSD (2014) Urban infrastructures. Final report. Geneva

Selected Websites and Social Media

#OneNYC
#artscienceproject
#EarthDay
www.arup.org
www.atkearney.com/global-cities
www.citiesofscience.co.uk
www.cite-des-sciences.fr
www.coe.int/en/web/culture-and-heritage/intercultural-cities
www.cultureactioneurope.org
www.culturemontreal.ca
www.ec.europa.eu/culture
www.ec.europa.eu/futurium/en
www.eit.eu
www.europanostra.org
www.fondationjacquesrougerie.squarespace.com/
www.futurium.org
www.galwayscience.ie
www.guangzhouaward.org/
www.hel.fi
www.icomos.org
www.ilab.harvard.edu
www.innovationpolicyplatform.org/ocean-economy-and-innovation
www.intelligentcommunity.org/
www.jpi-culturalheritage.eu
www.koshland-science-museum.org/explore-the-science/earth-lab/responses

www.levoyageanantes.fr
www.msp-platform.eu/
www.oecd.org/governance/observatory-public-sector-innovation/
www.seasteading.org/floating-city-project/
www.stationf.co
www.tallinn.ee/icapital/#together-city
www.unesco.org
www.worldcitiesculturereport.com/cities

Watercolour 10
Singapore, Most Technologically Ready City in the World (PwC 2016)

Chapter 8
Open Cities and the Ocean: Local and Global Engagement

Abstract Cities have withstood time by promoting local democracies, which are reinvented every day. Democracy implies the power of citizens and the risk of entrusting citizens with the governance of all. As new models of deliberative democracy emerge, this chapter illumines the importance of principled governance, leadership at all levels, and citizenship in cities and links with oceans and the world. Citizen education, empowerment, and participation are cornerstones in democratic policy making. In an increasingly digital world, open data cities multiply, often in alliance with scientist citizens, offering invaluable insights for enlightened actions, including for the global ocean. Transparent institutional and informal partnerships, enriched with nudge interventions and quintuple helices, can maximize the potential of synergies and serve as catalysts of transformational change. Coalitions and cross-networks allow overcoming national frontiers and horizons. Interconnected alliances of democratic cities can lead to a global solidarity bond involving coalitions and networks of cities and seas, with a major role in the world's most noble partnership of the future.

8.1 Citizen Education and Democracy for Achieving the SDG 2030

8.1.1 Education to Multiply and Fructify Choices

Education is the best way to explore many different possibilities, expand the spectrum of choices, push the threshold of understanding, and invest creativity to achieve objectives. Inclusive skills and awareness, and equitable quality education for all, require increasing efforts, especially for vulnerable populations including citizens with disabilities, indigenous populations, and poor children in rural areas. The education of women in Africa has long been considered the most productive world investment for sustainable development. It is crucial that education goes beyond the formal requirements and includes civic education and sustainable development options.

V. P. Mega, *Eco-Responsible Cities and the Global Ocean*,
https://doi.org/10.1007/978-3-319-93680-2_8

In most countries with available data, urban children score higher in reading than rural children. At the world level and despite considerable gains in education enrolment during the past 15 years, about 263 million children and youths, mostly in Sub-Saharan Africa and Southern Asia, were out of school in 2017, including 61 million children of primary school age. Even though more children than ever are going to school, many do not acquire basic skills in reading and mathematics. Recent studies show that in 9 of 24 Sub-Saharan African countries and 6 of 15 Latin American countries with data, fewer than half the students at the end of primary education had attained even minimum proficiency levels in mathematics. Concerning the acquisition of digital skills, data from 65 developing countries suggest that the average percentage of schools with access to computers and the Internet for teaching purposes is more than 60% in both primary and secondary education, but less than 40% in more than half of Sub-Saharan countries with data (UN 2017).

Equity is a major challenge in education, and digital skills are considered both objectives and the means of a potentially more democratic society. In all countries with data, children from the richest 20% of households achieved greater proficiency in reading at the end of their primary and lower secondary education than children from the poorest 20% of households. Gender gaps still persist even in the wealthier countries. The Mayor of Paris declared in 2017 that the €1 million prize from the EU Innovation capital award should best be invested in developing the digital skills of girls.

Investment in the education of urban youths and in blue–green domains not only allows succeeding towards sustainable development goals (SDGs) 4, 11, and 14 but also brings families precious knowledge and awareness about the blue planet. Marine subjects can greatly expand children's imagination and nurture plans for the future. In Ireland, the Marine Explorers' Education Programme has been running for many years in Galway, Mayo, Dublin, and Cork. The Programme aims to raise awareness and interest in the marine environment and promotes subjects in primary school that relate to marine species, ocean technology, and the shared seas.

At the frontier of creative industries, museums of natural history, floating museums, and aquariums are precious services for ocean literacy and awareness about marine biodiversity. In Washington, DC, the Ocean Hall, the National Museum of Natural History's largest attraction, offer visitors a unique initiation to the mysteries of the oceans. Exhibits include 674 marine specimens from starfish to corals, multi-media activities, and the newest technology allowing visitors and especially children to explore the ocean's past, present, and future. In Boston, the New England Aquarium presents a unique view on the world of jellyfish. The Shanghai Ocean Aquarium brings together representations of the world's seas, hosting an array of endangered Chinese marine species, most of which are rare breeds from the Yangtze River.

Perched at the edge of the Pacific Ocean, the Monterey Bay Aquarium hosts 600 species and two masterpieces, the Ocean's Edge Wing and the Outer Bay Wing. The Ocean's Edge Wing is a 10-m-high tank offering a privileged view into coastal marine life, and the Outer Bay Wing displays life in the open expanses of the Outer Bay. Monterey Bay is part of the world's largest protected marine sanctuary and is rich in diverse marine life, home to unimaginable numbers of fish, sea animals, and plant species and habitats.

European seas are endowed with some instructive educational services. In France, Boulogne-Sur-Mer hosts Nausicaá, a huge aquarium embarking on a new chapter in its history with an increase in its size and ambition. From May 2018, a remarkable new 10,000-m^3 tank invites us for ‘a voyage on the high seas,’ vast and fragile. Nearly 60,000 creatures will live all along the 15,000-m^2 space of the exhibition, and 22,000 creatures will be added to the iconic species that are already in place. Valencia, the marine complex located within the City of Arts and Sciences, features the largest aquarium tank in Europe. More than 45,000 marine creatures freely swim throughout nine underwater towers, structured on two levels. Each tower is built to represent different ecosystems from across the world. In the Greek Seas, the Neraida Floating Museum of Natsis Foundation seeks to raise awareness about Greek maritime history, develop academic research and infrastructure, and reward community development.

Some museums partner with universities to extend the limits of education and imagination. In 2017, the International University of Monaco, Monaco Impact, and the Oceanographic Museum have strengthened their collaboration by signing an official partnership agreement. Work on the project has been integrated in five different courses including Sustainable Development and Green Business.

Education at all levels is essential for cities and the seas. The World Maritime University (WMU) in Malmö, a postgraduate maritime university founded in 1983 by the International Maritime Organization (IMO), aims at achieving the objectives of safe, secure, and efficient shipping on healthy oceans, through education, skills, research, and innovation. The University operates on the basis of a charter adopted by the IMO Assembly as a centre of excellence for postgraduate maritime education, research, and global maritime capacity building. Working closely with all stakeholders, including IMO, the maritime industry, national administrations, and NGOs, the university is positioned as an international forum for networking and collaboration on critical issues and as a centre educating future global maritime leaders.

The citizens of the future should be given particular attention, as the well-being of children is a critical test for the present and future of society. It seems that children in coastal cities tend to develop a particular relationship with the sea. They love the blue horizons, the ever-changing weather patterns, and watching boats rising and falling on the waves. The Irish Marine Institute organised poetry competitions on marine life and the ocean. The marine environment can stimulate imagination and civic awareness on issues of sustainability and engage children as the citizens of the future (Children’s Society 2014).

The citizens of the future should be given the place that they deserve in all forums for Sustainable Development. In preparing for the Rio + 20, three NGOs, Oikos, Terre des hommes, and the World Future Council (WFC), launched an online petition for signatures demanding world leaders to support the establishment of Ombudspersons for Future Generations. Such initiatives can inspire further action for sharing and protecting the unknown interests of future generations. In 2012, the Future Policy Award celebrated the world’s most inspiring, innovative, and influential policies on the protection of oceans and coasts. Together, the WFC and Hamburg HafenCity University founded the international Expert Commission on Cities and Climate Change in 2008. The commission comprises 20 experts in the field of regenerative city development, including architects, city planners, and representatives of the United Nations Human Settlements Programme.

8.1.2 Deliberative Governance and Citizen Consensus

Urban democracy, representative and direct, is vital for the permanence of cities and their capacity for continuous renewal. Representative democracy has to address the challenge of the duly constituted authorities, linked to the proportional representation of all legitimate groups. Participative democracy can lead to truly publicly owned policies and facilitate their respectful implementation. Cities such as Madrid deployed exemplary efforts to make the best of the involvement of all stakeholders after hard crises. Citizens have been gradually invited to broadening the visions about the future and act as strategic partners rather than reactive protesters. Empowerment has become ethically and politically correct and has been recognised as a powerful driver of transformative change.

Urban democracies depend on the interactions among the many diverse stakeholders and actors of the cities and multi-level governance. In the increasingly multipolar and interconnected world, with the progressive diffusion of power, urban leaders often compete with national leaders in proposing innovative policies and initiating and scaling progress towards sustainable development. The role of businesses and NGOs is paramount, especially in coastal cities with maritime industry and advocacy groups striving for a healthier marine environment. Public–public multi-stakeholder partnerships can fill a decisive role for better urban politics.

Governance is the science and art of co-governing societies with the participation of all actors having a stake in the future. Thinking ahead and together with citizens can spectacularly open the spectrum of optimal futures and help co-evaluate the drivers, the barriers, and the conditions for change. The transition from government to governance implied the recognition that visions, strategies, and policy options and decisions should actively involve all those directly and indirectly concerned, at present and in the future. In many cities, new civic bonds have been sought with maritime business, environmental advocacy groups, and civil society, expected to invigorate the debate between governments and the constituencies they represent, increase public transparency and accountability, and enhance collective capacity for reflection, judgement, and decision.

Twenty-five centuries after the Athenian laboratory of urban democracy and the development of city-states, which often expanded as maritime powers and created colonies beyond the seas, the governance of cities is linked to global governance. Cities have evolved much as prominent political entities, but citizens remain the key political stakeholders and their participation is the common denominator of mobilising transformative projects. The array of stakeholders has been enlarged, and private local actors and civil society organizations have an expanding role. The ways and methods of active involvement have been multiplied through digital technology, and the joint actions by interconnected local and global partnerships are only limited by human imagination.

Eco-responsible effective governance is essential for cities to become global democratic leaders for sustainable development. Democracy is always reputed to be the less imperfect form of government for open societies. A governance framework to enhance marine capital should embrace vertical governance between different levels of government, as well as horizontal governance across multiple sectors, including engagement with nongovernmental actors, and governance across and between cities or coastal territories. A key issue for national policies is the empowerment of cities and citizens to become more effective in the design and implementation of actions driven from the bottom up and innovations that could subsequently be scaled up. A hybrid of the two frameworks provides opportunities for participative leadership towards a mutual trust paradigm.

At the dawn of the civilisation of sustainability and with the advent of participative and interactive 2.0 models, there has been a clear shift from indirect representational systems of democracy to a more interactive and direct democracy. Favouring interactivity, interoperability, and stakeholder-centred design, 2.0 has been grafted into diverse forms of governance for a continuous reinforcement of the civic bond and the incessant confirmation of sustainability values.

Anticipatory democracy can increase considerably the political capital of a city and become enriched with instant direct democracy, enabled by 2.0 processes. Much depends on the quality and commitment of the human and social capital and the political leadership. Governments should give unlimited opportunities to the concepts and ideas proposed by stakeholders, including those usually without a voice. Horizon scanning and anticipation may be decisive and mediation at an early stage critical to build trust. Leadership at all levels is a sine qua non condition and has to define territorial needs, propose visions, coordinate actions, mobilise public and private funds, and develop a permanent dialogue with the broad array of stakeholders and across the territorial levels of governance (EC 2017a, b).

At the heart of the preferred shared future is a common vision and consensual actions to make the vision come true. Civic Agreements signal the engagement by cities conceiving and forging new partnerships with their multiple and diverse stakeholders, far beyond the exchange of information and views. Citizens should become more than voters or customers; they should be engaged as co-designers and co-deciders in the increasingly complex policy-making process. The transient population should also commit and contribute in a spirit of openness and shared awareness. Manifold partnerships with civil society organizations and citizen consultation on the gestation of visions and plans expanded as a precondition for civic bonds and bold collective action.

Eco-responsible leadership can bring great impetus. Participatory and shared leadership has become a subject of attention among political analysts, social experts, and behavioural scientists. Shared leadership, occurring when the role of leader is actively and intentionally shifted, depending on the issues or the context, could raise ownership and increase excellence. It empowers all citizens to experience the responsibilities of the leaders and enables the gradual building of consensual mobilising visions.

Democracy has always to provide real and virtual forums to exert sound judgement and help citizens to be transformed from mere consumers and users of urban infrastructures and services into responsible city actors, sharing values, embodying visions, and performing actions. Their representation and participation can be extended through action planning and schemes, citizen and stakeholder forums, dialogue and consensus workshops, bringing together often opposed actors on neutral grounds and on equal terms (Abbott 1996).

Formal and informal mechanisms of horizontal and vertical cooperation between government bodies and partnerships with nongovernmental actors, mainly industry and civil society organizations, are necessary for weaving the democratic web of a city. Governments have, in some cases, begun to promote the formation of new spatial structures, such as inter-communal frameworks, regional platforms, territorial pacts, and sustainability areas. These processes, beyond traditional administrative boundaries, promise more coordinated policies and more coherent allocation of public resources, as well as greater transparency, visibility, and accountability.

In all spheres of government, it can occur that good decisions are not being made for fear of the political cost. Citizen juries, composed of randomly selected citizens not affiliated to a party, not up for re-election, and not linked to lobbyists have the power to act as a final filter to ensure worthwhile policy options have the chance to be publicly considered. This is the essence of the deliberative democracy, which should not be confused with groupthink, but serves as a way of finding common ground through nudge and progressive consensus building. In April 2013, the new Democracy Foundation was appointed by the South Australian Premier to conduct a Citizens Jury linked to alcohol-related violence.

Since 2016, the City of Athens Democracy Award has marked the events for the UN International Day of Democracy each September 15. It is associated with the New York Times Athens Democracy Forum, aspiring to become a modern-day version of the ancient agoras, gathering members of society for the debate of important civil matters. The *New York Times* brings together an impressive array of experts to identify and respond to the important questions surrounding liberal democratic rule, its evolution and new models, as well as the opportunities presented by the transformational forces of globalization and technology. The Mayor of Athens established the City of Athens Democracy Award to be bestowed on an individual or organization in recognition of exceptional achievements that demonstrate a lasting and compelling commitment to the advancement of democracy and to its enhancement on the international scene. The City of Athens also wished to honour all active citizens defending democratic principles and substantively contributing to the promotion of democracy as a system of ethical values. It reminded all world citizens that the concept of democracy in the contemporary world should be defined in terms of three inextricably linked and mutually reinforcing elements, including free and fair elections, adherence to the rule of law, and respect for human rights.

Boston won a 2016 Guangzhou Award with the project "Youth Lead the Change: Youth Participatory Budgeting." The initiative empowers youth to decide on the most beneficial use of $1 million of the municipal budget each year. Youth come up with ideas, propose priorities for investment, and decide in a democratic manner which projects to implement. This initiative engages youth in policy dialogue, to increase youth engagement in civic affairs, and above all to cultivate lifelong commitment in sound local politics.

8.2 Participative Cities, Institutions, Stakeholders, and Quintuple Helices

8.2.1 Public Goods and Commons Belong to All

Cities have to decide together with their citizens where they want to be in the sustainability journey and understand the significant trends that may influence the direction in which the future unfolds. Striving in the global economy for a sustainable future requires a dynamic local balance among economic, social, and environmental objectives and demands that the appropriate people, skills, and capabilities are developed, and that city leaders demonstrate their abilities to appreciate these assets and invest in the ways in which they can be enhanced and allowed to prosper. Transparency of public administrations, citizen participation, and cooperation processes form a triad bottom line of requirements which can be supported by social media and egovernment (Parliament of Finland 2011).

Coastal regional and local authorities should be endowed with the resources to manage the multiple functions leading to marine and coastal sustainability, fulfil new mandates, and ensure policy coordination, implementation, and evaluation. Fiscal federalism, based on the search of a balance between distribution of powers and allocation of resources, may be instrumental for sustainability. Responsible monitoring and evaluation may also be facilitated by 2.0 schemes, and crowdsourcing schemes can promote marine stewardship.

Leaders have to interpret the needs, values, and preferences of the citizens in relationship to the sea, elaborate plans towards sustainable development, facilitate and stimulate the interaction of the different actors and partners, and guarantee the coherence between options, means, and decisions. Stockholders, the owners of the physical assets, stakeholders, having particular interests in local life, and outside partners, such as shipping lines and offshore multi-functional platforms, could invest great energy in developing marine resources in an integrated and responsible way (CERES 2014). Coastal cities investing in ecological urban renaissance take special care of their urban and marine environment, their functions, and their evolution in time. Coastal and marine protected areas should be well integrated with urban and regional plans.

The renaissance of the planet cannot happen without healthy high seas. From the polar seas to the underwater abysses, the high seas are no longer the inaccessible waters beyond the national waters, the exclusive economic zone (EEZ), bound by state, regional, and international laws, although these are often flawed and ineffective. The immense development of technology and growth in demand for natural resources has allowed human exploitation to reach further and further beyond the EEZs into the high seas. This new territory covers, however, almost half the Earth's surface and it is virtually free of effective governance. This fact epitomises what is known as the tragedy of the commons: a resource used freely by all to the point of overuse, but owned and protected by no one. This is a great opportunity for cities to democratically lead coalitions of the willing to prevent irreversible tragedy (GOC 2016; Gore 2013).

Blue cities have to reinvent the art of negotiation for the brokerage between diverse industrial and organisational interests, including the protection of the high seas. Sustainable development ethics would hold cities responsible for the solid and chemical wastes that they produce and their wastewaters that should never impact water quality. Coastal cities should take the lead in preventing waste streams and supporting efforts to clean up the oceans. San Francisco, for example, through its Zero Waste policy, has been pioneer in banning plastic shopping bags at large supermarkets and has dramatically reduced the use of chemical pesticides in managing parks and urban infrastructures.

Academia created various centres of excellence to better nurture helixes of excellence. The University of Cambridge Institute for Sustainability Leadership helps companies with a global reach to derive value from embedding innovative approaches to sustainability across their organisation and supply chain. Assets include certified courses, online flexible learning, and access to the Institute's alumni network, consisting of nearly 8000 senior leaders and cutting-edge practitioners from the business, public sector, and civil society spheres in every industry.

The Stockholm Resilience Centre is another international centre of excellence focusing on sustainability science. Founded in 2007, as a joint initiative between Stockholm University and the Beijer Institute of Ecological Economics at the Royal Swedish Academy of Sciences, the centre invested in world-leading research for the reconnection of human and natural systems and the sustainable co-evolution of human civilizations with the biosphere. Its most outstanding works encompass the planetary boundaries wheel and foresight for the Arctic.

An #ArtScience project of the Stockholm Resilience Centre, the Radical Ocean Futures developed four scenarios and narratives built on a robust foundation of scientific knowledge, including on technological frontiers, marine ecology, ocean and fisheries science, the global fishing and seafood industry, and marine management, governance, and socio-economic shifts. The scenarios were developed following the method of science fiction prototyping, developed by B.D. Johnson in 2017, the futurist in residence at Arizona State University, Centre for Science and the Imagination. The heart of the project involves two axes, one on sustainability and one on connectivity with the four possible futures on each of the four quadrants, the 'collapsed to sustained' on the horizontal axis and the 'fragmented to connected' on the vertical one. The purpose is to help citizens think imaginatively, think about consequences, and proactively impact their future.

8.2.2 Volunteering for Interconnected Cities and Seas

Voluntary and active participation of all stakeholders, shareholders, and stockholders of the present and the future are most important for sustainable development. The plethora of diverse voluntary pledges for cities and seas has been one of the most encouraging signs of the first 2 years of the implementation of SDGs. Many cities develop particular approaches to attract voluntary commitment on various fronts of the sustainable development agenda. Cities such as Helsinki extend the city's reach through voluntary agreements with the private sector to strengthen contributions to sustainable development goals.

From voluntary schemes to volunteering for the city, the leap is not without meaning. Citizens regularly offering their time, energy, and skills pro bono in their city constitute a precious urban human resource that must be appreciated and enhanced. Exceptional events often served as catalysts of commitment. In preparing for the 1992 Olympic Games, for example, Barcelona created and trained a body of 40,000 volunteers. After the Olympic Games, this body was considered a living asset and offered new opportunities for local action. The municipality helped creating the association "Volunteers 2000" and availed its assistance in policies and projects.

In Athens, the Olympics of the Sea were a great part of the 2004 Olympic Games, which attracted 55,000 volunteers, Greek and foreign, carefully selected and trained to offer their services during the Games. The municipality of Athens organised a dedicated body of volunteers to help visitors discover another face of the Greek capital city and the coastal landscapes beyond tourist stereotypes. The next step would be for Athens to send its volunteers to other cities of its networks or/and have all their volunteers working together for the common public goods, including the oceans.

Volunteers are an invaluable resource for beach cleanups in coastal cities; they become true advocates for clean and healthy seas. The International Coastal Cleanup is the world's largest volunteer effort to clean up beaches and waterways, with partners around the globe. During Ocean Conservancy's 2014 International Coastal Cleanup, 648,015 volunteers collected more than 12 million pounds of trash around the shores of the world. The organisation makes accessible public data that help citizens better understand the problem of ocean trash, cooperates on innovative solutions with leaders from the private, public, and academic world, and tries to build a movement to decisively address the issue. In 2011, the Ocean Conservancy founded the Trash Free Seas Alliance to assemble leaders from industry, conservation, and academia and create pragmatic, real-world solutions to manage marine debris. Building upon the growing body of science on plastic marine debris, the Alliance aims to reduce the amount of plastic waste leaking into the ocean (Ocean Conservancy and Trash Free Seas Alliance 2017).

Committed citizens can act as ambassadors of their relationships with the seas. Many citizens have often acted as ambassadors of their cities and contributed to city twinning schemes. The earliest examples of twinning cities include the treaties between ancient city-states designed to protect each other's interests in times of hostilities. Most recent twinning schemes in Europe have their origins in the hope of peace and the unprecedented involvement of the citizenry in world conflicts.

Citizen diplomacy is vital for open cities. The choice of twin cities may be based on various geographic, industrial, or cultural characteristics and factors, growing from long-standing traditions rooted in past or recent links prompted by political solidarity. Old port or maritime cities have close links, as, for instance, Marseilles twinned with Piraeus. Sustainability ethics create new forms of global solidarity among local communities increasingly aware of their common destiny.

During the past decades, city twinning became frequent and active in Europe, including not only cities from countries with long-standing cooperative experience but also from those states recently opening to the world. Twinning among coastal cities, especially ports or tourist resorts, is viewed by many municipalities as an instrument for both addressing local problems and ensuring meaningful connexion with other world cities and shared projection to the future (Joenniemi and Sergunin 2011).

Volunteering for the planet is a valuable choice by many world citizens. For more than 30 years, the Ocean Conservancy has brought together more than 11.5 million volunteers in the annual International Coastal Cleanup that collectively removes 100 million kg of trash from the planet's coasts and waterways. In 2011, the Ocean Conservancy founded the Trash Free Seas Alliance to assemble leaders from industry, conservation, and academia and bring pragmatic, real-world solutions to marine debris. Building upon the growing body of science on plastic marine debris, the Alliance aims to drastically reduce the amount of plastic waste leaking into the ocean.

Without immediate intervention, 250 million metric tonnes of plastic waste could be in the ocean in less than 10 years. The lack of effective waste management is one of the greatest challenges and essential for a future in which waste can be recovered and repurposed. The aim of the Trash Free Seas Alliance is to dramatically reduce the waste by key countries in Asia Pacific, especially Indonesia, Vietnam, and Philippines. The Alliance highlights that it is possible to reverse trends if all stakeholders, government, the private sector, funders and investors, academia and civil society and community organisations work effectively together.

In adopting the Sustainable Development Agenda 2030, world leaders recognised that the work of volunteers will be central to its success. The Agenda explicitly mentions Volunteer Groups as a stakeholder working closely with Governments and public institutions, and the Addis Ababa Action Agenda on Financing for Development recognises that the resources, knowledge, and ingenuity of volunteers will be important in complementing the efforts of governments for the achievement of the Sustainable Development Goals. Volunteer Groups bringing together international networks and organisations represent the nearly 1 billion world citizens who offer their time and resource through public, non-profit, or for-profit organizations, and constitute 44% of the global non-profit workforce. The Volunteer Groups Alliance, a UN focal point for the active participation of volunteers in processes related to the implementation, follow-up, and review of the SDGs 2030, reflects the diversity of volunteer models, including local, national, and international volunteers engaged in online, employee, and diaspora volunteering.

8.3 Global Governance and Urban and Ocean Accountability

8.3.1 Responsibly Eco-Governing Cities and Seas

Citizens are the political stakeholders and society is the ultimate frontier for all policies. If governments wish their policies to be owned by citizens, they should allow as many voices as possible to be heard, and as many values as possible to be represented during the consideration of possible futures, the vision of the desired futures, and the elaboration of consensual plans for the concretisation of the vision. Residents, users of public infrastructures and services, and port and coastal communities have to be properly informed on important emerging issues, and duly involved in the formulation of possible policies. Decision makers should invest in a better understanding of public opinion and the preferred options. Making the community, especially the underrepresented social groups, better informed and more aware and willing to take part in a shared future is a noble challenge. Actions must be not only scientifically robust but also socially acceptable.

Partnerships are linked to the shift in public policies from direct interference to indirect or conditional policies, such as incubation, mediation, and nudge. They should enhance the capacity, contribution, and commitment of the public, private, and community sectors and improve the ability of society to act proactively and drive change. Public–private partnerships should work like an orchestra under a public conductor for the overall improvement of urban functions executed by private actors.

Strategic public–private, multi-stakeholder partnerships have a great potential in balancing objectives of competitive strength, social justice, and environmental well-being. These groups have the potential to take more risks, reduce the social costs of projects, and lead to enhanced outcome from public and private investment: they offer ample grounds for coalitions to overcome sector and institutional dissonance and take a critical role in the implementation of sustainable development policies. A clear vision and structure, a consistent and dynamic, strategic, and tactic approach, a critical mass, assertive leadership, flexible adjustments, and continued monitoring and assessment are often suggested as the success factors of the partnerships (Hall 2013).

Accountability is a sine qua non condition in the journey to sustainability. Progress against goals should always be closely monitored and assessed and publicly disclosed. In New York City, the OneNYC 2016 Progress report signalled an accomplished record total of nearly 4.3 million jobs, financing of more than 40,000 affordable apartments, which can accommodate 100,000 New Yorkers, since 2014, increase of the minimum wage for New York City workers to $15 by 2019, tripled solar energy installations citywide since 2014, expanded curbside organics collection for more than 700,000 New Yorkers, commitment to add 2000 electric vehicles to the City fleet by 2025, phasing out of the dirtiest heating oil in city buildings, and securing of another $300 million for new coastal resiliency investments. More than 95% of OneNYC's 202 initiatives were already launched and are underway (NYC 2016).

Sometimes the role of the orchestra conductor can be assumed by other citizen groups. The Committee for Sydney, an independent leadership group bringing together all sectors that share a passion for Sydney, created, in 2014, a Task Force to prepare the Sydney 2054 strategy in collaboration with government. The Committee favoured a "One Sydney approach" and enhanced coordination across the city's numerous councils. One can remark that the same unifying approach was also favoured in the City of New York and the OneNYC plan. Contribution from citizens and especially young leaders through the Sydney's Future Leaders sounding board has been essential.

Citizens, co-creators of visions and initiators of actions, may contribute decisively in creating a collective momentum for better public policies. Structured collaborative events, which unbound creative individuals and articulate a sense of vision, can create a thrust for the future. Processes such as the charette are being used to bring together the richness of diverse opinions and ideas and build consent on possibly controversial projects at the earlier possible stage. Citizen platforms can provide more permanent and effective interfaces among experts, policy makers, and citizens (Mega 2012, 2013, 2015).

The well-being of marine and coastal ecosystems has been at the heart of some citizen consultation schemes. Amsterdam introduced referenda to learn the residents' opinion on the extension of the city on artificial land and supporting infrastructures on sensitive coastal areas. Sustainable development public debates offer an excellent opportunity for raising awareness on threats to unique marine and land interactions and strengthening the urban social fabric.

Mayors have the reputation to be the most concrete results-oriented leaders, the ones most anchored in local realities (Barber 2013). Eco-responsible, accountable, and effective local government is the foundation of any nation that strives to provide its citizens with happiness, security, and sustainable prosperity. The City Mayors Foundation, an international think tank for local government, proposed the World Mayor Project and every 2 years recognises a mayor with an outstanding sustainability vision for urban communities. Conversely, incompetence, corruption, and misconduct in local government threaten the fundamental decency in a society and undermine the essentials of democracy.

The Code of Ethics underlines the Mayors' pledge to execute their office respecting the laws for the common good of their communities while refraining from actions that discriminate against individuals or groups because of their politics, communities, race, religion, gender, disability, or sexual orientation. It states their full responsibility for any acts performed by themselves or their administrations and emphasises that mayors should manage public resources for the public benefit of their communities and not for privileges or advantages for themselves and their close circle.

According to the code of ethics, mayors have to prove independence of judgment and demand the same degree of impartiality from their administrations. Mayors are subject to public scrutiny and should report any improper actions they may witness. They should work to strengthen civil society by raising public awareness of, and trust in, their city government's activities. Last but not least, Mayors committed to use their influence to promote cooperation and good will between cities, nationally and internationally.

Stewardship for the global ocean could be part of awards honouring mayors with the vision, passion, and skills to make the world a better place to live in and to work in, leading to a better planet. Various prizes highlight the results of city leaders with knowledge, enthusiasm and integrity, social and economic awareness, ability to provide security and to protect the environment, as well as the capacity to foster good relations between communities from different cultural, racial, and social backgrounds.

Ombudspersons for Future Generations could act as a driving force behind the sustainable development agenda while ensuring accountability and accessibility. They can bring long-term interests to the heart of decision making at all levels, offering an opportunity to break with the immediate concerns and stand up for future generations (WFC 2012).

Advantageous port conditions, adequate defence infrastructure, and good inland connections are the principal factors that guided the evolution of harbours and their complex relationships with the rest of the world. Ancient harbours were usually more exposed to the sea than the treasured agoras, at the very heart of the city. Many ancient port cities comprise civil and military harbours protected by towers and fortifications. Gates ensured the entrance into the town and the ancient agoras. Many harbours throughout history were later turned into commercial ports enhancing the valuable port conditions, defensive infrastructure, and inland connections.

Public places to be accessible to all and include everybody are essential in the era of sustainability. From ancient Greece to the civilisation of sustainability, the design of cities and urban spaces, such as open and light-filled assemblies, often reflect the democratic values and aspirations of citizens. The Agora, the focus of civic life in the archetypal city-state, constitutes a powerful public model place embracing civic and commercial spaces and orchestrating all urban functions. Citizenship, justice, culture, and exchanges were well anchored in these spaces, where the assembly, the theatre, the stadium, the marketplace, and the sanctuaries had the noble aim of promoting the physical and mental well-being of citizens. Environmental and cultural landscaping of public spaces can help forge urban identity. Qualitative recommendations for the functional and aesthetic character of squares, seaside plantations, and public lighting have been developed and implemented in many cities.

Historic public spaces can invite to endless journeys into ancient times. The Sacred Street in Athens goes deep into place and time. The city which invented theatre, philosophy, and democracy reorganised its ancient heritage surrounding the Acropolis, crowned by the Parthenon, into an urban archaeological park to be best enjoyed by citizens and visitors as a backdrop of the everyday environment. The park incorporates also green spaces, micro-squares, and pedestrian paths and bicycle routes providing a particular essence distilled from the magic of the ages.

Digital public spaces offer good access to cities. Assembling city data, structuring it, and publishing it without charge put the Helsinki region at the forefront of the open-data movement cities across the world. The Helsinki Region Infoshare service provides regional information quickly and easily accessible to all citizens, businesses, universities, research facilities, or municipal administration. The regional body works as a clearinghouse that confirms the data and offers training, support, and technical knowledge. The aim is to improve the compatibility of the data with other city data, including financial and location information, and encourage the development of innovative services for citizens. Helsinki Region Infoshare brought transparency in government and created a new digital commons.

The concept led to inspiring creations such as Ahjo, the Helsinki's city council paperless system, handling agenda items, minutes, and exhibits that accompany council debates. The city built a web-based interface for browsing the documents, and every citizen wishing to examine any issue brought to the attention of Helsinki's leaders can do so easily. Citizen with disabilities can enjoy particular applications adapted to their needs. The BlindSquare, a smartphone app that helps blind people navigate the city, has been made by an app developer who brought together the Helsinki region's data on public transport and services, location data from the social networking app Foursquare, as well as mapping and artificial voice capabilities of smartphones.

Helsinki Region Infoshare boosted cooperation in the metropolitan region and created new partnerships and alliances. The regional effort received funding from Sitra, the Finnish Innovation Fund, and Forum Virium Helsinki, a public–private partnership that helps develop new digital services in cooperation with the city. From the very beginning, the project was equally infused with regional inclusiveness and openness to the world. Helsinki Region Infoshare has capitalised on the open-data initiatives of New York City, Chicago, and London. Engaging the software developer community is an important part of the model. Forum Virium organizes monthly "Helsinki Loves Developers" events providing developers a place to network with each other and city officials to discuss projects that matter.

8.3.2 Inclusive Sustainability Indicators and Indexes

Better planning should lead to better policies and ultimately better lives. Well-being and quality of life have been approached through various metrics and qualitative indicators and have been the subject of many debates. Questions have been raised whether traditional measures of national and urban welfare built around macroeconomic statistics are able to capture the reality of citizens' lives in a complex globalised world. Many efforts have focussed since the turn of the century on sustainability indicators, often linked to the greening of national accounts systems and aggregating and comparing data in monetary forms. Genuine saving indicators or gross welfare product indices attempt to broaden the usual measure of saving to account for the cost of environmental depletion and degradation and investment in human capital or welfare.

Urban or ocean strategies should include a monitoring framework and a relevant number of thematic indicators assessing performance in all fields contributing to sustainable development and according to the strategic directions. Aggregate indexes such as the human development indicator, the genuine savings, or the ecological footprint and happy planet index, can shed light on the overall performance of cities. The development of a sustainability index after thematic policy indicators is a complex task, because indicators have to be weighted according to their contribution to sustainability levels and all the levels of aggregation have to be taken into due account. Finally, it is important to highlight that no indicator can show if a city integrates socioeconomic and environmental policy objectives, but this is at the heart of sustainable development practices (OECD 2011).

Many cities use indicators to measure their performance towards sustainable development, but there has not been a comparable standard for use by local authorities. Existing indicators are often not standardised, consistent, or comparable over time or across cities. Indicators are traditionally limited by national differences in data definitions and collection and composition methods. Systematic territorial indicators are necessary complements to national indicators serving for international comparisons. Regularly reporting on territorial progress towards international targets and commitments can promote policy coherence and accountability of national decision makers at the local and international level.

European efforts for creating a common set of urban indicators have multiplied over the last decade of the twentieth century. In 1997, the European Union (EU) launched the Urban Atlas, which initiated various waves of urban statistics progressively extending to the cities of the latest Member States of the Union. The sustainability monitoring initiative "Towards a local sustainability profile – EU common indicators," launched in 2000, has been developed as a bottom-up approach in close collaboration with local authorities participating on a voluntary basis. The set includes five core indicators on citizen satisfaction with the local community: contribution to global warming, mobility patterns, green spaces, local services, and air quality. Five additional indicators have been suggested focussing on children's commuting patterns: local management, noise pollution, sustainable land use, and eco-labelled organic and fair trade products.

Sustainable development is a binding principle of the European Union. Indicators should capture critical features of a city's endeavours towards sustainability and contribute to making it more visible and transparent, enhance structure and harmonize data banks, provide decision making with relevant and timely information, assist appraisals, comparisons, and prediction, stimulate communication. and promote citizen empowerment and participation. They should embrace all sectors and neighbourhoods and help tracing progress in the co-evolutionary process towards sustainable development. An indicators assessment board should validate the set of indicators and ensure that the framework is regularly updated and validated.

Many cities tested and introduced frameworks of monitoring indicators. In the US, Seattle is often quoted as a classic example of an open-minded dynamic city, a breeding ground of successful startups and businesses, with a coherent set of award-winning indicators. The Seattle framework demonstrates that indicators should reinforce local communication processes and promote shared values and common understanding. The Sustainable Seattle Indicators of Sustainable Community includes 40 indicators selected for tracking, reporting, and educating: these invite us to "move from being what we measure to measure what we want to be." The quality of the aquatic environment and marine ecosystems are key indicators for the well-being of the city and the citizens (Sustainable Seattle 2008).

Qualitative indicators and reports are also very important. The State of the City report, published by Oslo or Amsterdam, can be an inspiring monitoring tool. It synthesises information after a public survey and analysis of data from municipal statistics and other sources. In Amsterdam, results were compared to the national situation as well as to other Dutch cities. Such a State of the City report could be inspired by the great tradition of forward-looking "State of the Union" addresses on both sides of the Atlantic. They offer an assessment of the past and a compass for the future in a form to be understood by all and invite all to action.

The EU published a State of Cities report in 2016, the third perception study conducted in European cities (EC 2016). In 2017, the third edition of the Oslo State of the City report suggests that the city is improving its position and visibility despite the breadth and depth of world competition. As most cities, Oslo is competing for talent, visitors, investors, and attention. The Norwegian capital is becoming an internationally important location for entrepreneurship and innovation but must make more efforts to attract important businesses. Oslo is 4th of 50 global cities for climate change leadership, and the city's social and environmental model together with its maritime strength are considered to be a comparative advantage. Hospitality is the area that has improved most since 2015, but there is room to continue improving.

Boston is an extraordinary city in constant transformation. The Boston Indicators Project, initiated in 2000 with the goal of assessing and presenting progress towards 2030, Boston's 400th anniversary, offered new ways to understand the city and its neighbourhoods in a broader context. Benchmarks helped democratize access to information, foster insightful public debate, evaluate progress on shared civic goals, and build trust on essential achievements in the domains of Civic Vitality, Cultural Life and the Arts, Economy, Education, Environment, Health, and Housing, and the topics of Children and Youth, Competitive Edge, Fiscal Health, Race/Ethnicity, and Sustainable Development.

A series of imaginative reports underlined the emerging dynamics and identified prospects for Boston. A first report on "The Wisdom of Our Choices: Measures of Progress, Change and Sustainability" introduced the framework of indicators through a rigorous process involving more than 300 experts and stakeholders. The report noted that the booming knowledge economy had created an "education divide." A second report highlighted Boston's institutional, physical, and cultural assets, but alerted about a worrying brain drain resulting from the move of youths away from Boston and Massachusetts, mainly because of the high cost of living (Boston Foundation 2001, 2003).

Among the following reports, the "Thinking Globally, Acting Locally: A Regional Wake-Up Call" noted that the region was suddenly competing for jobs and talent not only with other US states, but with China, India, and other emerging economies in the multipolar world. This report called for a coherent, collaborative response and a civic agenda. In the journey towards 2030, Boston could be described as a resilient twenty-first-century city striving to develop a more robust, sustainable local economy. The vibrant Innovation District tries to capture synergies and embrace collaboration. The city wants to reach up, and to achieve this, it tries to reach out. Greater Boston has ridden out the economic downturn better than most US cities. Engagement has been the key to prosperity and to sustainability, and equality has been given special attention, as the region created new wealth, in which it wished to involve everybody (Boston Foundation 2009, 2012).

Boston has been among the top cities of the SustainLane U.S. City Rankings of the 50 largest cities, an inspiring benchmark exercise on the unfolding efforts of cities towards sustainable development. A San Francisco green media company, SustainLane, did an examination of sustainability initiatives in US cities, including such indicators of quality of life as local food availability, air and water quality, pedestrian and park space, sustainable mobility services, clean technologies, developments in renewable energy, waste management, alternative fuels, and green buildings. Since the first SustainLane rankings in 2005, extreme weather phenomena and crises underlined vulnerability and the need to foster urban resilience.

Portland promoted participative leadership for sustainability, with values-driven leadership exercised at all levels. It is not only demonstrated by the local governments but also by all actors including the civil society and businesses. The municipal administration gives the example by promoting renewable power and green buildings. Citizens opt for the most sustainability-oriented options in everyday life. Portland enhances its sustainability ethos to attract businesses, residents, and tourists and lead to a better future.

International comparative analyses should always be regarded with a sound dose of scepticism, because national statistics are culturally dependent and the conception of indicators is limited by differences in definitions and data collection and estimation methods. Comparisons are meaningful when they refer to units that are truly comparative in size and function. Systematic territorial indicators are necessary complements to national indicators serving for international comparisons. Moreover, by reporting regularly on territorial progress towards international targets and commitments, policy coherence gets more attention and national decision makers become more accountable at the local and international level.

The cross-country analysis of subnational data has been advanced by work jointly undertaken by the European Commission (EC) and the OECD on a new harmonised definition of a city and its commuting zone. According to this definition, 828 cities with an urban core of at least 50,000 inhabitants have been identified in the EU, Switzerland, Iceland, and Norway. Half these European cities are relatively small with a centre between 50,000 and 100,000 inhabitants. Each city is part of its own commuting zone or a polycentric commuting zone covering multiple cities. These cities host about 40% of the EU population without their suburbs, which together with other towns cover another 30% of the EU population. This methodology also helped identifying 492 cities in Canada, Mexico, Japan, Korea, and the US. The cities and commuting zones together, the Larger Urban Zones, account for 60% of the EU population (Eurostat 2016; OECD 2012).

Perception studies are of capital importance and accompany the EU framework indicators for cities. Since 2007, a series of surveys has been conducted to measure the local perceptions of quality of life in selected cities. The 2015 survey was conducted in a total of 79 cities in the 28 Member States of the European Union, Iceland, Norway, Switzerland, and Turkey. In 4 of these, an additional sample provided extra data, which allowed analysis of the perception of quality of life in Greater Paris, Greater Lisbon, Greater Athens, and Greater Manchester. Citizens seem in general satisfied with the city in which they live. There are significant differences in the level of satisfaction regarding public transport and health services. In ten cities, mainly Belgian, German, and French cities, at least nine respondents in ten are satisfied with the healthcare services. A majority of respondents in all but one city (Valletta) are satisfied with their city's cultural facilities and with sports equipment. Conversely, satisfaction with schools and educational establishments is low in many EU capitals and ranks lowest in Sofia and Bucharest.

Air quality is the aspect on which views diverge the most. Satisfaction with public and green spaces such as parks, squares, and pedestrian zones, markets, and shops is generally high. Finding good housing at a reasonable price is perceived as a challenge by most respondents in more than half the cities surveyed, especially in capitals. Trust in fellow citizens is high in more than three quarters of cities, and foreigners are well accepted, especially in cities in the periphery. Among the 17 cities in which fewer than half of the respondents trust people in their city are 10 EU capitals, including 7 from the Member States that have joined the EU since 2004. There was a strong correlation between the proportion of respondents who agreed that most of their fellow citizens could be trusted and the proportion who always felt safe in their city. Respondents across all surveyed cities were more likely to say that they always felt safe in their neighbourhood than in their city (EC 2016).

All citizens can establish personal sets of criteria and indicators and evaluate quality of life in given places according to their priorities. The OECD "Better Life Initiative" and "Measuring Progress of Societies" projects foster the development of key comparable indicators to measure the well-being of societies. The Organisation has suggested 11 topics reflecting the most essential factors for well-being in terms of material living conditions (housing, income, jobs) and quality of life (community, education, environment, governance, health, life satisfaction, safety, and work–life balance). The "Your Better Life Index" can be composed by citizens, based on the principle that everybody can easily establish a personal ranking depending on individual values and priorities (OECD 2011).

Open standardised data focussed on improving municipal performance, economic development, and quality of life can help international comparisons. An initiative by the World Council on City Data (WCCD), involving Ontario municipalities, is leveraging the ISO 37120 standard as a key opportunity for cities of all sizes to build their engagement with standardized open data. The ISO 37120 standard, adopted in 2014, provides clearly defined indicators covering the key themes proposed, prioritized, and agreed to by municipalities themselves. This central, although not exclusive, focus on ISO 37120 indicators ensure that the initiative is flexible and focussed on open data that help drive real improvements. The WCCD initiative for

Ontario is building direct engagement with municipalities and stakeholders, complementing existing open and municipal data initiatives and activities.

Ocean health and sustainability indicators are important monitoring tools as highlighted by the Marine Strategy Framework Directive, introduced in the European Union in 2008 to promote sustainable use of Europe's seas and marine ecosystems. It requires EU countries to adopt the ecosystem approach in their marine strategies, set objectives and targets, identify policy options, and engage in continuous evaluation and adjustment. The main goal of the directive is to achieve Good Environmental Status in Europe's seas by 2020. Good Environmental Status was defined as "marine waters that provide ecologically diverse and dynamic oceans and seas which are clean, healthy and productive." Eleven consensually agreed descriptors for the evaluation of Good Environmental Status include biodiversity, eutrophication, seafloor integrity, non-indigenous species, populations of commercial fish/shellfish, elements of marine food webs, hydrological conditions, contaminants of seafood, marine litter, and underwater noise.

European legislation has been a major driver for active citizenship and the participation of the organised civil society. The implementation of the Marine Strategy Framework Directive gave rise to multiple forms of public participation in coastal environments. Nongovernmental organizations (NGOs) but also the general public are very active despite a frequent lack of knowledge about European legislation and the descriptors. Access to information is the first step of any consultation process, and active participation the most advanced step leading to co-ownership. Cities and NGOs are usually very active in the public debate and the consultation in marine issues, which are largely deconcentrated processes. A joint NGO paper brought together Seas At Risk, the Marine Conservation Society, the Coalition Clean Baltic, the Swedish Society for Nature Conservation, Oceana, BirdLife Europe, the Mediterranean Information Office for Environment, Culture and Sustainable Development, The Fisheries Secretariat, Black Sea NGO Network, France Nature Environnement, the North Sea Foundation, the Surfrider Foundation Europe, Coastwatch Europe, the Sustainable Water Network, the Mediterranean SOS Network, and the WWF (World Wildlife Fund). The diversity of organisations joining forces to bring forward directions for sustainable ocean developments is indicative of the interest of the debate (Seas at Risk et al. 2014).

8.4 Constellations of World Networks of Excellence

8.4.1 Aligning Policies at All Governance Levels: The Multi-governance EU Paradigm

Urban and ocean issues are increasingly transnational in nature, and local, regional, national, and international policies have to be aligned for the final impact to be strengthened. To fully realise, for example, the potential of cities, European policies and rules intend to be more in line with local practice in cities. Alignment should not only increase impact but also heighten the level of commitment and cooperation between Member States, cities, European institutions, and other stakeholders. This stage is the purpose of the Urban Pact, an approach of the Urban Agenda for the EU including the development of a range of European partnerships. Within these partnerships, the European Commission, Member States, and European cities work together to ensure that the urban dimension is strengthened in EU policies.

The key concern issues gave rise to the first priority partnerships of the Urban Agenda: these include air quality, coordinated by the Netherlands (Ministry of Infrastructure and Environment); housing, coordinated by Slovakia (Ministry of Transport, Construction and Regional Development); inclusion of migrants and refugees, coordinated by the city of Amsterdam and the European Commission; and urban poverty, coordinated by Belgium (Federal Urban Policy) and France (Commissariat General à l'Egalité des Territoires).

Urban issues are often impacted by national culture, particularities, and traditions. Alongside the Urban Agenda for the EU, a model national Urban Agenda has been developed in the Netherlands. The Dutch government set city deals between cities on strategic urban domains providing a method of alignment between national and EU policies. The Dutch Urban Agenda can provide examples of innovative cooperation for other Member States. Dutch cities also learn from other European cities about addressing urban challenges. At the same time, the national Urban Agenda pinpointed European framework conditions that can be more effective and help urban areas realise their competitive and innovative capacity. These cases nurtured the dialogue within the Urban Agenda for the EU, aimed at improving European rules and regulations (EC 2017b).

The 2016 Dutch Presidency of the European Union advanced the urban alignments among the various governance levels. European Ministers responsible for Urban Matters launched the Urban Agenda for the EU and the first partnerships through the Pact of Amsterdam. This definitely sustainable city (City of Amsterdam 2014) hosted a City Makers Summit on 30 May 2016 to enable European social innovators to come together to exchange experiences and contribute to the EU Urban Agenda. The EU Agenda was designed for cities to have their say in policy making, from design to implementation. The "equal partnerships approach" on priority themes is rooted in the cooperation culture of European cities. With its 12 priority themes, its multilevel governance, and its focus on peer-learning, the EU Agenda is considered to have served as a model for the Global New Urban Agenda and contributes to the implementation of the New Urban Agenda on the EU territory (UN-HABITAT 2016).

The 2018 World Urban Forum in Kuala Lumpur provided the opportunity for the EU to report on the first achievements against its pledges in Quito in 2016. Significant progress has been accomplished under the three commitments presented at the Habitat III conference in Quito (October 2016) to harness the power of rapid urbanisation. City-to-city cooperation is flourishing across continents, important steps have been taken towards a single definition of cities at global level, and the EU is ready to lead the way for clean, safe, and prosperous cities around the globe. Europe and its partners are very active on these three concrete forefronts, which contribute to shaping more sustainable cities.

The first priority partnerships have already prepared their action plans. They include policy recommendations, good practices, and projects to be replicated across the EU and in the world. Beyond the thematic action plans, the Urban Agenda for the EU can inspire reforms in the way cities are governed across the globe and the will to bring together, on equal terms, cities, businesses, NGOs, and representatives from Member States and EU Institutions, for an integrated and balanced approach to sustainable urban development.

In the post-truth era, it is crucial to develop a global, harmonised definition of cities to facilitate monitoring, benchmarking, and eventually policy making at global and regional levels. A survey being carried out in 20 countries is gathering feedback on the global definition. Pilot projects are ongoing in 12 countries to compare the global definition with the national ones and assess the differences. The Commission and its partners work on a free online tool to help countries test this definition on their territories. The EU has been working on such a definition, which will be presented to the UN in 2019, in partnership with the OECD, FAO, and the World Bank. The European Commission (Joint Research Centre) has synthesised estimates of the level of urbanisation of each country in the world, and published the open access global city database, for all 10,000 urban centres across the globe.

The commitment to enhance cooperation between cities for sustainable urban development gave rise to the EU's International Urban Cooperation launched in 2016 to develop city-to-city cooperation around the globe. There are currently 35 twinnings under the programme, involving 70 cities, 35 from the European Union and 35 from outside the Union. All partnerships are working on local action plans on joint urban priorities, such as access to water, transport, or health, sharing knowledge, and best practices to reach their common goals. Pairings could also give rise to shared action for the protection and the sustainable use of the ocean.

Bringing together seas can also be extremely beneficial. In the EU, macro-regional strategies provide integrated frameworks, which may be supported by the European Structural and Investment Funds among others, to address common challenges faced by a defined geographic area relating to Member States and third countries. Strengthened cooperation contributing to economic, social, and territorial cohesion often goes through and is supported by cities. The Baltic Sea regional strategy and the Adriatic and Ionian Sea offer multiple examples.

8.4.2 Multi-cross-networks of Networks and Principled Connectivity

Sustainable Development Goal 17, with the aim to "Strengthen the means of implementation and revitalize the Global Partnership for Sustainable Development," recognizes multi-stakeholder partnerships as important means for sharing knowledge, expertise, technologies, and resources to raise awareness and amplify action towards the achievement of the sustainable development goals. Networks of public, public–private and civil society cross-sector, cross-frontier partnerships, building on the skilful strategies of all helixes, have to nurture the global orchestra.

Joint ocean-minded urban activities, for example, blue–green public procurement, can increase the direct impact of municipal governments on the global market for blue–green technologies. Cities purchase goods and services for thousands of activities, such as schools, hospitals, administrative offices, and police stations, port functions, and public lighting. Cities also buy and operate municipal fleets of vehicles and manage their water and waste systems. Through the cross-multi-network partnerships, cities are better able to pioneer energy-efficient and cleaner energy products and technologies. Their collective effort can significantly amplify results, raise awareness on the interconnected risks, prevent further deterioration of the ecosystems, and reduce pressures on the global ocean.

Oceans may have no voices, but city networks can plant the seeds towards the creation of influential world coalition of networks. A strong global partnership of cities could yield sufficient collective power and further encourage the sustainable development global agenda. Cities often have more freedom than countries to implement progressive strategies that are changing citizens' lives. The C40 alliance, formed in 2005, when representatives from 18 megacities came together to cooperate on reducing climate emissions through decisive and immediate action on procurement policies and the uptake of climate-friendly technologies and market dynamics, led to the C40 Climate Leadership Group, composed of cities in diverse stages of the development arc. Its partnerships, as for example with ICLEI, can provide an example in strategic networking. A Cities and Ocean Alliance and Leadership Group, broad coalition of networks with a consensual front, can have a major world impact on the global ocean and the future of life in cities.

As the leadership role of cities on the world scene expands, cities and citizens and networks of cities could organise ocean marches in various world cities, coastal and landlocked, for example, during the UN Ocean day of 8 June, to increase awareness. They can also create global platforms to present new and cutting-edge solutions for smart nations and urban environments. With more than 1000 participating companies in 2016 coming from 125 countries and regions, City Solutions Singapore (CSS) offers new partnerships and business opportunities for stakeholders across the value chain of sustainable development. The UN Sustainable Development Solutions Network (UNSDSN) can bring together and coordinate such activities.

City networks can have a great impact on mobilising the cities of the emerging world. Schemes of cooperation among blue cities could, for example, facilitate the transition to clean ocean energy options and promote sound marine and cruise tourism models. Cities from the EU, the world's largest donor of development aid, have a critical role, particularly for the sharing of knowledge and the transfer of cleaner technologies. Advanced marine technology projects, carried out in cooperation with emerging cities, could greatly expand the development of urban services based on offshore renewable sources. Furthermore, European cities could help developing cities to invest in state-of-the-art infrastructures and institutions, strengthen democratic governance structures and balance, and integrate their sustainable development agendas.

Multi-governance networks and compacts should support the commitment of cities and citizens to set ambitious voluntary sustainability targets, evaluate progress against objectives, strive for rigorous and consistent reporting standards, and make information publically available on recognized responsible platforms. A cross-sectoral, cross-stakeholder city–ocean compact could help establish robust and transparent data collection standards, and invest in common, consistent, and reliable processes for action. The commitment to contribute to more ambitious, transparent, and trustworthy targets could encourage more enabling policy environments and direct resources to improving cities in balance with nearby and faraway seas.

References

Abbott J (1996) Sharing the city: community participation in urban management. Earthscan, London
Barber BR (2013) If mayors ruled the world: dysfunctional nations, rising cities. Yale University Press, New Haven
Boston Foundation (2001) The wisdom of our choices: measures of progress, change and sustainability, Boston
Boston Foundation (2003) Creativity and innovation: a bridge to the future, Boston
Boston Foundation (2009) A great reckoning: healing a growing divide. Boston Foundation, Boston
Boston Foundation (2012) City of ideas: reinventing Boston's innovation economy, Boston
Children's Society (The) (2014) Ways to well-being: Exploring the links between children's activities and their subjective well-being, London
City of Amsterdam (2014) Amsterdam definitely sustainable 2011–2014, Amsterdam
City of New York (NYC) (The) (2016) OneNYC 2016. Progress Report, New York
Coalition of Environmentally Responsible Economies and Societies (CERES) (2014) Investing in the clean trillion: closing the clean energy investment gap, Boston
European Commission (EC) (2016) Quality of life in European cities. Flash Eurobarometer 419. Luxembourg/Brussels: Publications Office of the European Union, cop. 2016
EC (2017a) Report on the blue growth strategy. Towards more sustainable growth and jobs in the blue economy, Brussels
EC (2017b) Report from the Commission to the Council on the Urban Agenda for the EU, Brussels
Eurostat (2016) Urban Europe. Statistics on cities, towns and suburbs. Publications Office of the European Union, Luxembourg
Global Ocean Commission (GOC) (2016) The future of our ocean. Next steps and priorities, Oxford
Gore A (2013) The future: six drivers of global change. Random House, London
Hall P (2013) Good cities, better lives: how Europe discovered the lost art of urbanism. Routledge, London
Joenniemi and Sergunin (2011) Another face of integration: city twinning in Europe. In Research Journal of International Studies Issue 22, Eurojournals Publishing
Mega V (2012) Bio diver-cités. Editions Universitaires Européennes, Saarbrucken
Mega V (2013) Quintessential cities, accountable to the future. Innovation, sustainability and leadership. Springer, New York
Mega V (2015) Conscious coastal cities. Sustainability, blue green growth and the politics of imagination. Springer, Cham

Ocean Conservancy and Trash Free Seas Alliance (2017) The next wave: investment strategies for plastic free seas, Washington, DC
Organisation for Economic Co-operation and Development (OECD) (2011) OECD forum. Better policies for better lives. Paris: OECD
OECD (2012) Redefining "urban": a new way to measure metropolitan. OECD, Paris
Parliament of Finland (2011) Crowd-sourcing for democracy. A new era in policy-making, Helsinki
Seas at Risk et al (2014) Priorities for MSFD programmes of measures, Brussels
Sustainable Seattle (2008) Indicators of sustainable community, Seattle
UN-HABITAT (2016) New Urban Agenda, Nairobi
United Nations (UN) (2017) Progress towards the sustainable development goals 2017. Report to the UN Secretary-General, New York
World Future Council (WFC) (2012) The 2012 future policy award, Hamburg

Selected Websites and Social Media

#OceanYouthLeaders
@EUUrbanAgenda
https://sustainabledevelopment.un.org/sdg4
www.adelphi.de/en/news/adelphi-supports-laboratory-democracy-wheels
www.athensdemocracyforum.com
www.bcn.es
www.biennialtownplanning.org
www.bostonindicators.org
www.c40cities.org
www.citymayors.com/
www.ec.europa.eu/citizenship
www.ec.europa.eu/regional_policy/urban
www.glasgow.gov.uk
www.globalcitiesinstitute.org
www.harbourbusinessforum.com
www.iuc.eu/
www.microlearning.en-marche.fr/module
www.newdemocracy.com.au
www.nyc.gov/onenyc
www.oceanconservancy.org
www.opensocietyfoundations.org
www.oslobusinessregion.no/oslo-state-city-2017/
www.sdg.iisd.org/
www.sustainableseattle.org
www.sustainlane.com/us-city-rankings
www.tempomat.it
www.twinning.org
www.unhabitat.org
www.unorg/democracyfund/
www.urban-villages-forum.org.uk
www.wiserearth.org
www.worldfuturecouncil.org
www.worldmayor.com/

Watercolour 11
Istanbul, Only Transcontinental World City

Chapter 9
Messages for Leaders Towards a World of Difference

Abstract Ideas are a vigorous global currency and matter most for challenges which are too big or too complex and have to be collectively envisaged and addressed. Cities and the global ocean hold many hopes but require empowering actions by leaders able to understand and influence the interdependency of systems, build consensus, and mobilise stakeholders. What would be the most incisive insights for leaders at all governance levels to become champions of change? What are the priorities for cross-sector, cross-level, cross-stakeholder engagement at a moment when humanity is on the verge of multiple changes? This final chapter brings together key insights and messages for the committed to lead change towards a radically better future. Purposeful cross-implementation of SDG11 and SDG14 can lead to multiple benefits at all levels. Sustainability asks for the local, regional, national, and international actions to be fully coherent and aligned and bring the higher multiplier effect and impact. If all cities were adopting and preserving vulnerable marine ecosystems or protected areas, the overall effect could be greatly amplified. Significant data and insights are needed to address systemic stalemates that obstacles are overcome, lofty opportunities are co-created, and new social contracts are sealed. Trust, resilience, and inclusiveness may have never been as important as in the post-truth era in this closer-knit world, regulated by the force of oceans and powered by the will of humans.

9.1 World Partnerships of Purpose for Cities and Oceans: Beacons of Change

9.1.1 Cross-Implementation of SDG11 and SDG14: Joint Action on Urban and Ocean Biocycles

Eco-responsible actions on the seas can have a great contribution to the implementation of the SDG11 on cities and the accomplishment of its ten targets. Eco-responsible actions on cities can have an equally great contribution to the implementation of

V. P. Mega, *Eco-Responsible Cities and the Global Ocean*,
https://doi.org/10.1007/978-3-319-93680-2_9

the SDG14 on oceans and the accomplishment of its ten targets (EC 2017a, b; UN 2017; UNEP 2016).

The following suggestions are only indicators of a great spectrum of ideas that can be further developed at various contexts and levels of governance by coalitions of the committed in creating blue–green opportunities (EC 2016a, b, c).

SDG11 "Make cities and human settlements inclusive, safe, resilient and sustainable" and the contribution of the Oceans

It is important to note that all targets for cities have the horizon 2030, which indicates different conditions and difficulties than the variable horizon of more urgent 2020 and 2025 for the ocean, shared and relied upon by all (OECD 2017c).

11.1: By 2030, ensure access for all to adequate, safe and affordable housing and basic services and upgrade slums

New medicines and health services could come from the ocean, which could also provide materials for housing. Following the example of the first house from algae in Hamburg in 2017, adaptable bio-houses from algae could become affordable and accessible and increase the resilience of the built environment, especially in vulnerable deltaic areas.

11.2: By 2030, provide access to safe, affordable, accessible and sustainable transport systems for all, improving road safety, notably by expanding public transport, with special attention to the needs of those in vulnerable situations, women, children, persons with disabilities and older persons

Sustainable waterborne transport including automated eco-boats should be part of the public system in all coastal and riverside cities. All urban transport systems should benefit from second-generation biofuels from marine waste and third-generation biofuels from algae.

11.3: By 2030, enhance inclusive and sustainable urbanization and capacity for participatory, integrated and sustainable human settlement planning and management in all countries

Ecosystem-based management and compact developments should be the new norm for all cities, which should also develop adaptation policies and protect their populations and assets from extreme sea level rise, storm surge, and other adverse effects. Citizen involvement for resilience should be particularly encouraged in small island developing states and vulnerable deltaic megacities.

11.4: Strengthen efforts to protect and safeguard the world's cultural and natural heritage

Marine Protected Areas contribute much to the world's heritage and the health of the blue planet. They remind us of all the importance for underwater natural (e.g., coral reefs) and cultural (e.g., submerged ancient cities) commons to be equally protected as the urban natural and cultural heritage (OECD 2017b; UNESCO World Heritage Centre 2013, 2014).

11.5: By 2030, significantly reduce the number of deaths and the number of people affected and substantially decrease the direct economic losses relative to global gross domestic product caused by disasters, including water-related disasters, with a focus on protecting the poor and people in vulnerable situations

Improving resilience to sea level rise and storm surge is capital for protecting people and assets. Adaptation strategies and actions of some coastal areas and the 100RC are inspiring efforts to be deployed to protect populations, assets, and ecosystems, as also the UNISDR handbook for local government leaders (UNISDR 2012).

11.6: By 2030, reduce the adverse per capita environmental impact of cities, including paying special attention to air quality and municipal and other waste management

Proactive circular economy solutions to prevent plastics reaching the oceans improve environmental quality in cities, particularly in developing world megacities with open-air landfills and chaotic footprints. Integrated resource management with increasingly eliminated waste contributes to better air quality and less greenhouse gas (GHG) emissions, and prevents plastic pollution of the oceans.

11.7: By 2030, provide universal access to safe, inclusive and accessible, green and public spaces, in particular for women and children, older persons and persons with disabilities

Water commons, ocean gardens, and public waterfronts are essential for the well-being of cities and citizens, and many of them can be designed to become transformed into protective reservoirs at need.

11.A: Support positive economic, social and environmental links between urban, peri-urban and rural areas by strengthening national and regional development planning

Alignment of policies and ecosystem-based plans and generation and capture of synergies at all scales and governance levels on land and in the sea are critical for reinforcing mutual and overall benefits and preventing conflicts among value chains.

11.B: By 2020, substantially increase the number of cities and human settlements adopting and implementing integrated policies and plans towards inclusion, resource efficiency, mitigation and adaptation to climate change, resilience to disasters, and develop and implement, in line with the Sendai Framework for Disaster Risk Reduction 2015–2030, holistic disaster risk management at all levels

Integrated ocean policies and plans aim at addressing climate change and related extreme effects, and reinforcing resilience and security of strategic resources (FAO 2016; IEA 2017b). Foresight can serve as a compass and engage with the complex and uncertain future (Millennium Project 2017; NIC 2017), and ocean science can contribute to better evidence and policies (UNESCO/IOC 2017).

11.C: Support least developed countries, including through financial and technical assistance, in building sustainable and resilient buildings utilizing local materials

The resilience of particularly vulnerable Small Island Developing States (SIDS), totally dependent on the ocean, is essential for humankind and has to be assisted especially by those responsible for the greatest part of GHG emissions. Built environments have to use the most accessible local material, for example, from marine waste and algae, and the best possible technologies, on land and the sea, to withstand extreme effects and develop sustainable blue economies. The assistance of advanced cities can lead to rewarding results for all.

SDG 14 "Conserve and sustainably use the oceans, seas and marine resources" and the contribution of Cities

Interconnected cities can be great guardians for the ocean, coasts, and marine ecosystems inherited jointly and freely and held in trust for future generations.

14.1: By 2025, prevent and significantly reduce marine pollution of all kinds, in particular from land-based activities, including marine debris and nutrient pollution

Efficient waste prevention and management in cities is critical for clean oceans. The success of plastics in cities has come at the price of the oceans. As 80% of marine litter originates on land and especially in cities, where also are concentrated the companies involved in plastics, packaging, consumer products, and resource management, cities must and can work for a difference for the ocean (City of Amsterdam 2015).

14.2: By 2020, sustainably manage and protect marine and coastal ecosystems to avoid significant adverse impacts, including by strengthening their resilience, and take action for their restoration in order to achieve healthy and productive oceans

Ecosystem-based management of coastal and ocean-minded cities is protecting coastal ecosystems and the global ocean. Powerful cities and organisations could adopt vulnerable marine ecosystems such as coral reefs and actively assist their restoration and recovery.

14.3: Minimize and address the impacts of ocean acidification, including through enhanced scientific cooperation at all levels

Effective action on cities, which produce 70% of global emissions, can be instrumental for limiting the adverse effects of climate change on the ocean. The powerful knowledge of cities can promote international scientific cooperation on global commons phenomena such as ocean acidification, or adopt research and observation stations.

14.4: By 2020, effectively regulate harvesting and end overfishing, IUUF and destructive fishing practices and implement science-based management plans, in order to restore fish stocks in the shortest time feasible, at least to levels that can produce MSY as determined by their biological characteristics

Closing seas, ports, and markets to illegal, unreported and unregulated (IUU) fishing is crucial. Many cities joined the Port State Measures Agreement, which entered into force in 2016 as a binding, international treaty designed to combat IUU fishing. By denying unscrupulous fishers safe haven and access to markets, cities can promote fairer trade and sustainability. The EU Fair Trade City award could inspire empowering measures.

14.5: By 2020, conserve at least 10% of coastal and marine areas, consistent with national and international law and based on the best available scientific information.

Cities can influence and support the strategic designation and the management of protected areas on land or under the waters, especially in high seas. They can adopt Marine Protected Areas (MPAs) in the high seas for exemplary preservation.

14.6: By 2020, prohibit certain forms of fisheries subsidies which contribute to overcapacity and overfishing, eliminate subsidies that contribute to IUUF and refrain from introducing new such subsidies, recognizing that appropriate and effective special and differential treatment for developing and least developed countries should be an integral part of the WTO fisheries subsidies negotiation

In the era of urban geopolitics, powerful cities can impact equitable global governance and the end of the fisheries subsidies crime, through arrangements reflecting differential responsibilities.

14.7: By 2030, increase the economic benefits to SIDS and LDC from the sustainable use of marine resources, including through sustainable management of fisheries, aquaculture and tourism

Given the strong challenges facing SIDS and coastal LDC, cities can help with capacity building and partnerships for a sustainable blue economy, from eco-responsible aquaculture to marine tourism, creating new assets and sustainable wealth from the ocean.

14.A: Increase scientific knowledge, develop research capacity and transfer marine technology, taking into account the IOC Criteria and Guidelines on the Transfer of Marine Technology, in order to improve ocean health

Knowledge and technology sharing between resilient powerful cities and vulnerable ocean and island cities are essential for awareness raising, and the expansion of cleaner, cutting-edge technologies for healthier cities and seas.

14.B: Provide access for small-scale artisanal fishers to marine resources and markets

Urban markets in line with sustainability should provide a good place for nutritious food and ethical products from small-scale artisanal fishers and provide incentives for blue–green goods.

14.C: Enhance the conservation and sustainable use of oceans and their resources by implementing international law as reflected in UNCLOS, which provides the legal framework for the conservation and sustainable use of oceans and their resources, as recalled in paragraph 158 of The Future We Want

Coalitions of willing powerful world cities and multi-stakeholder forums could overview the implementation of international law as reflected in UNCLOS and act as caring custodians of the oceans in a global democratic governance framework.

9.1.2 Strategic Common Denominators for Eco-responsible Action

The possible pathways towards sustainable cities and seas depend on multiple priorities and goals by various actors at all levels UN-HABITAT (2016). Nevertheless, some common denominators for all actions and all those involved can act as a compass and only produce results with no regrets.

Citizen-oriented and ocean-minded co-action cannot cross-pollinate without true citizen involvement and engagement, starting with co-design as a pillar for co-creation. Local, regional, macro-regional, national, and supranational policies and investments should be well aligned and horizontal and vertical dimensions well integrated. A city is far more than the sum of its people and buildings. The ocean is far more than water and living resources. The ingenuity of nature for a more sustainable and harmonious approach to the ocean should be reintroduced and strengthened in cities. Greening walls and roofs, creating and maintaining wetlands and ocean gardens, and managing and caring for the soil–water–vegetation interface in cities can greatly amplify results for a healthier ocean.

Ocean and urban policies have many common challenges to address, including global climate, the sustainable management of strategic resources, but also education and empowerment, economy and jobs, governance, accountability, and leadership. Acting on the cities–ocean nexus can be critical for shared systemic benefits. Solidarity between the less and the most inhabited places on earth can take many expressions, foster links between urban and ocean socioeconomic agendas, and open the imaginary of landlocked cities and citizens to the vastness of the oceans.

Monitoring of progress is essential and cities can provide citizen-centred measures. Cities could take as a reference the global indicator set endorsed by the UN Statistical Commission at its 47th session, or work after the OECD set proposing 131 indicators covering 98 targets spanning all 17 SDGs, which enable only 57% of all the SDG targets to be evaluated. The OECD suggested that, on average, even OECD countries, the most advanced economies, have some distance to travel to reach the SDGs targets, particularly on the goals related to gender equality, education, the economy and jobs, and institutions. Countries that mark their highest score in relationship to oceans, on the basis of this very restricted indicators set, include Denmark, Finland, Italy, Netherlands, Norway, Slovenia, and Sweden, which contain some of the most sustainable cities of the planet (OECD 2017c).

If all cities were closing their markets to boats with unsustainable practices, and adopting and protecting vulnerable marine ecosystems or protected areas, the overall effect could be greatly amplified. No longer can any resource be wasted: everything has to be valued and enhanced through responsible innovation and concerted action. Significant data and insights are needed to address systemic stalemates, create and grasp lofty opportunities, and seal new sustainability contracts.

9.2 Shared Commitment, Data, and Trust for the Future of Cities and Oceans

9.2.1 For Cities and the Oceans, the Fourth Industrial Revolution Is First and Foremost a Data Revolution

From the facial recognition of humans and fishes to blockchains, the fourth industrial revolution holds many innovations for the better governance of cities, oceans, and the planet. Connected technology and innovations can foster the sustainable management of public goods. The heart of the matter is good governance to ensure the best possible use of digital technologies for the benefit of all (WEF 2016c, 2018).

The innovations of the fourth industrial revolution create new possibilities for all stakeholders, including governments, citizens, organisations, universities, and markets, to better understand and manage the resources upon which they collectively depend. Streams of information on cities and oceans can be translated into new understandings and insights and lead to new discoveries and achievements, building better sustainable value chains (WEF 2016c, 2017).

For cities and the oceans, the future depends on the use of data and how ecosystems and environments are protected and enhanced through the merging of digital, physical, and biological realms. Powerful new tools can ensure that everything can be traced at each stage. Blockchains allow stakeholders who do not know each other to organise into networks and keep trusted records, inviting transparency and accountability.

Traceable fisheries can be a prime example. Market incentives for fish that comes from sustainable fisheries can be very influential and make traceability very potent. Large ships are required to carry automatic identification system transponders designed to prevent accidents. Their signals are tracked by satellites and, combined with other data, can help monitoring fishing, shipping, mining, or other activities. Satellite tracking of marine endangered species can help preservation of ocean wildlife and facilitate accountability and involvement.

Streams of information on public activities or marine ecosystems can inform the design of cities and the creation of marine protected areas. Drones are being developed to help restore degraded coastal ecosystems, for example, mangroves. Flying 100 m above the ground, drones can take detailed pictures, process them through a machine learning algorithm, and help decide on the best species to plant according to the characteristics of each location.

Capturing the systemic and cumulative impacts of activities and investments, across value chains, is essential for optimising impact. Success required investments in building resilience by reducing stressors and identifying and protecting areas that are affected by increasing threats. More secure cities could partner with the most vulnerable ocean environments and deltaic megacities. Transparent sharing of urban and ocean knowledge, including potential impacts, beyond the immediately concerned communities, best practices, lessons learned, and ideas, is important for the overall 2030 Sustainable Development Agenda.

9.2.2 From Dystopia to Eutopia: Integrity and Inclusiveness Never Mattered That Much

Technology is becoming increasingly connected and allows for more agile management in very complex environments, such as cities and the ocean. Innovations are becoming faster, more efficient, and more widely accessible than before. But trust in government, essential for the success of a wide range of public policies that depend on behavioural responses, is deteriorating in many OECD countries (OECD 2017a). The following principles try to encapsulate some key requirements.

Purpose and Leadership
Achieving the Sustainable Development Goals 11 and 14 and other SDGs, especially those which contribute to good governance of cities and the ocean, must be recognised by all, and especially young citizens, as being of the highest importance for humanity's common journey, an opportunity to offer time, resources, work, ideas, perspectives, and enlightened leadership. Interconnected cities should act as an orchestra which has no other choice but produce music directed by a great conductor (Attali 2017).

Democratic Anticipation and Trust
Trust in institutions is important for the success of many government policies, programmes, and regulations that depend on the cooperation and confidence of investors and consumers. Through foresight and democratic anticipation, governments have an obligation to minimise uncertainty and extend the possibilities for engagement in the complex economic, social, and political environment (Camdessus 2017; EEA 2015a, b; OECD 2017a).

Integrity and Fairness
Integrity is a crucial determinant of trust and is essential if governments want to be recognised as forward looking, fair, and open. It is a crucial and invaluable resource for cross-sectoral, multi-stakeholder cooperation and value chains (WEF 2016b; WIR 2016).

Openness and Principled Connectivity
Open government policies that concentrate on citizen engagement and access to data and information can increase mutual trust. Openness to the world and principled connectivity with the global society are essential in this shared journey with uncertain winds. (WEF 2016a, c, 2017)

Inclusiveness and Participation
Participative policies are important for social justice, fairness, and the rule of law and also public services that anticipate needs and are co-created with stakeholders. Informed and proactive participation in the design and implementation of policies should empower all stakeholders, strengthen institutions, and promote trust between government and citizens (EC 2017a, b).

Assessment and Impact
Investment decisions should be based upon holistic and long-term assessments that account for economic, social, and environmental values, quantified risks and systemic impacts, and be adapted to reflect new knowledge of the associated potential risks and opportunities (Mori Memorial Foundation 2017).

Science Foundations and the Precautionary Principle
All activities should be based on knowledge and data on the potential risks and impacts associated with investments, share scientific information and data on cities and the marine environment, and encourage sustainable investment opportunities. The precautionary principle should always prevail, especially when scientific data are not available (OECD 2016; UNESCO/IOC 2017).

References

Attali J (2017) Histoire des Mers. Fayard, Paris
Camdessus M (2017) Le Monde vers 2050. Flammarion, Paris
City of Amsterdam (2015) Circular Amsterdam. A vision and action agenda for the city and metropolitan area. Amsterdam
European Commission (EC) (2016a) Opportunity Now! Europe's mission to innovate. Brussels
EC (2016b) Sustainability Now! A European vision for sustainability. Brussels
EC (2016c) Shaping the future. Thoughts on the future of society and governance. European Strategy and Policy Analysis System, Brussels
EC (2017a) Report on the blue growth strategy. Towards more sustainable growth and jobs in the blue economy. Brussels
EC (2017b) Report from the commission to the council on the Urban Agenda for the EU. Brussels
EEA (2015a) The European Environment. State and Outlook 2015. Synthesis report. Copenhagen
EEA (2015b) The European Environment. Assessment of Global Megatrends. Copenhagen
Food and Agriculture Organisation (FAO) (2016) The state of world fisheries and aquaculture, Rome
IEA (2017b) World energy outlook, Paris
Millennium Project (The) (2017) 2017 State of the future report and index. Washington DC
Mori Memorial Foundation (The) (2017) Global Power City Index. 2017. Tokyo Institute for Urban Strategies
National Intelligence Council (NIC) (2017) The paradox of progress: Global Trends 2035. Washington, DC
Organisation for Economic Co-operation and Development (OECD) (2016) Making cities work for all, data and actions for inclusive growth. OECD, Paris
OECD (2017a) Trust and public policy: how better governance can help rebuild public trust. OECD, Paris
OECD (2017b) Marine protected areas: economics, management and effective policy mixes. OECD, Paris
OECD (2017c) Measuring distance to the SDGs targets; an assessment of where OECD countries stand. OECD, Paris
UN Environmental Programme (UNEP) (2016) Global Environmental Outlook (GEO-6) (6th edition). Paris
UNESCO World Heritage Centre (2013) Safeguarding the crown jewels of the ocean. Paris
UNESCO World Heritage Centre (2014) The List. Global Strategy, Paris
UNESCO/IOC (2017) Global Ocean Science Report. Paris
UN-HABITAT (2016) The New Urbana Agenda. Nairobi
UN Office for Disaster Risk Reduction (UNISDR) (The) (2012) How to make cities more resilient. A handbook for local government leaders. UNISDR, Geneva
World Economic Forum (WEF) (2016a) The Global Competitiveness Report (36th edition). Geneva
WEF (2016b) The Global Gender Gap Report (11th edition). Geneva
WEF (2016c) Fourth Industrial Revolution. Geneva
WEF (2017) Harnessing the Fourth Industrial Revolution for Oceans. Geneva
WEF (2018) The Global Risks Report (13th edition). Geneva
World Resources Institute (WIR) (2016) Towards a More Equal City: Framing the Challenges and Opportunities. Washington, DC

Selected Websites and Social Media

#ocean literacy
@UNDP_SDGs
www.panda.org
www.c40cities.org
www.citymayors.com
www.csiro.au
www.deltacities.com
www.eea.org
www.espas.eu
www.europarl.europa.eu/stoa/
www.globalcovenantofmayors.org
www.goc.org
www.habitat3.org/
www.iclei.org
www.iom.int
www.metropolis.org
www.millennium-project.org
www.myworld2030.org/
www.nesta.org.uk
www.nexcities.org
www.nyc.gov/forward
www.oecd.org
www.ourocean2017.org/commitments
www.sdg.iisd.org/
www.un/sustainable-development-goals www.un.org/sustainabledevelopment
www.undp.org
www.unesco.org/new/en/natural-sciences/ioc-oceans
www.unfpa.org
www.unhabitat.org
www.wef.org
www.wri.org

Watercolour 12
Jellyfishes at the New England Aquarium, Boston

Annex: Concert of Inspiring Initiatives and Paradigm Shifts

1. Planet Ocean, a World of Cities: A Journey Around the Shores of the Anthropocene

1.1 Cities and the Seas: Megafacts, Trends, and Drivers of Change

The Millennium Project, State of the Future Report and Index
OECD International Futures Programme
Club of Rome
ESPAS, European Union
STOA, European Parliament
Sustainable Cities Index 2016

1.2 The Urban and Ocean Capital in the Digital Age of Overexploitation

World Economic Forum
World Resources Institute
European Environment Agency

1.3 Cities and Oceans in the Aftermath of Historic Commitments

Global Covenant of Mayors
Ocean for Climate Platform
UNISDR work on Cities

V. P. Mega, *Eco-Responsible Cities and the Global Ocean*,
https://doi.org/10.1007/978-3-319-93680-2

1.4 The Bond Between Cities and the Oceans: Expectations from the Sustainable Development Goals 2030

UN SDGs Knowledge Platform, UN Sustainable Development Solutions Network
OECD Distance to travel to SDGs targets
UNDP, IDDR work on SDGs

2. Threatened Urban and Ocean Biodiversity: The Imperative of Resilience

2.1 The Impacts of Excessive Consumption, Pollution, and Overfishing

Sunken Billions studies by the World Bank
OECD-FAO Outlooks
WRI work on coral reefs

2.2 Towards Resourceful Urban and Marine Ecosystems

UN-HABITAT, UNEP, and EEA work
Vancouver Greenest City
San Francisco Zero Waste
Port of Antwerp Sustainability Report

2.3 The Overstressed Geostrategic Nexus "Water–Food–Energy"

Food Cities (Eurocities)
Milan Food Policy Pact
Closed-Loop Oceans
Singapore Four National Taps

2.4 Ecological Cities and Protected Areas Under the Water

SymbioCities (Sweden)
Dutch Compact City and Natura 2000 Sites
Danish Model of Decentralised Concentration
OECD work on MPAs

3. The Future of Climate Will Be Decided in Cities and Oceans

3.1 Cities, Seas, and Climate: Highly Interlinked and in Still to Explore Ways

IPCC and IUCN reports
UNEP Global Emission Gap reports

C40 Cities Climate Leadership Group
Ocean Climate Platform
Ocean Climate Nexus Consensus

3.2 Ocean Warming and Acidification: Grieves for the Reefs and Not Only

NOOA State of the Coast
World Futures Council 2012 Awards
Global Mangrove Alliance
WRI work on coral reefs

3.3 Cities, Climate, and Intertwined Extreme Events

100 Resilient Cities (100RC) Rockefeller Foundation
No Regrets Charter (Metropolis principles for climate adaptation in cities)
ICLEI / UNISDR handbook for local government leaders

3.4 Strengthening Resilience in Riverside, Coastal, and Deltaic Cities

Schoolyards to Playgrounds Programme, New York City
Copenhagen Adaptation Strategy
Rotterdam Adaptation Strategy and Port Vision
Amsterdam Climate Strategy, Lagos, Makoko School

4. The Paths to Decarbonisation Through Cities and Seas

4.1 Renewables and Efficiency as Cornerstones of the Energy Transition

EDF Lab Singapore on Sustainable Cities
Amsterdam hydrogen-fuelled public transport including buses and boats
BIQ House, Hamburg, first algae-powered building, 2017

4.2 Urban and Local Excellence for the Post-carbon World

U.S. Energy Stars
Be exemplary (Brussels)
Zero Emission Cities (WBCSD)
European Sustainable Energy Week and Mobility Week Awards (EU)
Global Covenant of Mayors for Climate & Energy

4.3 The Sustainability Promises of the Maritime Paths of the Planet

Environmental Ship Index
World Port Climate Initiative

4.4 Smarter Mobility for Cleaner and More Inclusive Cities

SiMPlify (WBCSD and ECF)
Electricity cities (Oslo, Amsterdam)
Safer City Streets (International Transport Forum
Las Zebras (La Paz)
Cycle superhighways (Copenhagen, bike capital)
Hangzhou bike-sharing perception survey

5. Towards Purposeful, Circular, Sustainable Economies

5.1 Eco-Creative Cities at the Crossroads of Sustainable Shared Values

WEF Future of Urban Development and Services Initiative
Circular Amsterdam
Laudato Si′ Challenge
Ellen McArthur Foundation work on Biocycles and the New Plastics Economy

5.2 Established, Reinvented, and Burgeoning Blue–Greener Activities

OECD Ocean Economy 2030
BlueCity (Rotterdam)
Shanghai, Singapore, Rotterdam: First World port cities (since 2016, 1986–2016, up to 1986 respectively)

5.3 Responsible Corporate Conduct and Co-optimised Investments

CERES Principles and Campaigns
UN Global Compact Pioneers
WWF Principles for the Blue Economy
Corporate Social Responsibility in the Maritime Industry (WMU)

5.4 Prosperous and Sustainable City Rankings and Indexes

EIU Liveability Index 2012, 2017
Global Power City Index, Annually since 2008 by the Mori Memoriam Foundation, Institute for Urban Strategies, Tokyo
T. Kearney Cities Index 2014, 2017
Cities of Opportunity (PwC) 2011, 2014, 2016
Quality and Cost of living surveys (Mercer) 2017
Mastercard Global Destination Cities Index 2017

6. An Inclusive, Just Society as the Ultimate Frontier on Land and the Sea

6.1 Cities and Seas Without Borders: Welcoming and Sharing Places
Cities of Migration Initiative
SUCCESS Vancouver

6.2 Universal Values, Local Communities, and Intercultural Dialogue
Barcelona Inteculturality Plan
Anna Lindh Intercultural Dialogue
Intercultural Cities Programme (ICC)

6.3 Housing and Integration, Diversity and Solidarity
Stockholm Royal Seaport
Copenhagen North Harbour
Coalition for the Homeless (USA)

6.4 Healthy and Safe Cities and Seas, Youth and Citi-Zen
Healthy Cities (WHO)
EIU Safe Cities Index

7. Cities and Oceans at the Crossroads of Knowledge, Ingenuity, and Co-creation

7.1 Forward Cities: Investing in Science, Technology, and Sustainability
Smart Cities Awards (Vienna, New York)
EU Innovation Capitals (Barcelona, Amsterdam, Paris)
SparkLabs, Songdo, Seoul

7.2 Outstanding Urban and Underwater Heritage Commons
UNESCO Heritage (and heritage to watch) lists
Networks of marine protected areas

7.3 Tangible and Intangible Co-created Assets and Innovations

Hamburg HafenCity
Bilbao Design Cities
Museo Atlantico Lanzarote
Cultural Capitals of Europe (EU)

7.4 The Science and Art of Ecosystem-based Strategic Planning

OECD compact cities
OneNYC, OneSydney
Kobé & WBCSD Urban Infrastructures
Ecosystem-based MSP (EU)

8. Open Cities and the Ocean: Local and Global Engagement

8.1 Citizen Education and Democracy for Achieving the SDG 2030

Athens Democracy Award
Citizen Jury (newDemocracy Foundation)

8.2 Participative Cities, Institutions, Stakeholders, and Quintuple Helices

World Futures Council
International Coastal Cleanup (Ocean Conservancy)
Trash Free Seas Alliance

8.3 Global Governance and Urban and Ocean Accountability

Helsinki Region Infoshare
OECD Better Life Initiative
Oslo State of the City report
UN Volunteer Groups Alliance

8.4 Constellations of World Networks of Excellence

EU Urban Agenda
Our Ocean Conferences and Commitments
EU-OECD-FAO-World Bank initiative for a global harmonised definition of cities
International Urban Cooperation Initiative (EU)
UN Sustainable Development Solutions Network

9. Messages for Leaders Towards a World of Difference

9.1 World Partnerships of Purpose for Cities and Oceans: Beacons of Change

Proposed World Cities and Oceans University
Proposed World Council on City and Ocean Data

9.2 Shared Commitment, Data, and Trust for the Future of Cities and Oceans

Proposed World Urban and Ocean *Summits* and Forums for Young Leaders
Proposed Ocean and Urban Conferences and Commitments
Proposed State of Cities and Ocean reports and Outlooks

Watercolour 13
Bangkok, First Global Destination City
(Mastercard 2017)

About the Writer *(and Self-portrait)*

Dr Voula P. Mega graduated as an engineer from the National Technical University of Athens and had her first DEA Diploma at the National Geographical Institute in Paris. She continued with a DEA at the French Institute of City Planning, where she also received her PhD on City and Regional Planning and Policy. Her post-doctorate studies include regional policy at the School of Planning, Oxford Brooks University, and environmental economics and policy analysis at the Harvard Institute for

V. P. Mega, *Eco-Responsible Cities and the Global Ocean*,
https://doi.org/10.1007/978-3-319-93680-2

International Development. She also received a scholarship from MIT and attended a summer programme on innovation and the dynamics of technology and organizations. Engaged with thought leadership and lifelong learning, she was selected to attend the "Leadership Decision Making: Optimizing Organizational Performance" executive programme at the Harvard Kennedy School in 2016.

She is pursuing an international career on science and innovation-based strategies for sustainable development. She has offered special advice to governments, been research manager and leader at the EU level, and provided policy development and advice to the OECD and the European Commission. In 2001, she joined the European Commission in Brussels, where she works as senior policy analyst at the Directorate-General for Research and Innovation. Her strategic policy analysis and advice, research management and publications address the scientific foundations, values, culture and innovations for sustainable development policies and the cities of the future. She firmly believes that Science and the Arts can herald a better world and that concepts, ideas, foresight, visions, and actions for desired futures can far extend the limits of the possible.

In continuous search of excellence, she has published and co-published many books and hundreds of articles with the European Union, OECD, UN, UNESCO, and international publishers and magazines, in Greek, English, French, Spanish, and Italian. Most recent titles include "Conscious Coastal Cities. Sustainability, Blue Green Growth and the Politics of Imagination" (Springer, 2015), "Quintessential Cities, Accountable to the Future. Sustainability, Innovation and Citizenship" (Springer, New York, 2013), "Bio-diver-cités" (EUE, Saarbrucken, 2012), "The Desirable Future of Innovative Cities" (LAP, Saarbrucken, 2011), "Sustainable Cities for the Third Millennium. The Odyssey of Urban Excellence" (Springer, New York, 2010), "Modèles pour les villes d' avenir" (L'Harmattan, Paris, 2008), and "Sustainable Development, Energy and the City" (Springer Science+Business Media, New York, 2005).

Citizen of the world and passionate European, Voula was nurtured by many cultures, including French, Spanish, American, Thai, and Japanese. Her academic activities include invited Master Lectures at the University of California at Berkeley, Tokyo Institute of Technology, Columbia University, Prince of Wales Institute of Architecture, Institut Français d'Architecture, the University College of London, Catalan Institute of Technology, and the Universities of Athens, Bologna, Trieste, and Parma. She has also been Associate Professor at the National School of Public Administration, in Athens.

She actively promotes the dialogue between Science and the Arts and among various forms of arts, especially poetry, design, and painting. She published the poetry collections "Siren Cities" (in Greek, English, and French) accompanied with drawings (Exantas, Athens, 1997), "Dawns and Souls for Europe" (in English and French, by Persée, Paris, 2008), official selection for the European Book Prize 2008, "Héliopolis. Sourires de Villes Solaires" (HD, Paris, 2013), and "The Invincible Journey" (Iolkos, Athens, 2016). An exhibition of watercolours, drawings, diptychs (poetry/design), and photographs at the European Parliament brought together her artistic work on cities under the title "The song of siren cities."

Watercolour 15
Rising Arc of Life (Inspired by Ephemeral Sculptures in New Zealand)

Index

V. P. Mega, *Eco-Responsible Cities and the Global Ocean*,
https://doi.org/10.1007/978-3-319-93680-2

C

D

Q

R

MIX
Papier aus verantwortungsvollen Quellen
Paper from responsible sources
FSC® C105338

If you have any concerns about our products,
you can contact us on
ProductSafety@springernature.com

In case Publisher is established outside the EU,
the EU authorized representative is:
Springer Nature Customer Service Center GmbH
Europaplatz 3, 69115 Heidelberg, Germany

Printed by Libri Plureos GmbH
in Hamburg, Germany